Experimental General Chemistry

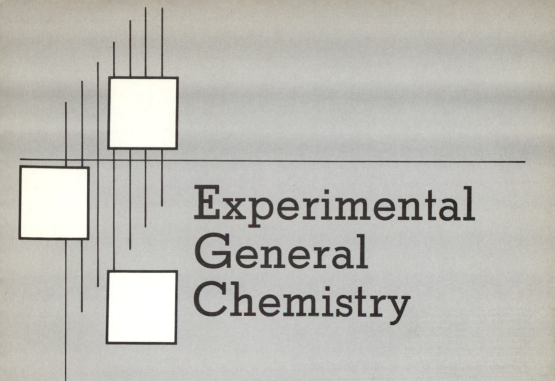

Experimental General Chemistry

Stanley Marcus
Cornell University

The Late Michell J. Sienko
Cornell University

Robert A. Plane
Cornell University

McGraw-Hill Book Company
New York   St. Louis   San Francisco   Auckland   Bogotá   Hamburg
London   Madrid   Mexico   Milan   Montreal   New Delhi
Panama   Paris   São Paulo   Singapore   Sydney   Tokyo   Toronto

Experimental General Chemistry

234567890   DOC DOC   89210987

ISBN   0-07-054420-4

This book was set in Times Roman by Automated Composition Service, Inc.
The editors were Irene M. Nunes, Karen S. Misler, and Annette Bodzin;
the designer was Caliber Design Planning, Inc.;
the production supervisors were Diane Renda and Louise Karam.
The cover designer was Charles A. Carson.
The cover photograph is by Ken Karp.
The drawings were done by Fine Line Illustrations, Inc.
Project supervision was done by Total Book.
R. R. Donnelley and Sons Company was printer and binder.

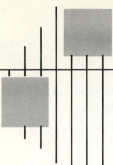

Contents

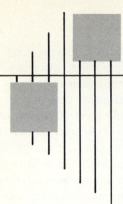

Preface

This laboratory manual can be used with any general chemistry textbook, including books designed for courses that cover qualitative analysis and/or a substantial amount of organic chemistry. Most of what is contained in this manual has been drawn from the laboratory manual *Experimental Chemistry, Sixth Edition*, by Sienko, Plane, and Marcus. The experiments all have been tested by thousands of students, both at Cornell University and at many other schools.

The two main differences between this manual and *Experimental Chemistry* are

1. The numbering of the experiments has been changed in order to group experiments dealing with similar conceptual material together and to present them in the order followed by most general chemistry courses.

2. The number of organic chemistry experiments has been substantially increased.

The approach in this manual is *guided discovery* rather than simple verification. Students are taught to observe and to report what actually happens, not what they think should happen. Nearly all of the experiments are designed to be completed in a single two- or three-hour laboratory session, but there are also open-ended experiments that require several laboratory periods for completion. Experiments Q.1 through Q.10 (qualitative analysis), for example, require ten or more laboratory sessions. Experiments E.21 (Preparation of a Complex Iron Salt) and E.22 (Analysis of a Complex Iron Salt) could be done either as separate experiments or as a single combined experiment. Experiment E.34 (Separation and Analysis of a Mixture), which also uses procedures described in Experiments E.35 through E.37, could take as many as five periods to complete.

The manual has four major parts. Part A describes techniques of experimental chemistry. As is the case throughout the manual, the primary emphasis is on safety. Also discussed in this section are balances, volumetric measuring devices, filtration techniques, the gas burner, glassworking, the centrifuge, determination of melting point, spectrophotometers, and proper methods of recording and treating data.

Part B presents 52 experiments designed to teach the important general principles and facts of general chemistry. For each major conceptual area of chemistry, there are several experiments that are at different levels of complexity. There are, for example, five experiments relating to properties of matter and Dalton's elementary atomic theory, five on the gas laws, five on atomic and molecular structure and bonding, more than fifteen on stoichiometry (synthesis and analysis), more than ten on solutions, three on kinetics, more than six on aqueous equilibrium, seven on thermodynamics and electrochemistry, and more than ten on descriptive chemistry.

Part C consists of eleven organic chemistry experiments. There are two molecular model exercises designed to familiarize the student with various types of isomerism and to provide some practice in naming organic compounds. Important classes of organic compounds (hydrocarbons, alkyl halides and alcohols, aldehydes and ketones, carboxylic acids and esters) are studied in a series of experiments by examining characteristic reactions of each functional group. There are also experiments on the synthesis of aspirin and sulfanilamide and on the paper chromatography of amino acids.

Part D is an abbreviated scheme of qualitative analysis in which groups of elements are added to the scheme not in the order in which they appear in the classical scheme of qualitative analysis but in the order in which they appear in the periodic table. As each element is added, the student carries out experiments to find out how that element fits into the scheme. The experiments are designed to allow the student to discover what eventually turns out to be the classical scheme of analysis. The main purpose of working through the series of experiments is to learn some of the chemistry of individual elements and some important principles of aqueous solution chemistry.

There are also five appendixes containing such general reference material as constants and conversion factors, equilibrium constants, and atomic weights, which will be of use to students in making computations.

Detailed information on the preparation of reagents for the experiments and on answers to the questions is available in an instructor's manual, which can be obtained by writing to the publisher.

Most of the credit for this laboratory manual must be given to Michell J. Sienko, late professor of chemistry at Cornell University, who developed the majority of the experiments. By his teaching, example, and writings, Dr. Sienko, together with his colleague Robert A. Plane, made an impact on chemical education in the United States and many other nations that may never be duplicated.

Stanley T. Marcus

typical chemistry laboratory equipment

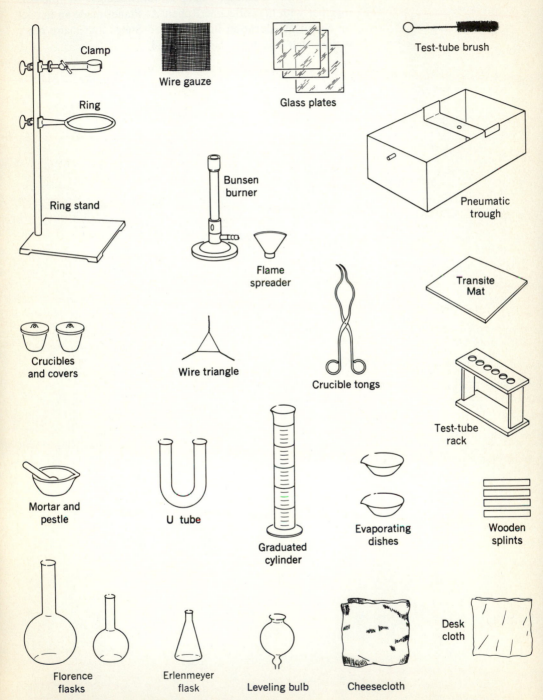

Clamp

Ring

Ring stand

Wire gauze

Glass plates

Test-tube brush

Bunsen burner

Flame spreader

Pneumatic trough

Transite Mat

Crucibles and covers

Wire triangle

Crucible tongs

Test-tube rack

Mortar and pestle

U tube

Graduated cylinder

Evaporating dishes

Wooden splints

Florence flasks

Erlenmeyer flask

Leveling bulb

Cheesecloth

Desk cloth

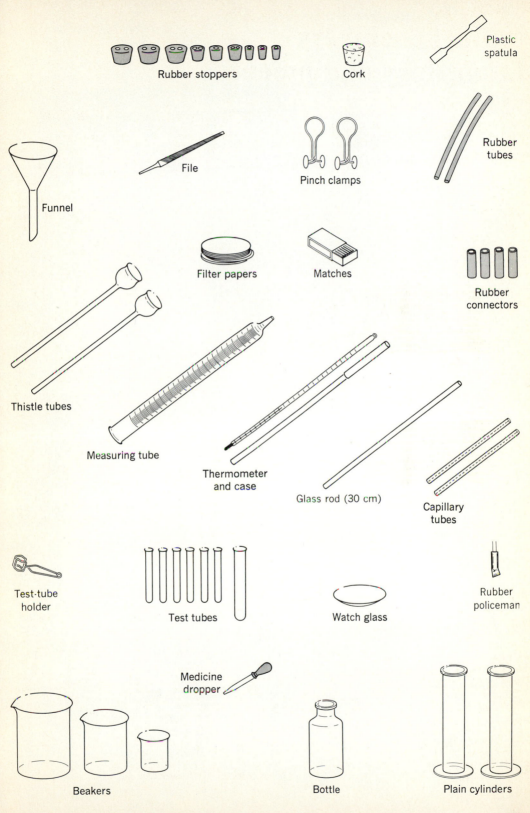

Rubber stoppers

Cork

Plastic spatula

Funnel

File

Pinch clamps

Rubber tubes

Filter papers

Matches

Rubber connectors

Thistle tubes

Measuring tube

Thermometer and case

Glass rod (30 cm)

Capillary tubes

Test-tube holder

Test tubes

Watch glass

Rubber policeman

Medicine dropper

Beakers

Bottle

Plain cylinders

Typical Inventory

1 transite board
5 beakers: 30, 50, 100, 200, 600 mL
1 bottle, wide-mouth: $\frac{1}{3}$ liter
1 bunsen burner with flame spreader
1 capillary: 25 cm
1 clamp
1 cheesecloth
2 crucibles and covers: No. 00
1 crucible tongs
1 cylinder, graduated: 25 mL
3 cylinders, plain: 125 mL
1 evaporating dish: No. 00
1 file
25 filter papers: 9 cm
3 flasks, flat-bottom: 125, 300, 500 mL
1 funnel: 65 mm
1 gas-measuring tube: 50 mL†
3 glass plates: 10 × 10 cm
1 leveling bulb
matches
1 medicine dropper
2 pinch clamps
1 pneumatic trough
1 ring: 8 cm

1 ring stand: 45 cm
4 rubber connectors: 5 cm
1 rubber policeman
9 rubber stoppers:
 3 No. 8, 2-hole
 1 No. 5, 2-hole
 1 No. 4, 2-hole
 1 No. 4, 1-hole
 3 No. 1, 1-hole
60-cm rubber tubing: 7 mm
60-cm rubber tubing: 5 mm
15-cm ruler
10-cm spatula
10 splints
1 stirring rod: 5 mm × 30 cm
6 test tubes: 16 × 150 mm
1 test tube: 25 × 200 mm
1 test-tube brush
1 test-tube holder
1 test-tube rack
1 thermometer: −10 to 110°C
1 thistle tube
1 triangle
1 U tube
1 washcloth
1 watch glass: 80 mm
1 wire gauze: 10 × 10 cm

If qualitative experiments are included, each student should be provided with an individual set of reagents in 60-mL dropping bottles. These could profitably include: 6 M HCl, 15 M NH_3, 1.7 M thioacetamide, 3 M H_2SO_4, 3 M $NH_4C_2H_3O_2$, 2 M NaOH, 6 M NH_4Cl, saturated Na_2SO_4, 2 M $NaHSO_4$, 6 M HNO_3. Students also need a 15-cm piece of nichrome wire and a double thickness of cobalt glass. It is also helpful to have 12 additional test tubes (13 × 100 mm).

†Purchasable as "Dennis buret" or "Cornell buret" from Ace Glass Co., Vineland, N.J.

Experimental General Chemistry

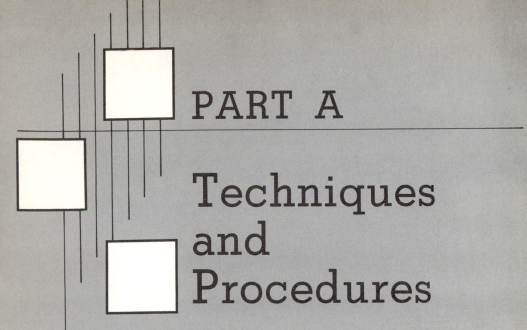

PART A

Techniques and Procedures

The entire structure of chemistry rests on a foundation of empirical evidence. To maintain the integrity of that foundation, chemistry insists that its facts be reproducible and independent of the observer.

In performing experiments, there are two absolute requirements: first, that the experiment not expose the experimenter or others to danger; second, that the experiment be designed and carried out such that valid information is generated.

The purpose of this part of the manual is to instruct you in practices and techniques that will lead to a safe working environment as well as to the generation of valid data.

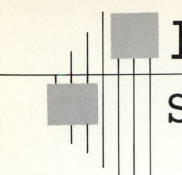

1

Safety

The chemistry laboratory is really not a dangerous place, but it demands reasonable prudence on the part of an experimenter to keep it safe. In the following paragraphs, the more important precautions are discussed.

Eye Protection

The eyes are particularly susceptible to permanent damage by corrosive chemicals as well as by flying fragments.

It is required that each student wear safety goggles at all times in the laboratory.

In doing hazardous experiments, follow all directions carefully and, in particular, take care not to endanger your neighbor. For example, when heating a test tube, do not point its mouth toward anyone. Report any accident immediately to your instructor. In case of injury to the eye, *immediately* flood the eye with lots of water, and continue to rinse for at least 10 min. If an eyewash fountain is not immediately available, use a rubber tube connected to a faucet. *All* injuries involving the eyes should be referred to a physician at once.

Cuts and Burns

The great majority of laboratory injuries are cuts and burns. Virtually all of these can be prevented by following a few simple rules:

1   In case of any injury, report it at once to your instructor for treatment.
2   Do not insert glass tubing, especially thistle tubes, into rubber stoppers without first moistening the tubing and the hole with water or glycerin. Also, it is a wise precaution to shield the hands by use of cloth, as shown in Figure A.1. To reduce leverage on the glass, hold the hands close together. While twisting the stopper back and forth, gradually work the glass

3

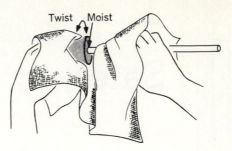

Figure A.1

tubing through the hole. It may be necessary to moisten the stem several times while the operation progresses.

3   When trying to remove glass tubing from a stopper or from a rubber fitting, lubricate it by prying a bit of the rubber away from the glass and dribbling in a bit of water or glycerine. Then try to twist off the stopper. It may be necessary to re-lubricate. If the joint is really stuck, do not force it. Cut off the rubber with a razor blade.

4   Never set heated glass on the bench top. Place it on a transite board to cool.

5   Never pick up a heated piece of glass until it has had time to cool. Unfortunately, hot glass looks just like cold glass; when in doubt, use tongs.

6   Use of glass equipment that is cracked or broken may ruin an experiment. Worse, it may cause an injury. Since damaged equipment must be replaced eventually, discard it immediately. Place it in a "waste glass" container, not in the wastepaper basket.

Poisonous Chemicals

Most of the chemicals you will work with are poisonous to some degree. It is obvious that you should never taste a chemical unless specifically directed to do so. However, there are more subtle ways of being poisoned. One of these is by breathing toxic vapors. Be careful to work in a ventilating hood whenever instructed to do so. Even such common substances as carbon tetrachloride, benzene, and mercury are poisonous and potentially dangerous. Avoid prolonged exposure to these liquids or the accompanying vapors. Since heating favors the vapor state, these and other poisonous liquids should be heated only in a hood.

Occasionally, you will be directed to test the odor of a substance. The proper way to do this is to waft a bit of the

Figure A.2

vapor toward your nose as shown in Figure A.2. Do not stick your nose in and inhale vapor directly from the test tube.

A possible poisoning hazard, frequently overlooked, is contamination through the hands. Some poisons—e.g., benzene—are rapidly absorbed through the skin. All poisons can stick to the hands and eventually end up in the mouth. Immediately scrub your hands thoroughly after exposure to hazardous chemicals, and get into the habit of always washing your hands before leaving the laboratory.

Food should not be brought into the laboratory.

Corrosive Chemicals on Skin or Clothing

If you should accidently spill a corrosive chemical, such as a concentrated acid or base, on yourself, quickly remove any contaminated clothing, flush the affected skin for at least 10 min with water, and notify your instructor. Do *NOT* neutralize chemicals on the skin by adding other chemicals.

Footwear

Shoes, not sandals, should be worn in the laboratory. Preferably, the shoes should be of the hard-toe variety.

Essential Precautions

Follow all directions with utmost care, especially those having to do with hazardous conditions. Do not perform any unauthorized experiment. If you want to change or supplement the assigned material, first consult your instructor and get his or her permission. Irresponsible behavior will result in immediate expulsion from the laboratory.

In using chemical reagents, double-check the label to make sure you are not using the wrong chemical. Serious explosions have frequently resulted from such errors.

Smoking is not permitted in the laboratory.

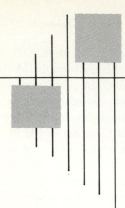

2

Operating Common Laboratory Equipment

a. Balances

One of the most common operations in experimental chemistry is the determination of mass or weight. The mass of an unknown is usually determined by comparing it with a known mass. For making the comparisons, there are three kinds of balances in common use.

Platform Balance   One of these is the platform balance, two models of which are shown in Figure A.3. When the type shown on the left is used, the unknown on the left pan is balanced by placing large weights on the right pan and then sliding the weights attached to the crossbeams until the pointer makes equidistant swings to the right and left of the center point on the scale. When the type on the right is used, balance is achieved by sliding the weights along the three beams until

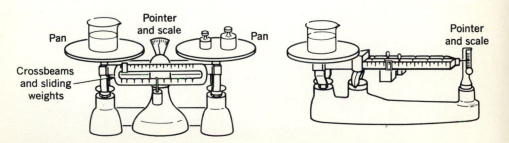

Crossbeams and sliding weights
Pan
Pointer and scale
Pan
Pointer and scale

Figure A.3

the pointer at the extreme right makes equal swings above and below the zero point. Platform balances, when used properly, are capable of measuring mass to within ±0.05 g. To prevent corrosion of the pans, chemicals should never be placed directly on the metal surface. Always use a container or, for solids, a piece of weighing paper. In using weighing paper, it helps to have the paper creased upward along the two diagonals in order to prevent spillage.

Two-Pan Analytical Balances   The second kind of balance in common use is the two-pan analytical balance, illustrated in Figure A.4. In principle it is the same as the two-pan platform balance. There is a horizontal beam with two pans suspended from the ends. The beam pivots about its midpoint, where there is a knife-edge resting against a flat, hard surface. This contact point about which the beam turns is one of the most delicate parts of the balance and must be treated with care to prevent permanent, costly damage. Similar delicate contact points are found where each pan hangs from the beam. To prevent injury to all three contacts, the beam and pans are supported from below in a locked position except when the balance swings are being observed.

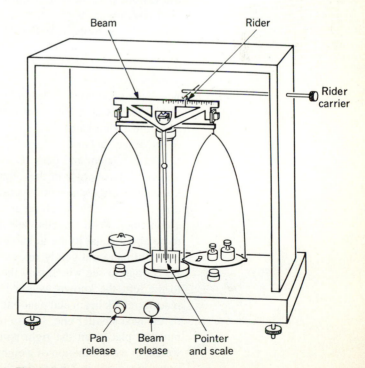

Figure A.4

■ CAUTION

Keep beam and pans locked at all times when changing loads on the pans, when opening or closing the balance case, and when leaving the balance after use.

The beam support is controlled by an external knob extending from the center of the floor of the case; the pan support is frequently controlled by a button next to the beam-support control, although sometimes the two controls operate from the same knob. Because beam deflections are small, to facilitate observation there is a long pointer rigidly attached to the beam and extending toward the bottom of the case, where it sweeps across a ruled scale. Rough balance is attained by placing known weights on the right-hand pan; fine balance is then achieved by sliding the wire rider along the beam. The rider is manipulated by a hook fastened to the rod that protrudes from the top right of the case.

The rider weighs 10 mg but exerts its full value only when at position 10 on the beam. As the rider is moved toward the center of the beam, its effective value is reduced to the value shown by the graduations on the beam.

Procedure for Two-Pan Balance   Familiarize yourself with the various parts of your balance. Check to see that the rider is in position at the zero line on the beam. Gently release the pan and beam supports so that the pointer swings freely across its scale. Note the number of divisions the pointer moves right and left of the center line. Locate the midpoint of the swings. This is called the ''rest point'' and represents the point at which the pointer will eventually stop. If the rest point is more than three divisions from the center line, call your instructor to readjust your balance. Record the position of the rest point of the empty balance.

Lock beam and pans. Open the case, and place an unknown (e.g., a coin) in the center of the left pan. If the unknown is a single piece of noncorrosive solid, it may be placed directly on the pan; in all other cases, weigh the substance in a container. With forceps, select a weight that you estimate to be about equal to the unknown. Place it near the center of the right pan, and gently release pan and beam supports just far enough to see which way the pointer swings. If it swings to the left, the known weight is too heavy and must be replaced (relock beam and pans each time you change a weight!) by the next smaller one. If it swings to the right, more weight must be placed on the right pan. Try substituting a heavier weight or a combination of weights until the unknown is found to be no more than 1 g heavier than the known weights. Using the same procedure, add milligram weights until the unknown is found to be no more than 10 mg heavier than the known weights. Close the case. Keeping the beam locked, shift the rider to about position 5. Release pans and beam, and locate the midpoint of the swings. If the midpoint lies to the right of the rest point of the empty balance, then the combination of known weights including the rider is too small. Relock the beam, and move the rider farther to the right. If the midpoint of the swings lies to the left of the rest point of the empty balance, relock the beam and move the rider to the left. Adjust the rider until the midpoint of the swings is within a division or so of the rest point of the empty balance. You have now determined the weight of your unknown to the nearest milligram.

When finished with the balance, check to make sure that beam and pan supports are locked. Have all weights properly put away in the box. Brush up any spilled chemicals, and close the balance-case door. Set the rider at its zero point, and push the rider-carrier arm all the way in. The latter is important to prevent damage to the balance from accidental brushing against the rod.

Single-Pan Analytical Balance   A third type of laboratory balance is the single-pan analytical balance, illustrated in Figure A.5. The principle of operation is based on substitution

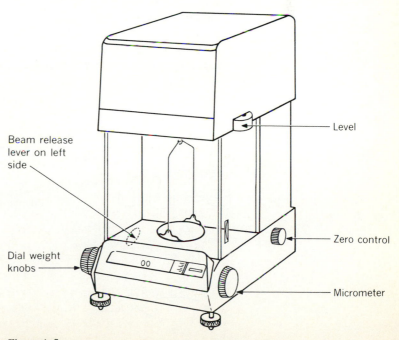

Figure A.5

weighing—i.e., replacing built-in known masses by the unknown object and removing an amount necessary to restore balance. We can see how this is done by referring to Figure A.6, which shows schematically the internal workings of a typical single-pan balance. Initially, with pan empty and all "removable known masses" in place, there is balance (as shown on the optical scale) because of the counterweight. The object to be weighed is placed on the pan, and some removable known masses are lifted off the beam by internal levers so as to restore approximate balance. Any remaining imbalance tilts the beam slightly, and the amount of tilt is projected on the optical scale. The total mass of the unknown is then equal to the masses removed plus the reading of the optical scale. Finally, it might be noted that the counterweight is constructed to act also as a damper that prevents oscillation of the beam.

Procedure for Single-Pan Balance   Detailed procedure depends on the particular make of balance used. Typical steps are as follows:

1   *Check zero.* Set all controls to zero. Turn beam release lever down to "fine weigh" position. If scale does not stop on zero, adjust with zero control knob until it does.

2   *Weigh the unknown.* Arrest the beam. Place object on pan. Turn beam release lever up to "coarse weigh" position. Turn the knob of the largest weight decade until the scale swings back below "zero." Back off one unit. Then proceed with the next lower series of weights (the next knob to the right) and repeat the procedure. Arrest the beam and then, after a pause,

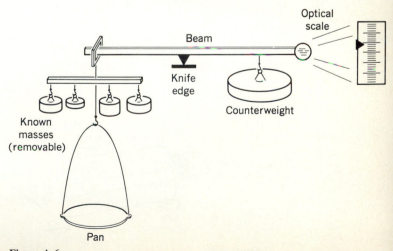

Figure A.6

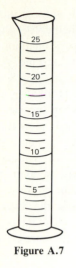

Figure A.7

turn the beam release lever down to the fine weigh position. When the optical scale has come to rest, adjust the micrometer knob until the reading indicator is superimposed on the next lower division of the scale.

3   *Read the result.* Read total weight from left to right: decade knob—unit knob—decimal—scale indication—fine adjustment. After completing the weighing, arrest the beam, and return all weight knobs and the micrometer drum to zero.

The main precaution in using single-pan analytical balances is to minimize vibration. Do not bump the case or the bench at any time, and during final readings in steps 1 and 3, *do not touch any part* of the balance or the bench. Care and cleanliness are, of course, just as important for single-pan as for double-pan balances. Brush up any spilled chemicals, and close the balance-case door when done.

b. Volumetric Measuring Devices

Graduated Cylinder   To measure the amount of a liquid sample, it is usually more convenient to determine its volume than its mass. One of the simplest devices for doing this is the graduated cylinder (Figure A.7). The correct way to read the volume of liquid is to hold the cylinder vertical at eye level and look at the top surface (meniscus) of the liquid. The meniscus is curved, with a rather flat part in the center. By noting the position of this flat part relative to the calibration marks on the cylinder, you can, with the cylinder shown, determine the volume of a liquid to approximately ± 0.1 mL. It is important to keep graduated cylinders clean for at least two reasons: Dirt may chemically contaminate an experiment, and the presence of dirt in the cylinder may throw off accurate determination of volume. (Glassware that is dirty does not drain properly, so the delivered volume is not equal to that indicated by the calibration marks.)

Cleaning Glassware   Despite the fact that there are elaborate recipes for making cleaning solutions, the most effective way to clean glassware is to use soap, water, and a brush. Get into the habit of washing your equipment after each use to have it ready for the next experiment. Do not forget to rinse off the soap with lots of water. Invert the glassware so that it will drain dry.

Burette   Another common device for measuring liquid volume is the burette (Figure A.8). In general, burettes are con-

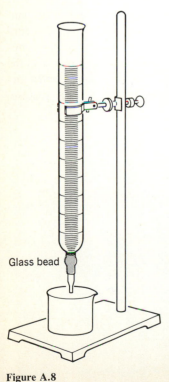

Glass bead

Figure A.8

structed so that it is possible to measure volume more precisely than with a graduated cylinder. The burette illustrated can be made by attaching a rubber connector containing a glass bead to the constricted tip of a gas-measuring tube. A nozzle made by drawing down a piece of glass tubing (see Section e) is inserted at the lower end. The burette is filled by pouring liquid through a funnel into the top. When the side of the rubber connector at the glass bead is pinched, some liquid can be drained out to free the tip of all air. It is not necessary for the meniscus to be at the zero mark, since the volume delivered from a burette can be obtained simply by taking the difference between the initial and final readings. In reading a burette, it is imperative that the eye be at the level of the meniscus. For colorless liquids like water, read the bottom of the meniscus; for dark liquids, it is generally more convenient to read the top (i.e., the place where the liquid contacts the glass). Keep your burette cleaned; otherwise, it will not drain properly—as you can easily tell by noting whether drops of liquid are left hanging from the walls. Burettes equipped with stopcocks (Figure A.9) should be rinsed with water and filled

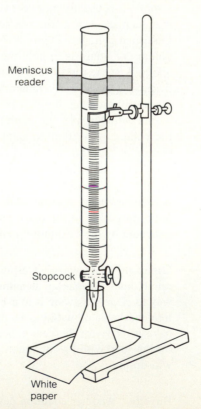

Meniscus reader

Stopcock

White paper

Figure A.9

with distilled water for storage. Rinsing is especially important if a strong base has been used. For accurate use of a burette, it is necessary that the volume of liquid held below the calibration marks be the same for both the initial and the final readings. To achieve this, make sure that there is no air trapped in the connector or tip. A good test is to twist the tip upward so that a trapped air bubble can rise and then be squeezed out. Furthermore, make sure that no drop is left hanging from the burette tip either before or after delivering liquid. After delivering a volume, always touch the tip to the wall of the receiver.

Pipette   Sometimes it is necessary to obtain several equal samples of an accurately determined volume of liquid. This can be done by means of a pipette. As shown in Figure A.10, a pipette consists of an elongated glass bulb having a delivery tip at the lower end and an opening for suction filling at the upper end. A ring mark engraved on the stem indicates the height to which the liquid level must be drawn to get exactly the volume of liquid marked on the bulb. The proper way to use a pipette is to draw the liquid up past the ring mark using a rubber suction bulb, detach the bulb, and quickly place a fingertip over the suction end to hold the suction, and then, by slightly rolling the fingertip aside, allow excess liquid to drain out of the delivery tip until the meniscus descends to the ring mark. The tip of the pipette is then transferred to a receiving vessel, and the liquid is allowed to drain out. Usually 10 sec more are counted off to permit the walls to drain properly, and then the tip is touched to the receiver wall to get off the last drop. Pipettes are available in various sizes—1, 5, 10, 25, 50 mL—and are generally calibrated to deliver that volume at a standard temperature of 25°C.

Handling Reagent Bottles   When adding liquids directly from bottles, you should observe certain elementary precautions. First, double-check the label to make sure that you have the right reagent. Second, never set the stopper down so as to contaminate the inner surface. The proper way to hold one type of stopper and pour from a reagent bottle is shown in Figure A.11. Note that the stopper is equipped with a projection that is clasped between the forefinger and middle finger. Note also that a stirring rod is held against the mouth of the bottle so as to permit liquid to flow down along it without splashing into the receiver. Finally, when done, replace the stopper, and *set the bottle back on the reagent shelf*. Under no circumstances should you pour chemicals back into a re-

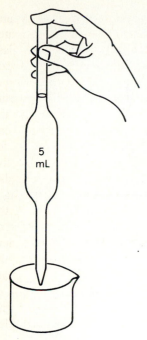

Figure A.10

5
mL

Figure A.11

agent bottle or, for that matter, place anything—e.g., a medicine dropper—in it.

c. Decantation and Filtration

A common problem in the laboratory is that of separating a liquid from a solid. This should be done in two stages: First, decant, i.e., let the solid settle, and carefully pour off most of the liquid down a stirring rod. Second, proceed to filter. When only a few milliliters of liquid remain in the beaker, swirl it gently and pour the mixture into the funnel. Rinse down the sides of the beaker with a stream of distilled water from your wash bottle, swirl, and pour into the funnel. Repeat until beaker and stirring rod are clean. Allow filter to drain.

Prepare the filter paper for insertion into the funnel by folding it in half along one diameter and then in quarters, as shown in Figure A.12. For rapid filtration, the filter paper should fit snugly in the funnel so that air cannot be drawn down between the paper and the funnel. Tear off a corner of the folded paper as shown in the third sketch of Figure A.12 to get a tighter seal. Insert the paper into the funnel, and pour some solvent (usually water) through it. It is then possible to press the wet paper gently against the funnel wall. When the funnel is properly sealed, liquid will be retained in the funnel stem, and its weight will help pull liquid through the paper. If liquid is not retained in the stem, either the paper is not sealed or the stem is dirty, in which case it should be cleaned (use a pipe cleaner or a twisted strip of cheesecloth).

When pouring a solid-liquid mixture into a filter, it is a

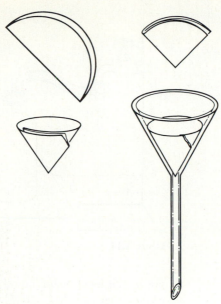

Figure A.12

good idea to pour down a stirring rod, as shown in Figure A.13. However, be careful not to punch a hole in the filter paper with the end of the glass rod. To prevent splashing, have the bottom tip of the funnel touch the receiver wall. Do

Figure A.13

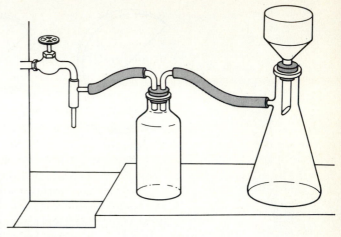

Figure A.14

not let the level of the liquid in the funnel go above the top of the paper.

In quantitative separations you may find that it is necessary to loosen the filter paper in the funnel so as to let the last liquid trapped in the stem flow out to a receiver.

Suction Filtration   Filtration is frequently a slow operation. It can be speeded up by use of suction filtration. Suction can be applied by means of a water aspirator (Figure A.14), which is a gadget that fits on a water faucet and uses a fast stream of water to suck in air through a side arm. Because the suction is quite strong, it is necessary to use a special funnel, called a *Büchner funnel*, to support the filter paper; otherwise, the paper might break through. Put the paper in the funnel, and wet it first with solvent. Then decant most of the liquid into the center of the filter paper. Pour it through slowly so the precipitate does not get a chance to seep under the filter paper.

d. Gas Burner

The classic device for providing heat in a laboratory experiment is the bunsen burner or some modification of it. The essential construction of a bunsen burner is shown in Figure A.15. The gas enters the burner at the base, and its supply is regulated externally by the gas cock. As the gas streams upward through a jet inside the base, air is pulled in through the air-intake holes just above the base. The amount of air can be controlled by rotating a sleeve that fits over the holes in the

Figure A.15

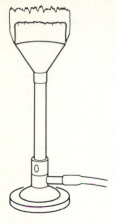

Figure A.16

barrel of the burner. Maximum air enters when the holes in the sleeve match those in the barrel. To keep the flame from blowing out, special tips are frequently provided to fit over the top end of the barrel.

The proper method for lighting a gas burner is to close off the air supply, turn on the gas, and light. The flame will be large and yellow. Gradually open the air intake until the flame takes on a blue color. For maximum heat, open further so that two distinct zones appear, the inner of which is light blue and cone-shaped. The hottest point of the flame is just above the top of this inner zone.

In some instances, when the air supply to the burner is increased, the flame blows out or detaches itself from the burner tip. This means that the gas flow is too great and needs to be cut down. When relighting, remember to shut off the air supply first.

The normal tip shown in Figure A.15 gives a concentrated flame. If you wish to spread out the flame, as is necessary for glass bending, replace that tip with a flame spreader, or wing tip (Figure A.16). Again, there should be two zones, the inner one having a level top as shown. If the inner zone is jagged, part of the wing-tip opening is constricted and needs to be opened up. This can be done by inserting the edge of a triangular file and rocking it back and forth along the entire length of the slot. If you have trouble, consult your instructor.

e. Glassworking

Glass is usually supplied in standard forms, such as straight tubing, which frequently must be modified before use. Glass becomes plastic when hot, and the essential secret of successful glassworking is to get the glass sufficiently hot. Most student trouble in glassworking stems from failure to get the glass hot enough. Keep this in mind when going through the following operations.

1 *Cutting and fire-polishing.* Lay the tubing flat on the bench. Holding it down firmly with the left hand, make a single deep scratch across the tubing with the edge of a triangular file. (Press the file down firmly but not so much as to shatter the glass. Do not saw back and forth. If a single stroke does not produce a deep cut, your file is no good and should be replaced.) Holding the glass as shown in Figure A.17, with the thumbs opposite the scratch, pull and bend outward from the scratch. The glass should break cleanly with only moderate

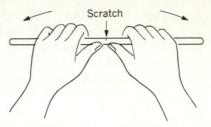

Figure A.17

pressure. The cut ends must now be fire-polished by being heated with a rotary motion in the edge of a burner flame. Continue heating until the sharp edges become rounded; do not heat so long that the end becomes constricted. It is wise to get into the habit of considering that the cutting operation is not finished until the fire-polishing is done.

2   *Constructing a nozzle.* Cut and fire-polish a foot length of tubing as directed above. When the polished ends are cool, heat the center of this tube in a normal burner flame, rotating the tube all the while to heat uniformly. Continue heating and rotating (Figure A.18) even after the heated portion softens. Do not pull the ends apart while the tube is in the flame; if anything, slightly push the ends together to thicken the molten glass. Remove from the flame, continue rotating for a few seconds, and then gradually pull the ends apart until the diameter of the hot portion is reduced to the dimension required. Place on a transite board until cool. Make a file scratch at the point chosen for the tip end of the nozzle, and break gently. Fire-polish the end, being very careful not to seal it. Cut and fire-polish the other end.

3   *Sealing a tube.* Suppose that you wish to make a very small test tube from glass tubing. Select a piece of tubing about 20 cm long and heat its center while rotating as described above

Figure A.18

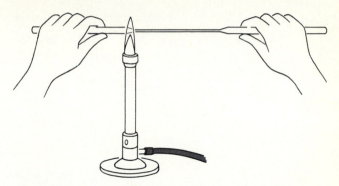

Figure A.19

in operation 2. When the glass is soft, remove it from the flame and quickly pull the ends apart to form a constriction. Heat the shoulder of the constriction (Figure A.19) very hot and immediately pull off the excess glass to the right. If you want to be fancy and make a rounded end on the tube, return the sealed end to the flame, heat it up while rotating, remove from the flame, and blow gently into the cool end. This operation may have to be repeated.

4   *Bending tubing.* For making a glass bend, it is necessary to heat a rather extended portion of the tubing. To do this, replace the normal burner tip (probably hot! use crucible tongs, and don't forget to turn off the gas!) with a flame spreader. Relight the burner. Heat the tubing at the place where you want to make the bend, while rotating the tube in the flame as above. Continue to heat and rotate until the glass is quite soft. This occurs well after the flame has taken a yellow coloration from the hot glass. Remove from the flame, and while gently pressing the glass against a transite board on your bench, bend the glass to the desired angle (Figure A.20). The purpose of

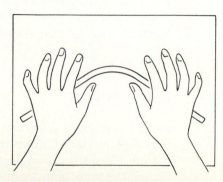

Figure A.20

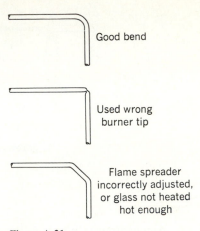

Good bend

Used wrong burner tip

Flame spreader incorrectly adjusted, or glass not heated hot enough

Figure A.21

the transite board is to ensure that the entire bend is in one plane. If you wish, you can draw the desired angle on the transite board before making the bend. Good and bad bends are illustrated in Figure A.21. Practice making right-angle bends until one satisfies you and your instructor. Don't be impatient; glassworking is an art.

5   *Constructing a wash bottle.* It is convenient to have a supply of distilled water at your bench. A handy container is a wash bottle such as shown in Figure A.22. The bends are numbered in the order in which they are made. Bend 1 has an interior angle of about 45°; bends 2 and 3, about 135°. Bend 2 can be made while bend 1 cools. Before making bend 3, hold the tube with bend 1 next to your 500-mL flask to estimate the position required for bend 3. In making bend 3, make sure that bend 1 lies flat against the transite board. See to it that all ends are fire-polished. Take great care when inserting bend 3 through the rubber stopper so that you do not injure yourself. Protect your hands with cheesecloth (Figure A.1), lubricate the glass and rubber well, and keep your hands close together. Have the final assembly approved by your instructor.

f. Centrifuge

Rapid separation of solids and liquids can also be done in a centrifuge, an example of which is shown in Figure A.23. The centrifuge consists of two parts: a fixed base that houses an electric motor and a freely rotating head that accommodates test tubes in an inclined position. When the head is rotated at

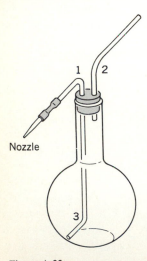

1   2

Nozzle

3

Figure A.22

Figure A.23

high speed, the denser phase in the test tubes (usually the solid phase) separates to the part of the test tube farthest from the axis of rotation. The liquid phase, which corresponds to the filtrate, can then be poured off, or it can be sucked up into a collector, e.g., medicine dropper or pipette. Precipitates can be washed by adding wash solution to the solid, stirring the mixture with a glass rod, and repeating the centrifugation.

When operating a centrifuge, the following rules should be observed:

1   The weight load in the rotating head must be symmetrically distributed about the axis of rotation. If you put your mixture in one slot, you must place an identical test tube with an approximately equal volume of water in the slot diametrically opposite.

2   Allow the centrifuge to rotate about one minute for most efficient separation. The optimum time is shorter for centrifuges rotating faster than 1700 rpm.

■ CAUTION

Keep your hand away from the top of the centrifuge while it is rotating.

3   Allow the rotating head to come to a full stop before trying to remove your sample. You can help slow the rotation by gently pressing your hands to the sloping sides of the cone. Do not brake too hard, however, because if you do you will stir up the sediment.

g. Melting Point

If a material is *pure*, melting occurs over a very narrow temperature range, and the solid is said to have a "sharp" melting point. This melting point is of considerable value in identifying a compound, because it is unlikely that other compounds will melt over exactly the same temperature range. An *impure* sample of the same compound melts at a lower temperature and over a wider temperature range due to the well-known phenomenon of freezing-point depression of a solvent by a solute.

A melting-point apparatus can be constructed as follows: A heating bath is made of a large test tube supported in a clamp and containing mineral oil to a height of 5 cm. A stopper assembly, such as the one shown in Figure E.37.1 on page 220, contains a stirrer and a 250° thermometer.

Introduce a *very small* quantity of powdered sample into a melting-point capillary tube by pressing the open end of the tube vertically into the solid. Gently tap the tube on a hard surface to bring the solid to the sealed end of the tube, and repeat the process until a column of approximately 1 cm has been introduced.

Attach the melting-point tube to the thermometer using a small rubber band near the top of the tube. Position the tube along the stem so that the column of solid is adjacent to the thermometer bulb. Insert the thermometer into the stopper groove, and carefully place the assembly in the heating bath with the bulb well immersed in the oil. (*Note:* Water must be kept out of the oil bath; even one drop will lead to serious spattering when the test tube is heated.)

Heat the test tube with a *small* gas flame. If the melting point is not known, a rapid preliminary heating that overruns the melting point gives a clue to the correct value. After the bath cools slightly (remove the burner and cool by fanning), repeat the determination with a *fresh* sample and tube. This time, heat until the temperature is within $15°$ of the approximate melting point. Continue heating intermittently so the temperature rise is no more than $3°$ per min near the melting point. Stir while heating.

h. Spectrophotometers

A *spectrophotometer* is an instrument for measuring the fraction of an incident beam of light that is transmitted by a sample at each wavelength. In general, spectrophotometers consist of the five basic components shown in the block diagram of Figure A.24. A light *source*, which may be a tungsten lamp or a hydrogen discharge tube, radiates light over a broad range of wavelengths. The *monochromator*, which may be a grating or a prism, selects a narrow band of these wavelengths and sends it at an intensity I_0 to impinge on a sample in the *sample cell*. Absorption may occur here, in which case the intensity I coming out of the cell is less than what went in. A *detector*, such as a phototube, measures I and sends a proportional signal to an indicating device that tells on a *meter* how much light has been transmitted or, conversely, absorbed. Mirrors and lenses are used to regulate the path of the light through these various components.

Absorption spectra can be presented in various ways. One is to specify the *percentage transmittance* at each wavelength. This is defined by

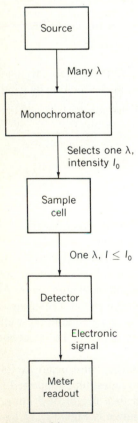

Figure A.24

$$T = 100 \times \frac{I}{I_0}$$

where T is the percentage transmittance, I_0 is the intensity of the incident beam on the sample, and I is the intensity transmitted by the sample. If the sample is a solution and is contained in a cell, then I is the intensity of light transmitted by the cell when it contains solution and I_0 is the intensity when the cell is filled with solvent alone. Another way of describing an absorption spectrum is in terms of *absorbance*. This is defined by

$$A = \log \frac{I_0}{I}$$

where A is the absorbance and $\log (I_0/I)$ is sometimes known as the *optical density*. A completely transparent sample has $T = 100$ percent and $A = 0$, whereas a completely absorbing sample has $T = 0$ percent and $A = \infty$. Absorbance is related to the path length d of the sample and the concentration c of the absorbing molecules by what is called *Beer's law:*

$$A = Ecd$$

where E is the *absorptivity* or *extinction coefficient*. When Beer's law is obeyed, there is a linear relation between the absorbance and the concentration. For analytical applications of spectrophotometry, however, one must check the validity of Beer's law for the particular system considered, since deviations occasionally occur, in which case a calibration curve of absorbance vs. concentration will be required.

The Spectronic 20 is a small, single-beam spectrophotometer for measuring light absorption in liquid samples. It generally operates over the wavelength range of 375 to 650 nm, but by simple addition of a red filter and exchange of phototubes the range can be extended to 950 nm. Figure A.25 shows a schematic of the various components. (The Spectronic 21 is an instrument similar to the Spectronic 20, except that it has a digital readout, a solid-state photodetector, and a range of 340 to 1000 nm.) White light from a lamp is focused by lens A onto the entrance slit. Lens B collects the light and sends it to a grating, where it is reflected and dispersed into its various wavelengths. To obtain the various wavelengths the grating is rotated by means of an arm that is positioned by a cam attached to the same shaft as the wavelength-control knob. Selected light of a single wavelength goes through the exit slit, passes through the sample, and finally falls on the sensing surface of a phototube. The phototube output is read

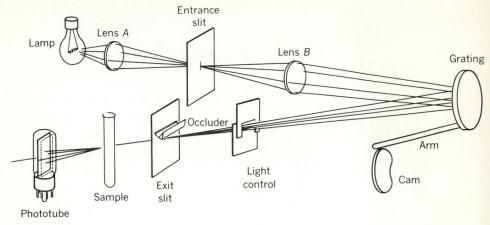

Figure A.25

directly either as optical density or as percentage transmittance on the double scale of a meter on the face of the instrument. When there is no sample in place, an occluder (opaque screen) intercepts the light beam so zero transmittance may be set. A light-intensity control is provided to permit setting the meter at full-scale deflection when there is a reference cell in the sample compartment.

Procedure

Operating Instructions for Spectronic 20   (Refer to Figure A.26.)

1   Turn on by rotating power-control knob clockwise. Allow 5 min for instrument to warm up. Adjust meter to 0 percent transmittance by turning the zero-adjust knob. Select the desired wavelength by adjusting the wavelength control. During these operations no sample should be in the holder, and the cover should be closed.

2   To standardize the instrument, use the pure solvent as a reference. Fill the special test tube (or cuvette) about halfway with liquid; insert it in the sample holder so that the line on the test tube is aligned with that of the holder. This is to make sure that the test tube is always placed in the holder in exactly the same position so the light path through the tube is exactly the same. Close the holder cover and adjust the light-control knob so zero absorbance (100 percent transmittance) is read on the scale.

3   To measure the absorbance of a solution, fill an identical tube

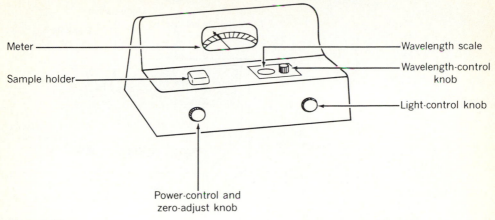

Meter

Sample holder

Wavelength scale

Wavelength-control knob

Light-control knob

Power-control and
zero-adjust knob

Figure A.26

(or the other side of the double cuvette) halfway with the solution, and place it in the sample holder. Again match up the line on the tube with that on the holder. Read the absorbance (or percentage transmittance) directly from the scale.

4   To change to a new wavelength, remove the sample from the holder. Turn the light control counterclockwise. (This protects the phototube, which may be more sensitive to the new wavelength.) Repeat standardization at the new wavelength, since the solvent may absorb more or less light. Now measure the absorbance of the sample. Be careful not to get fingerprints on the test tubes or cuvettes, and after each sample rinse two or three times with the new sample to be used.

Operating Instructions for Spectronic 21—Model DV   (Refer to Figure A.27.)

1   Turn on with the power switch in the lower right corner. Allow 30 min for instrument to warm up. Set the mode switch to absorption. Set the wavelength selector to the wavelength of maximum absorption for the substance being measured. Set the sensitivity to low.

2   To standardize the instrument, use a blank (solution containing solvents and all other substances in the same proportions as in the sample solution but without the substance being measured). Rinse the special test tube (or cuvette) several times with distilled water and then with the blank solution. Fill the cuvette about halfway with blank solution. Wipe all fingerprints and dirt off the outside of the cuvette, rinse with distilled water, and wipe dry with a soft tissue (e.g., Kim-Wipe). Insert the cuvette into the sample holder so that the line on the test tube is aligned with that of the holder. Close the holder

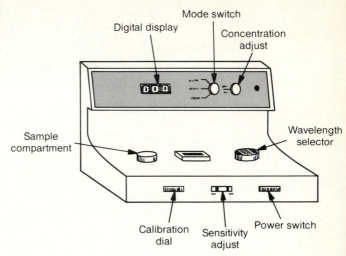

Figure A.27

cover, and turn the calibration dial until the digital display reads zero absorbance. If zero cannot be reached, increase sensitivity to medium and try again.

3   To measure the absorbance of a solution, use a cuvette identical to that used for the blank. Clean, rinse with distilled water, and dry the cuvette with a soft tissue (do *not* use paper towel). Rinse the cuvette with one or more portions of the sample, fill it halfway with sample, and align the tube in the sample holder so that the lines match up as before. Close the cover. Read the absorbance from the digital display. If more than one sample is being studied, it is advisable to check with the blank and check for any drift in the zero value between samples.

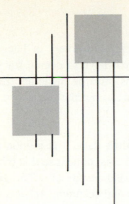

3

Recording and Treating Data

Research chemists consider their notebook one of their most valuable possessions. It summarizes the work they have done and the results they have obtained. Loss of a notebook can be catastrophic. Furthermore, the notebook record should be kept so that it will be meaningful at a later date both to its author and to other chemists.

Data should be recorded as soon as they are obtained. Consult your instructor about choosing between pen or pencil. Writing should be concise and legible. Qualitative data, which are just as important as quantitative data, should not be omitted. A few words, perhaps with a sketch, will often suffice. In recording quantitative data, make sure the numbers are labeled with both units and property being measured. Pay strict attention to significant figures so that the numbers give maximum information.

In calculating results from experimental data, it is good practice to show specifically the method of calculation. This can be done either by giving a sample calculation using numbers labeled with units or by writing out a general expression indicating how the experimental data are mathematically combined. For example, in an experiment to determine density, we can write either

$$\text{Density of unknown} = \frac{27.1 \text{ g}}{22.6 \text{ mL}} = 1.20 \text{ g/mL}$$

or

$$\text{Density of unknown} = \frac{\text{mass of unknown}}{\text{volume of unknown}}$$

In neither case is it useful to show the detailed arithmetic, which is better done on scratch paper or in some inconspicuous place in the record book.

You will note that for the earlier experiments in this manual labeled blanks are provided to indicate the particular data needed for the experiment. However, in later experiments this is not done. You will have to decide for yourself what experimental observations must be made and recorded. If in doubt, consult your instructor in order not to overlook an essential measurement.

Reliability of Data   In recording data, it is a good practice to give the uncertainty of each measurement so that others will know how reliable your data and results are. The aim is to have data that are both precise and accurate. *Accuracy* means closeness of a measurement to the true value; *precision* means closeness of approach of a number of repeated measurements to a common value. To get high precision, measuring instruments are designed so that they can pinpoint measurements to within a very small range. Suppose that you have an analytical balance with a readability of 0.001 g. This means that the balance can be read to within ±0.0005 g. A reading of 5.142 g would mean that the mass was measured to be somewhere between 5.1415 and 5.1425 g, that is, 5.142 ± 0.0005 g. Just because the balance is precise, however, doesn't mean that it is accurate. To be accurate, the balance must be calibrated, that is, adjusted so that the readings are true. This is done by weighing standard masses that are known to be accurate to many decimal places. An analytical balance that has been properly calibrated will give readings that are not only precise but also accurate to ±0.0005 g.

Estimating Precision of Data   It is often the case that the precision of a measurement is unknown. A common way to estimate precision is to make several measurements of the same quantity and then find an *average deviation* of the measurements. To do this, you first find an average, or mean, value by adding up the measurements and dividing by the number of measurements. Then you go back to each item of data and compare it with the mean value. Without caring about whether the mean is larger or the item of data is larger, you calculate an absolute difference between each item of data and the mean. These differences are called *absolute deviations* because they have no plus or minus signs. To get the average deviation, you add up the absolute deviations and divide the total by the number of measurements. The result is expressed by giving the mean plus or minus the average deviation.

For example, suppose that you have measured the mass of a stopper five times on a triple-beam balance and got the

Trial	Measurement, g
1	2.00
2	2.03
3	2.02
4	2.01
5	2.04
Total	10.10

$$\text{Average (mean)} = \frac{10.10 \text{ g}}{5}$$
$$= 2.02 \text{ g}$$

following values: 2.00, 2.03, 2.02, 2.01, and 2.04 g. First you would calculate an average of the five measurements. The average, or mean, value would be 2.02 g. Then you would find the absolute difference between 2.02 g and each of the individual measurements. Next you would add up the absolute deviations and divide by 5 to get the average deviation.

Trial	Measurement, g	Absolute deviation
1	2.00	0.02
2	2.03	0.01
3	2.02	0.00
4	2.01	0.01
5	2.04	0.02
Total		0.06

$$\text{Average deviation} = \frac{0.06}{5} = 0.01$$

The result would be expressed as 2.02 ± 0.01 g.

Standard Deviation   A more reliable estimate of precision of data than average deviation is standard deviation. The main advantage of standard deviation is that the number of measurements N is taken into consideration in estimating the precision. The procedure given here is for calculating the *standard deviation s* for a finite number of measurements of a quantity the true value of which is unknown. This is the usual case in scientific work.

The first two steps in finding the standard deviation are the same as those used to find the average deviation. First you calculate an average, and then you find the absolute deviation of each of the measurements. Third, you square each of the absolute deviations. Fourth, you add up the squared absolute deviations and get a total. In the fifth step, you divide this total by $N - 1$. This gives a value called the *variance*. To get the standard deviation s, you take the square root of the variance. Suppose that you want to find the standard deviation for the same set of five ($N = 5$) measurements of the mass of a stopper that we used to illustrate how to calculate average deviation, i.e., 2.00, 2.03, 2.02, 2.01, and 2.04 g.

For the first two steps you calculate an average (2.02 g) and find the absolute deviations. Then, as shown below, you square each absolute deviation, add up the squared deviations, and divide the total by $N - 1$ to get the variance.

Trial	Measurement, g	Absolute deviation	Squared absolute deviation
1	2.00	0.02	0.0004
2	2.03	0.01	0.0001
3	2.02	0.00	0.0000
4	2.01	0.01	0.0001
5	2.04	0.02	0.0004
Total			0.0010

$$\text{Variance} = \frac{0.0010}{5 - 1} = \frac{0.0010}{4} = 0.00025$$

The standard deviation is the square root of the variance,

$$s = \sqrt{0.00025} = 0.016$$

The result would be expressed as 2.02, with a standard deviation of 0.016.

Propagation of Error   It is very likely that in an experiment you will have to combine several items of data mathematically. When you do this, the uncertainty in your result is based on the precision of each of the items of raw data.

In addition or subtraction, you simply add the uncertainties, so that the precision of the total is the sum of the individual uncertainties. For example, consider the following subtraction to give the mass of the contents of a crucible:

Crucible + contents	85.235 ± 0.0005 g
Empty crucible	26.202 ± 0.0005 g
Contents	59.033 ± 0.001 g

The result is precise to 0.002 g, or ± 0.001 g.

In multiplying data, you calculate a maximum and a minimum product based on the uncertainties in the data. The precision of the product is equal to the difference between the maximum and minimum products. For example, suppose that you want to find the precision of an area calculated from the following measurements:

Length = 8.422 ± 0.0008 cm
Width = 1.071 ± 0.0005 cm

Maximum area = $(8.4228 \text{ cm})(1.0715 \text{ cm}) = 9.0250 \text{ cm}^2$
Minimum area = $(8.4212 \text{ cm})(1.0705 \text{ cm}) = 9.0149 \text{ cm}^2$
Difference between maximum and minimum = 0.010 cm^2

The precision associated with the area would be 0.010, or ± 0.005 cm^2. You would express the result as the average of the maximum and minimum values, together with the uncertainty. In this case, the average would be 18.040 cm^2/2 = 9.020 cm^2, so the result would be given as 9.020 $\pm$ 0.005 cm^2.

In dividing items of data, the procedure is very similar to that used for multiplication, except that instead of products you find maximum and minimum quotients. The precision of the result is the difference between the maximum and minimum quotients.

Relative Error   Sometimes an experiment is designed to measure a quantity whose true value is known. The usual practice in such cases is to express the accuracy of your measurement as *relative error*, which is simply the difference between your value and the true value, divided by the true value.

Consider, for example, an experiment in which you were supposed to find the atomic weight of aluminum, which is known to be 26.98154 g per mol. Let's say your result turned out to be 26.92 g per mol.

The absolute error would be

$$(26.98154 - 26.92) \text{ g/mol} = 0.06 \text{ g/mol}$$

$$\text{Relative error} = \frac{\text{absolute error}}{\text{true value}} = \frac{0.06 \text{ g/mol}}{26.98154 \text{ g/mol}} = 0.002$$

Note that there are no units for relative error. It is usually expressed as percent (%), parts per thousand (ppt), or parts per million (ppm). A relative error of 0.002 could be given as

$$0.002 \times 100\% = 0.2\%$$

$$0.002 \times 1000 \text{ ppt} = 2 \text{ ppt}$$

$$0.002 \times 1,000,000 \text{ ppm} = 2000 \text{ ppm}$$

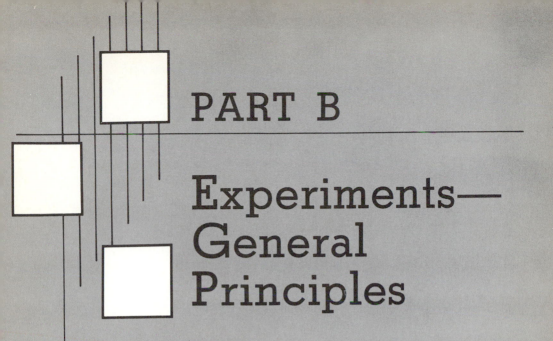

PART B

Experiments— General Principles

Included in this section of the manual are experiments representing physical chemistry, analytical chemistry, and inorganic chemistry.

Chemistry is an experimental science; that is, it is founded on empirical evidence gathered in the laboratory. One can learn a great deal *about* chemistry through reading or by watching others, but one cannot truly *understand* or appreciate chemistry without performing experiments.

Observe carefully and completely, and record what *actually* happens, not what you think *should* happen. Prepare well beforehand, so that your time spent working in the laboratory is most productive. Before anything else, practice *safety*. Know what you are doing, and observe every precaution.

Experiment E.1

Density

Special Items

Ruler calibrated in
  centimeters
Unknown specimens
Two 20-mL samples of
  unknown solutions
Burette (if not included in
  normal desk equipment)

TABLE E.1.1 Densities of some common metals, g/cm³

Aluminum	2.7
Lead	11.4
Magnesium	1.8
Monel metal (alloy of Ni, Cu, Fe)	8.9
Steel (Fe, 1% C)	7.8
Tin	7.3
Wood's metal (Bi, Pb, Cd, Sn)	9.7
Zinc	7.1

TABLE E.1.2 Volume formulas for regular shapes

Cube	$V = l^3$
Right-parallelepiped	$V = lwh$
Sphere	$V = \frac{4}{3}\pi r^3$
Cylinder	$V = \pi r^2 h$
Cone	$V = \frac{\pi}{3} r^2 h$
Tetrahedron	$V = 0.11785 l^3$
Octahedron	$V = 0.4714 l^3$

Density of Solids

The *density* of a substance is defined as its mass per unit volume. The most obvious way to determine the density of a solid is to weigh a sample of the solid and then find out the volume that the sample occupies. In this experiment, you will be supplied with variously shaped pieces of metal. You are asked to determine the density of each specimen and then by comparison with a table of known densities to identify the metal in each specimen. As shown in Table E.1.1, density is a characteristic property. Table E.1.2 gives the formulas for calculating volumes of various shapes.

Density of Liquids

The density (mass per unit volume) of liquids can be determined like that of solids by measuring both the mass and the volume of a given sample. Density is a characteristic property of a substance; it remains fixed unless the temperature or pressure is changed. For liquids, a relatively small change in temperature can affect the density appreciably, but a pressure change must be quite great to have a measurable effect. In this experiment, you will determine the density of liquid water. Once this density and its change with temperature are known, you can use the information to find out what volume should be occupied by a known mass of water at a given temperature.

Calibration is the process by which a stated measure, such as a mass or a volume, is checked for accuracy. In the second part of this experiment, you are asked to calibrate a burette. A precisely measured volume of water at a known temperature is drawn from a burette and weighed. By use of its mass and its density (from Table E.1.3), its volume can be calculated and compared with that measured by the graduations on the burette. The density data in the table are the most accurate available; the weights you get are quite reliable.

TABLE E.1.3   Density of water at various temperatures*

Temp., °C	Density, g/mL	Temp., °C	Density, g/mL
15	0.9979	26	0.9959
16	0.9978	27	0.9957
17	0.9977	28	0.9955
18	0.9975	29	0.9952
19	0.9973	30	0.9949
20	0.9972	31	0.9946
21	0.9970	32	0.9944
22	0.9968	33	0.9941
23	0.9966	34	0.9938
24	0.9964	35	0.9935
25	0.9962		

*When weighed in air, an object is buoyed up by the weight of air displaced. The values in this table have been corrected for this effect.

Therefore, only the volume markings on the burette are in question.

Procedure

(a)   Obtain an unknown specimen from your instructor. Using the procedure described in Section 2 of Part A of this manual, weigh the sample accurately on an analytical balance. Pay attention to the precautions discussed there.

■ **WEAR YOUR SAFETY GOGGLES**

Determine the volume of your specimen by measuring appropriate dimensions. For example, for a cylindrical sample, measure the diameter and the length of the cylinder. Calculate the volume of the sample.

Determine the volume of your specimen directly by carefully sliding the specimen into a graduated cylinder containing a known volume of water. Make sure that the specimen is completely covered with water and that no air bubbles are trapped. Note the total volume of the water and specimen.

Repeat with other unknowns as directed by your instructor.

(b)   Thoroughly clean your graduated cylinder. Add about 20 mL of distilled water. Record the volume to the nearest 0.1 mL. Weigh the cylinder plus water to the nearest 0.1 g on the platform balance, and record. Measure the temperature of the water. Pour out the water, hold the cylinder inverted for at least 10 sec, and reweigh. Calculate the density.

■ **WEAR YOUR SAFETY GOGGLES**

In a similar fashion, determine the density of each un-known provided.

(c)   Thoroughly clean your gas-measuring tube or burette, follow-ing the directions given in Part A, Section 2. Fill the tube with distilled water. Weigh your 125-mL flask with a rubber stop-per on an analytical balance to the nearest 0.001 g. Measure the temperature of the water in the burette. Into a beaker drain out enough of the water so that the meniscus comes into the graduated range (make sure all air is out of the tip). Record the burette reading to the nearest 0.02 mL. Into the weighed flask drain about 40 mL of water. Stopper immediately, re-cord the burette reading, and weigh the stoppered flask plus water. Calculate the volume of water from its mass and table-given density. Compute a correction factor for the burette, a factor that can be used in the future to get true volume by multiplying the volume read by the factor.

(d)   In clean dry containers, obtain approximately 25 mL of each of four different sodium chloride solutions: 4, 8, 12, and 16% NaCl by weight. Rinse your graduated cylinder, used in part (b) above, with about 5 mL of the 4% solution, and then pour about 20 mL of the 4% solution into the graduated cylinder. Using the procedure described in part (b), determine the den-sity. Then determine the density of each of the other "known" solutions. Obtain a solution of unknown concentration of NaCl from your instructor and determine its density. Plot a graph showing density (on the vertical axis, the ordinate) vs. con-centration of NaCl (on the horizontal axis, the abscissa). Using the graph, obtain a value for the concentration of NaCl in your unknown and report it to your instructor.

Data

(a)

Unknown:	1	2	3
Mass of specimen	_____	_____	_____
Dimensions of specimen (label the length, diameter, etc., of each)	_____	_____	_____
	_____	_____	_____
	_____	_____	_____
Volume of water in cylinder	_____	_____	_____
Volume of water plus specimen	_____	_____	_____

(b)   Mass of graduated cylinder plus water   _____

Mass of graduated cylinder   _____

Volume of water in cylinder   _____

Temperature of water   _____

Mass of graduated cylinder plus
unknown   _____   _____

Volume of unknown in cylinder   _____   _____

Temperature of unknown   _____   _____

(c)   Mass of flask plus water   _____

Mass of flask   _____

Initial burette reading   _____

Final burette reading   _____

Temperature of water   _____

(d)

	1	2	3	4
Mass of graduated cylinder plus known NaCl solution	_____	_____	_____	_____
Volume of NaCl solution in cylinder	_____	_____	_____	_____
Mass of graduated cylinder plus unknown NaCl solution	_____			
Volume of unknown NaCl solution	_____			

Results

(a)   Calculated volume of
specimen   _____   _____   _____

Density (calculated from
dimensions)   _____   _____   _____

Measured volume of
specimen   _____   _____   _____

Density (calculated from
measured volume)   _____   _____   _____

Identity of specimen   _____   _____   _____

(b)   Density of water   _____

Density of unknown   _____   _____

(c)   Measured volume of water   _____

Calculated volume of water   _____

Correction factor for burette (calculated
volume per measured volume)   _____

(d)   1   2   3   4

Density of known
NaCl solution   _____ _____ _____ _____

Density of unknown NaCl solution   _____

Questions

E.1.1   Which method used in part (a) of this experiment—the one in
which the dimensions of the specimen are measured or the one
in which the volume is measured by water displacement—
should give the more precise estimate of volume?

E.1.2   How can you establish the precision of your different measurements in part (a)?

E.1.3   Design and sketch an apparatus that would permit more precise determinations of the volume of a solid by water displacement than you were able to obtain using a graduated cylinder.

E.1.4   Find the minimum error in density that would result in part
(a) if the analytical balance were inaccurate by 0.005 g for a
mass of 1.525 g and a volume of *exactly* 9.420 mL.

E.1.5   Obtain from your classmates as many values that you can for
the density of one of the solids (e.g., all the samples of aluminum) you analyzed. Determine an average (mean) value for
the density, and calculate an average deviation.

E.1.6   In part (b) of this experiment, you determined the density of
water using your graduated cylinder. Calculate the percent error in your determination by comparing your value with the
value given in Table E.1.3.

E.1.7   In part (c) of this experiment, suppose that you determined
the mass of 41.20 ± 0.01 mL of water at 24°C to be 41.052
± 0.0005 g. Calculate the density and its uncertainty.

E.1.8   Using information from Table E.1.3, find the minimum percent error in the density of water introduced by reading the
temperature as 20°C when it is actually 30°C.

E.1.9   In part (d) of this experiment, you are directed to pour "about
20 mL" of each solution into your graduated cylinder. What
advantage might there be in taking a more precise, constant
amount, such as 20.0 ± 0.1 mL, of each solution?

E.1.10   In part (d), what density value would you deduce for the solution with zero percent NaCl from the densities you actually
measured?

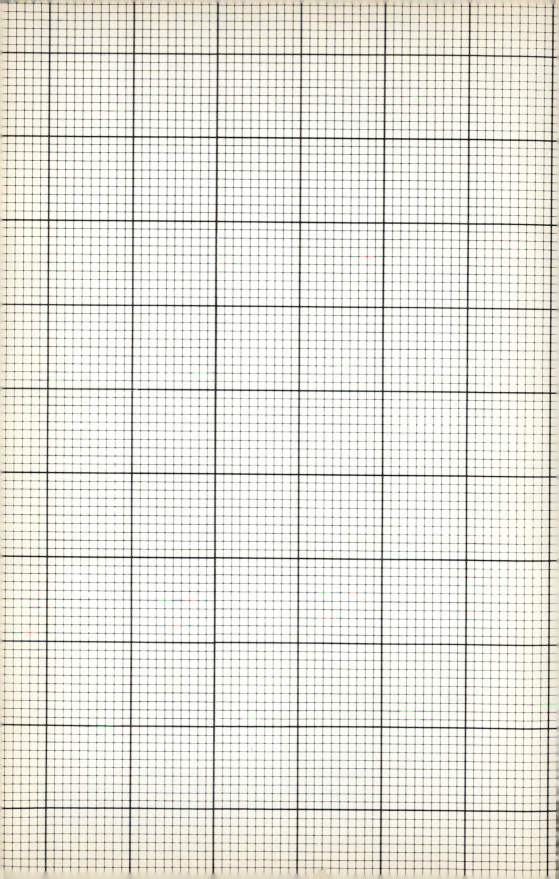

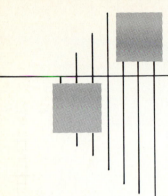

Experiment E.2

Conservation of Matter

Special Items

Two small vials with stoppers and labels
10 mL Na_2CO_3 soln (124 g $Na_2CO_3 \cdot H_2O$ per liter)
3 mL $CaCl_2$ soln (111 g $CaCl_2$ per liter)
3 mL 3 M H_2SO_4

In the study of mass changes in chemical reactions, it is of fundamental importance to find out if the total mass present is the same before and after reaction. To find out if mass is conserved, one could weigh the reactants and then weigh the products after reaction is finished. Needless to say, for such an experiment to be valid, *all* the reactants and *all* the products must be weighed.

In this experiment, you will study first the reaction between solutions of calcium chloride and sodium carbonate and then the reaction between the product obtained (calcium carbonate) and sulfuric acid.

Procedure

■ **WEAR YOUR SAFETY GOGGLES**

Into a 125-mL flask pour 10 mL of the solution of sodium carbonate. Stopper the flask. Pour 3 mL of the solution of calcium chloride into a small labeled vial; similarly, pour 3 mL of dilute sulfuric acid into a second labeled vial. Stopper both. Make sure that the exteriors of the flask and vials are dry. Place all three simultaneously on an analytical balance, and weigh to the nearest 0.001 g. Follow the directions and precautions given in Section 2 of Part A.

Removing the flask from the balance case to protect the balance, carefully pour the solution of calcium chloride into the flask. Swirl gently to mix. Note any changes that occur. Restopper the vial and flask, and set them back on the pan. Redetermine the total mass.

Again remove the flask from the balance case, and this time add the sulfuric acid carefully. With stopper off, swirl and note any evidence of reaction. Continue to swirl until all evidence of reaction has disappeared. If the flask is not at room temperature (which you can determine by touching it with the inside of your wrist), wait a few minutes until it is. Restopper the flask and vial, and redetermine the total mass.

Data

Total mass before first mixing _____

Total mass after first mixing _____

Total mass after second mixing _____

Results

What do your data indicate about the mass change in a chemical reaction?

Questions

E.2.1   Describe any changes of state (solid, liquid, gas, solution) you observed in the first reaction and in the second reaction. What evidence was there to show that a reaction was going on in each case?

E.2.2   Why are you directed in the second reaction to swirl the solution with the stopper off?

E.2.3   Suppose that you had by mistake added the sulfuric acid along with the calcium chloride in the first reaction. What do you think would have happened?

E.2.4   Why at the end of the experiment do you need to wait for the temperature of the flask to come back to room temperature?

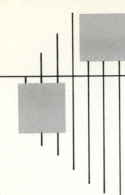

Experiment E.3

Definite Composition

Special Items

13-cm length of No. 18
  copper wire
10 g powdered sulfur

When a chemical reaction occurs, the total mass of material at the end of the reaction is observed to be equal to the total mass of material at the beginning of the reaction. Such is the gist of the law of conservation of matter. However, this does not mean that when element A is allowed to react with element B, all A and all B must necessarily react chemically with each other. If A and B react to form a compound, then the mass of A used up plus the mass of B used up is equal to the mass of compound formed; but there may be some A or some B left unreacted. In other words, there may be a limitation to the amount of one element that can be chemically combined with a given mass of another. In this experiment, we determine the maximum mass of sulfur that under specified conditions combines with a given mass of copper.

A weighed piece of copper is heated with some sulfur. Copper and sulfur react, and any excess sulfur burns off. The gain in mass allows us to calculate the mass of sulfur that has combined with the copper. More sulfur is added, and the heating is repeated to see how much additional sulfur will combine.

Procedure

**■ WEAR YOUR
SAFETY
GOGGLES**

Coil up a length of copper wire weighing about 1 g. Weigh it on an analytical balance to the nearest 0.001 g. Place the coil of copper in a porcelain crucible, and cover it completely with sulfur. Set the crucible on a triangle, and support it *in the hood* as shown in Figure E.3.1. Heat gently with the lid on the crucible. A light-blue flame around the edge of the lid is due to burning sulfur. When sulfur burns, irritating fumes of sulfur dioxide are produced—hence the necessity of using a hood. When the light-blue flame has died down, step up the heating

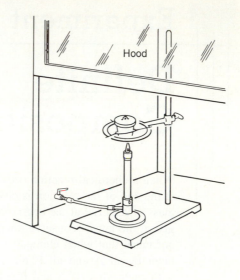

Figure E.3.1

by lowering the ring closer to the burner. Heat with full intensity for about 5 min. Cool to room temperature. Carry the crucible and its contents, held over a transite board, together with the crucible tongs to the analytical balance. Working gently so as not to crumble it, lift out the product with the crucible tongs, and place it on the balance pan. Weigh to the nearest 0.001 g.

Replace the product in the crucible, cover it with sulfur again, and repeat the heating in the hood as before. Reweigh the product.

Data

Mass of copper _____

Mass of product after first heating _____

Mass of product after second heating _____

Results

Calculate for both parts of the experiment the mass of sulfur that has combined with the fixed mass of copper. Calculate also the percentage of sulfur in the product after the first and after the second heating.

Questions

E.3.1.   Estimate the uncertainty (precision) in each item of data and in the results.

E.3.2.   Calculate the percentage of sulfur in (a) pure CuS, (b) pure Cu_2S. How do these values compare with your experimental result?

E.3.3.   Explain how each of the following would affect your calculated value for the percentage of sulfur in the product:

(a) Some of the weighed copper reacted with oxygen to give CuO.

(b) Some of the weighed copper reacted with oxygen to give Cu_2O.

(c) Some of the copper did not react but remained as the metal.

(d) Some of the wire assumed to be pure copper had oxidized to CuO before it was weighed as pure copper.

E.3.4.   There is some evidence that the compound produced in this experiment is somewhat deficient in copper from the value expected. What effect would this have on the percentage of sulfur in the product?

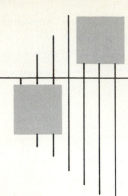

Experiment E.4

Heat and Temperature

Special Items

Boiling chips
Laboratory barometer
String
15-cm strip of nichrome wire
10-g piece of metal rod

Heat differs from temperature in that heat is a quantity of energy whereas temperature is the intensity of heat (or hotness) within a body. The *specific heat* of a substance is the amount of heat needed to raise the temperature of 1 g of the substance 1°C. In this experiment you will investigate (*a*) the calibration of a thermometer as a temperature-measuring device, (*b*) the relative temperatures of different portions of a gas burner flame, and (*c*) the determination of the specific heat of a given metal.

The thermometer can be calibrated by immersing it first in melting ice (0°C) and then in boiling water (100°C at standard atmospheric pressure). Since the boiling point of water varies with changing barometric pressure, Table E.4.1 will enable you to find the true boiling point of water under the conditions of the experiment.

Because the ordinary mercury thermometer cannot measure the high temperatures that exist in burner flames, you will have to use indirect methods to find out the temperatures of different portions of your flame. One possible method is to place a high-melting solid in various parts of the flame and note the color of the glowing solid. An approximate color scale of temperature is given in Table E.4.2.

An alternative method is to select solids of known melting point and to note in what regions of the flame they melt. Solids suitable for exploring a gas-burner flame are given in Table E.4.3.

To determine the specific heat of a metal, you will heat a weighed amount of metal to a known temperature (that of boiling water) and then place the hot metal in a known amount of water. From the temperature rise of the water, you can calculate the amount of heat transferred from the metal to the water. This can be done directly because it takes 4.2 joules (or 1 calorie) to raise the temperature of 1 g of water by 1°C. Since the density of water is almost exactly 1 g per mL (see Table E.4.1), 1 g of water is very nearly 1 mL. The specific heat of the metal is equal to the amount of heat liberated by the metal divided by its temperature drop times its mass.

TABLE E.4.1 Boiling point of water at various pressures

| Pressure | | Bp, |
mmHg	atm	°C
700	0.921	97.7
705	0.928	97.9
710	0.934	98.1
715	0.941	98.3
720	0.947	98.5
725	0.954	98.7
730	0.961	98.9
735	0.967	99.1
740	0.974	99.3
745	0.980	99.4
750	0.987	99.6
755	0.993	99.8
760	1.000	100.0
765	1.007	100.2
770	1.013	100.4

TABLE E.4.2 Color Scale of Temperature	
Incipient red	500
Dark red	700
Bright red	900
Orange	1100
Incipient white	1300
White	1500

TABLE E.4.3 Melting Points of Solids	
Solid	Mp, °C
Barium nitrate	592
Potassium iodide	723
Sodium chloride	801
Sodium carbonate	851
Brass	940
Copper	1083

Procedure

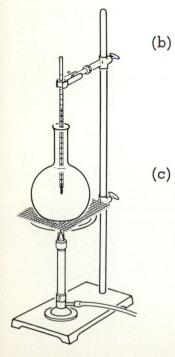

■ WEAR YOUR SAFETY GOGGLES

Figure E.4.1

(a) Wash off some crushed ice, and place it in a small beaker. Add distilled water until the ice is nearly covered. Immerse your thermometer in the ice-water mixture as deep as possible; stir gently; record the thermometer reading when it has become constant.

Place about 100 mL of distilled water in your 500-mL flask. Drop in a boiling chip to prevent bumping during boiling. Insert your thermometer into a stopper (lubricate well!) and clamp it just above the water level as shown in Figure E.4.1. Heat the water to boiling, and record the thermometer reading when it has become constant. Record the barometric pressure that you read from the laboratory barometer.

(b) Get a piece of nichrome wire about 15 cm long, and fashion the end of it into a tiny loop. Hold the loop in various portions of the burner flame, and note the color. Record your observations. Repeat your exploration of the flame by placing a tiny sample of each available solid (from Table E.4.1) on the wire loop and holding it in the flame. Note whether and where the solids melt. Record.

(c) Obtain a piece of metal (about 10 g) for which the specific heat is to be found. Weigh it on the platform balance to the nearest 0.1 g. Tie it with a string, and suspend it above boiling water in a flask as the thermometer is set in Figure E.4.1. Do not let the metal get wet. (If water condenses on the metal, pull the metal out, dry, and replace.) Allow sufficient time for the metal to reach the temperature of the vapor from boiling water. Record the barometric pressure. With a graduated cylinder, measure 10.0 mL of water into a test tube. Measure the temperature of the water to the nearest 0.1°. Withdraw the thermometer, and hold it directly over the test tube so that any drops of water will drain back into the test tube. Quickly remove the heated metal, and carefully lower it into the water in the test tube. Stir by raising and lowering the metal. Return the thermometer, and note the maximum temperature registered.

Data

(a)   Thermometer reading in ice-water mixture   _____

Thermometer reading at water boiling point   _____

Barometric pressure   _____

(b)   Map of burner flame, showing colors of glowing wire:

Map of burner flame, showing observed melting behavior of various solids in different portions of flame:

(c) Mass of metal   _____

   Barometric pressure   _____

   Volume of water in test tube   _____

   Initial temperature of water in test tube   _____

   Final temperature of water and metal   _____

Results

(a) Compare your thermometer reading at the ice point with the standard value.

   From the barometric pressure and Table E.4.1, find the boiling point of water under the prevailing conditions. Compare your thermometer reading with this value.

(b) Draw a map of the burner flame, indicating the approximate temperature at various points. Show also the range of uncertainty for each temperature.

(c) Temperature rise of water   _____

Mass of water   _____

Heat absorbed by water   _____

Heat lost by metal   _____

Temperature drop of metal   _____

Specific heat of metal (heat lost per gram per degree)   _____

Questions

E.4.1   Suppose that you had a mercury-in-glass thermometer marked and calibrated to measure temperatures from $-10°C$ to $+110°C$ and you wished to measure a temperature of approximately $115°C$ accurately. How might you use your thermometer to do it?

E.4.2   In calibrating your thermometer by setting only the freezing and boiling points, what assumptions are you making regarding the internal diameter of the thermometer and the expansion of mercury?

E.4.3   Is a calibrated thermometer more precise than one that has not been calibrated? What implications does this have for measuring temperature *changes* as contrasted with measuring individual temperatures?

E.4.4   Oxidation is favored in the hottest portions of the burner flame; reduction is favored in the coolest portions. Map the flame in terms of oxidation and reduction regions.

E.4.5   Identify the main sources of error in part (*c*) of the procedure. How might some of these errors be reduced?

E.4.6   An unknown metal having a mass of 14.00 g is dropped into 11.00 mL of water at $20.0°C$. The starting temperature of the chunk of metal is $100.0°C$. The final temperature of the water and metal is $32.5°C$. Find the specific heat of the metal.

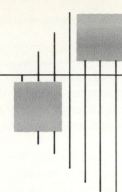

Experiment E.5

Atomic Weight of a Metal

Special Items
Two pieces of metal foil (ca. 1 g each)
10 mL concentrated nitric acid

The atomic weights of elements are assigned relative to the value 12, arbitrarily chosen for the most abundant isotope of carbon. If we know the relative number of atoms in a compound containing carbon and one other element, then from a weight analysis we can apportion the observed mass to the different elements and decide how much mass must be contributed by the atoms of the element other than carbon. An indirect way of determining the atomic weight of an unknown element is to combine it with an element whose atomic weight is known, such as oxygen.

In this experiment you will determine the atomic weight of a metal M by converting a weighed amount of M to an oxide. From the gain in mass you can calculate the mass of oxygen that has combined per gram of M and thus decide the relative contributions of oxygen and M to the mass of the compound. If there are equal numbers of M and O atoms in the compound, then the observed mass contributions simply reflect the relative masses of individual M and O atoms. If, however, there are twice as many M atoms as O atoms, each M atom must be made half as heavy. To decide the relative number of M and O atoms in the oxide we can use the law of Dulong and Petit, which states that the product of the atomic weight of a solid element and its specific heat expressed in joules is approximately 26. The specific heat of M can be measured as described in part (c) of Experiment E.4. If you do not have time to do that part, use the value given by your instructor.

Procedure

Mark two crucibles and covers so that they can easily be distinguished, e.g., by filing a notch in one rim and two notches in the other. Heat gently each of the crucibles with its cover ajar over a burner flame as shown in Figure E.5.1. To save time, with crucible tongs set the first crucible on a transite

52

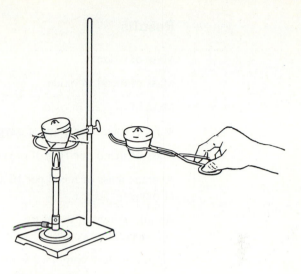

Figure E.5.1

board to cool while the second is being heated. When cool, transfer to an analytical balance, and weigh each crucible and cover to the nearest 0.001 g. Place about 1 g of metal in each crucible, and reweigh. Support one of the crucibles as in Figure E.5.1, and set the stand *in the hood*. Slowly and carefully add about 5 mL of concentrated nitric acid, set the cover ajar, and warm gently for 15 min over a small flame.

■CAUTION

There may be spattering. Do not spill on skin.

Stop heating if there is any evidence of spattering. After all the liquid is boiled off, heat strongly for about 10 min more. Set aside on a transite board to cool. While the first crucible and contents are cooling, set the second on the stand and repeat the nitric acid treatment and ignition. Set the second crucible aside to cool, and meanwhile weigh the first. Finally, weigh the second.

Data

	1	2
Mass of crucible and cover	_____	_____
Mass of crucible, cover, and metal	_____	_____
Mass of crucible, cover, and oxide	_____	_____

Results

Mass of oxide	_____	_____
Mass of metal in oxide	_____	_____
Mass of oxygen in oxide	_____	_____
Mass of metal per gram of oxygen	_____	_____
Mass of metal per 16 g of oxygen	_____	_____
Average mass of metal per 16 amu of oxygen	_____	_____

Atomic weight of metal if:
    1 metal atom combined with 1 oxygen atom   _____

    1 metal atom combined with 2 oxygen atoms   _____

    2 metal atoms combined with 1 oxygen atom   _____

Specific heat of metal   _____

Rough value of atomic weight of metal from
Dulong and Petit (26/sp ht)   _____

Probable exact value of atomic weight of metal   _____

Questions

E.5.1 Refer to a periodic table showing atomic weights, and select the element most likely to be metal M.

E.5.2 Assuming your answer to question E.5.1 is correct, calculate the percent accuracy of your result.

E.5.3 Identify two possible procedural errors that could make your determined value of the atomic weight too high, and identify two other errors that could make it too low. Explain.

E.5.4 Use the law of Dulong and Petit to estimate the atomic weight of the following metals, given their specific heats: copper (0.386 joule per g-deg); nickel (0.439 joule per g-deg); tin (0.222 joule per g-deg).

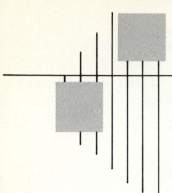

Experiment E.6

Hydration of Plaster of Paris

Special Items

7 g plaster of paris
115°C oven

Plaster of paris is made by partial dehydration of the mineral gypsum, $CaSO_4 \cdot 2H_2O$. In this experiment, by determining mass loss on dehydration to $CaSO_4$, you will find the formula of plaster of paris and study its hydration when it is mixed with water and allowed to set.

Procedure

(a) On a watch glass, mix about 5 g of plaster of paris with enough water (about 20 drops) to make a thick paste. Set it aside to harden and dry, either in an oven or on a radiator. [While waiting, you can go on to part (*b*).] In a weighed crucible, place about half of the set product. Weigh on an analytical balance. Cover the crucible and heat it, very gently at first and then with the full flame for about 10 min to dehydrate completely. Remove the cover, cool, and weigh the crucible and contents.

■ WEAR YOUR
SAFETY
GOGGLES

(b) To a weighed crucible, add about 2 g of fresh plaster of paris. Weigh on an analytical balance. Cover the crucible and heat it, very gently at first and then with the full flame for about 10 min. Remove the cover, cool, and weigh crucible and contents.

Data

Record weighings, clearly labeled.

Results

Calculate the formula of plaster of paris and the formula of the product that it forms when it sets.

Questions

E.6.1 How would your determination of the formula weight of the set product be affected by each of the following?

(*a*) The plaster of paris treated with water at the beginning of part (*a*) was lumpy, so that a portion of it was not exposed to the water.

(*b*) The set product in part (*a*) was not given sufficient time in the oven to dry thoroughly.

E.6.2 Suggest a reason for heating the crucible in parts (*a*) and (*b*) very gently at first.

E.6.3 Calculate the percent weight change in going from:

(*a*) Plaster of paris to the product it forms when it sets

(*b*) Plaster of paris to anhydrous $CaSO_4$

E.6.4 Strong heating of a hydrate of cupric nitrate converts it completely to anhydrous $Cu(NO_3)_2$. If the weight loss upon heating is 22.37 percent, find the number of moles of H_2O per mole of $Cu(NO_3)_2$ in the hydrate.

E.6.5 Assuming that the formula you determined for plaster of paris is correct, use an atomic weight table to find an accurate value for the formula weight. Then determine the percent error in your experimentally determined value for the formula weight of plaster of paris.

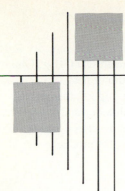

Experiment **E.7**

Analysis of a Mixture

Special Items

1 g potassium chlorate
Pinch of manganese dioxide
1 g unknown mixture

Stoichiometry concerns mass relations in chemical formulas and chemical reactions. In this experiment, you will investigate the stoichiometry of potassium chlorate and use the results to analyze a potassium chlorate–potassium chloride mixture. Potassium chlorate is a compound containing the elements potassium, chlorine, and oxygen. When heated strongly, the compound decomposes and all the oxygen can be driven off. The potassium and chlorine remain as the product potassium chloride, for which the formula is KCl. When a mixture of potassium chlorate and potassium chloride is heated, the mass loss due to oxygen evolved from the potassium chlorate enables you to calculate how much potassium chlorate was present in the original mixture. To speed the rate at which the decomposition occurs, manganese dioxide is used as a catalyst. The mass of the manganese dioxide remains constant.

■ CAUTION

Potassium chlorate is a powerful oxidizing agent. It must not be mixed with reducing agents such as sulfur or paper. Any spillage should be flushed down the drain with plenty of water.

Procedure

(a) To a clean, dry, small Pyrex test tube add a pinch of manganese dioxide, and weigh the tube plus manganese dioxide to the nearest 0.001 g on an analytical balance. Add about 1 g of dry potassium chlorate, and reweigh. Tap gently to mix. Clamp the tube to a ring stand at an angle of about 45°. Heat gently with a gas-burner flame. When the solid melts, increase the heating, and heat as strongly as possible for several minutes. Allow the tube to cool, and weigh.

■ WEAR YOUR
SAFETY
GOGGLES

(b) Add a pinch of manganese dioxide to a clean, dry Pyrex test tube, and weigh the total to the nearest 0.001 g. Obtain from your instructor an unknown mixture of potassium chlorate and potassium chloride. Add about 1 g of the mixture to the weighed tube plus manganese dioxide, and weigh the total. Tap to mix. Heat the mixture as described above, allow the tube to cool, and weigh the tube plus product.

Data

(a) Mass of test tube plus catalyst _____

Mass of test tube and catalyst plus potassium chlorate _____

Mass of test tube and catalyst plus potassium chloride residue _____

(b) Mass of test tube and catalyst _____

Mass of test tube and catalyst plus unknown mixture _____

Mass of test tube and catalyst plus residue _____

Results

(a) Mass of oxygen lost _____

Number of moles of oxygen atoms evolved _____

Mass of potassium chloride (residue) _____

Number of moles of potassium chloride _____

Number of moles of potassium atoms in original sample _____

Number of moles of chloride atoms in original sample _____

Number of moles of oxygen atoms in original sample _____

Simplest formula of potassium chlorate _____

(b)   Mass of unknown mixture   _____

Mass of oxygen lost   _____

Number of moles of oxygen atoms evolved   _____

Number of moles of potassium chlorate decomposed   _____

Mass of potassium chlorate in original sample   _____

Percentage by weight of potassium chlorate in unknown   _____

Questions

E.7.1   Suppose, in part (*a*), that your sample of potassium chlorate contained a small amount of water. How would this affect your calculated ratio of oxygen atoms to chlorine atoms in your sample?

E.7.2   This experiment could be carried out by decomposing either potassium chlorate or potassium perchlorate, $KClO_4$. Suppose that, in part (*a*), you heated a mixture of 0.001 g of manganese dioxide and 0.950 g of $KClO_4$. Predict the maximum number of grams of oxygen you could produce.

E.7.3   KCl(*s*) has some tendency to vaporize (sublime) when heated strongly. How would your results for part (*a*) be affected if some of the KCl vaporized?

E.7.4   Suppose that you had a sample, in part (*b*), containing 75.0 percent by weight potassium chlorate and 25.0 percent inert, nonvolatile material. What weight loss should you observe after intense heating, if the original sample weighed 1.100 g?

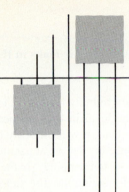

Experiment E.8

Gases and Absolute Zero

Special Items

Meter stick
40 cm capillary tubing
Laboratory barometer

NOTE: Students work in pairs.

TABLE E.8.1 Pressure of vapor above liquid water at various temperatures

Temp., °C	Pressure	
	atm	mmHg
15	0.0168	12.79
16	0.0179	13.63
17	0.0191	14.53
18	0.0204	15.48
19	0.0217	16.48
20	0.0231	17.54
21	0.0245	18.65
22	0.0261	19.83
23	0.0277	21.07
24	0.0294	22.38
25	0.0313	23.76
26	0.0332	25.21
27	0.0352	26.74
28	0.0373	28.35
29	0.0395	30.04
30	0.0419	31.82
31	0.0443	33.70
32	0.0469	35.66
33	0.0496	37.73
34	0.0525	39.90
35	0.0555	42.18

In this experiment, you will investigate the way in which the volume occupied by a gas sample at constant temperature changes with pressure. Then you will use this relation between volume and pressure to determine the pressure of a gas at various temperatures. The graph of pressure vs. temperature can be extended to the point of zero pressure. The temperature corresponding to zero pressure is called "absolute zero." By investigating the behavior of gases, you will thus determine the location of absolute zero.

A complicating feature of this experiment is that the gas sample will be in contact with water. Water tends to evaporate, so that any gas above it becomes contaminated with water vapor, which itself exerts a pressure. Fortunately, we can easily subtract out this water-vapor pressure, because it depends only on the temperature of the liquid water. Table E.8.1 gives the values of water-vapor pressure at common room temperatures.

Pressures are usually measured with mercury columns. In this experiment, some pressures are measured with water columns. Mercury has a density of 13.5 g per cm^3; so a column of water (of density 1.0 g per cm^3) is equivalent to a column of mercury $1/13.5$ as high.

Procedure

(Select a laboratory partner to work with in this experiment.)

■ WEAR YOUR SAFETY GOGGLES

(a)

Assemble the apparatus shown in Figure E.8.1. Trap about 40 mL of air in the gas-measuring tube, and set the clamp on the rubber connector. Raise and lower the leveling bulb to chase air bubbles out of the rubber tubes.

61

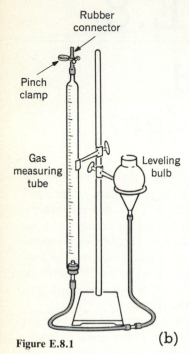

Rubber
connector

Pinch
clamp

Gas
measuring
tube

Leveling
bulb

Figure E.8.1

It is most important that there be no air leaks in the system.

Make sure that all joints are tight by lowering the leveling bulb and noting whether the level in the gas-measuring tube stays constant. If there is a leak, it will probably be necessary to replace the rubber connector. Set the apparatus aside to allow the water to come to room temperature. [In the meantime, start work on part (b) by following the directions given in the first paragraph of part (b).]

Adjust the height of the leveling bulb so that the water in it is at exactly the same level as the water in the gas-measuring tube. Record the volume of the gas sample to the nearest 0.1 mL. Have your partner raise the leveling bulb so that the difference in levels is 1 m. (Meter sticks are provided on the reagent table.) Record the gas volume. Lower the bulb below the desk level so that the difference in levels is 1 m. Record the volume. Measure the water temperature and barometric pressure.

(b) Get a 40-cm capillary tube from the stockroom (or use two 20-cm pieces). Make two right-angle bends about 8 cm from the ends. Fire-polish ends. (For bending capillary, it is not necessary to use the flame spreader on the burner.) Take your 125-mL flask, and heat it dry. Insert the capillary ends into No. 1 one-hole stoppers so that the capillary is flush with the rubber. When the flask is cool, insert one of the stoppers tightly into it. By clamping onto the stopper, suspend the flask inside a 600-mL beaker as shown in Figure E.8.2. Set aside a small beaker of water to come to room temperature. [Go back to part (a).]

Assemble the rest of the apparatus of Figure E.8.2. Take the gas-measuring tube and leveling bulb from part (a), leaving the water in it since it is already at room temperature. With the pinch clamp open and the bottom of the neck of the leveling bulb opposite the 40-mL mark, pour room-temperature water into the leveling bulb until it is filled to the neck. Measure the temperature of this water. Pour cold water from the tap into the beaker to the level of the bottom of the rubber stopper. Stir the water gently for about 1 min. Record the temperature of the water in the beaker. Being very careful not to force water over to the dry flask, insert the capillary stopper. Close the pinch clamp, and record the gas volume to the nearest 0.1 mL. Heat the water until its temperature has risen by about 10°. Remove the burner; stir the water; lower the gas-measuring tube until its water level matches that in the

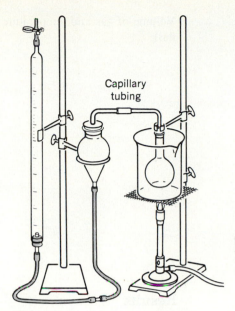

Figure E.8.2

leveling bulb; record the gas volume when it has become constant. Immediately note the temperature in the beaker. Resume heating, and in similar fashion take volume and temperature readings at 10° intervals up to the boiling point of water. To prevent heating the water in the leveling bulb, it is advisable to hold a transite board vertically between it and the flame. (See, for example, Figure E.9.1.)

Record the barometric pressure.

Data

(a)   Volume of gas (levels equal)   _____

Volume of gas (bulb 1.00 m above)   _____

Volume of gas (bulb 1.00 m below)   _____

Temperature of water   _____

Barometric pressure   _____

(b)   Temperature of water in leveling bulb   _____

Barometric pressure   _____

Volume of gas and temperature of
flask

    _____ _____

    _____ _____

    _____ _____

    _____ _____

    _____ _____

    _____ _____

    _____ _____

    _____ _____

    _____ _____

Results

(a) Levels Equal
Observed pressure of wet gas (barometric
pressure)   _____

Calculated pressure of dry gas (barometric
pressure minus vapor pressure of H_2O)   _____

Volume of dry gas ($=$ volume of wet gas)   _____

PV product for dry gas   _____

Bulb 1.00 m Above
Hg equivalent of 1.00-m H_2O column   _____

Observed pressure of wet gas (barometric
pressure plus Hg equivalent of 1.00-m H_2O
column)   _____

Calculated pressure of dry gas   _____

PV product for dry gas   _____

Bulb 1.00 m Below
Observed pressure of wet gas   _____

Calculated pressure of dry gas   _____

PV product for dry gas   _____

(b) To locate absolute zero, we must find how the pressure of a
fixed mass of gas changes with temperature if its volume stays

constant. In your experiment, the mass of gas in the flask changes slightly because some gas is transferred from the flask to the leveling bulb. This small increase of volume will have to be corrected for. However, before worrying about this, we will find out the pressure in the flask by using the observed volume of the gas in the gas-measuring tube. Presumably the air trapped in the measuring tube follows Boyle's law [as shown in part (a)]; so we can use the relation $PV = $ constant to find the pressure. Since the trapped gas is over water, it will be necessary to correct for the vapor pressure of water.

Proceed now to fill in the following table, using the method of calculation outlined.

Temp. in Flask	Press. of Dry Gas in Tube	Press. in Flask	Press. in Flask if V Const.
————			————
————	————	————	————
————	————	————	————
————	————	————	————
————	————	————	————
————	————	————	————
————	————	————	————
————	————	————	————
————	————	————	————
————	————	————	————

For the first entry, the "Temp. in flask" is assumed to be the same as the initial temperature of the beaker of water; the "Press. in flask if V const." is simply the barometric pressure. To calculate the pressure for subsequent entries, proceed as follows:

Original pressure P_0 of dry gas in measuring tube = barometric pressure

$\qquad$ − vapor pressure of H_2O at the temperature of the measuring tube

Pressure of dry gas in measuring tube

$$= \text{original pressure } P_0 \text{ of dry gas in measuring tube} \times \frac{\text{initial volume}}{\text{observed volume}}$$

Pressure in flask = pressure of wet gas in measuring tube

$\qquad$ = pressure of dry gas in measuring tube + vapor pressure of H_2O

Pressure in flask if V constant $=$ pressure in flask $\times \dfrac{\text{actual volume of gas sample}}{\text{initial volume of gas sample}}$

$= $ pressure in flask

$\times \dfrac{\text{volume of flask} + \text{volume of gas expelled}}{\text{initial volume of flask}}$

$= $ pressure in flask

$\times \dfrac{125 + \text{net } V \text{ change in measuring tube}}{125}$

(*Note:* The net V change is equal to the volume of water forced into the measuring tube. It is found by subtracting the observed reading on the measuring tube from the initial reading.)

Plot your results for the *P-T* relation, using "temperature in flask" in degrees Celsius along the horizontal axis and "pressure in flask if V constant" along the vertical axis. Draw the best straight line through the points, and extend it to zero pressure. Where does absolute zero lie on the Celsius scale?

Questions

E.8.1   In part (a), how would your calculated value for the PV product when the water level in the bulb is 1.00 m above the level in the gas-measuring tube be affected if there were a small air leak in the rubber connector at the top of the gas-measuring tube?

E.8.2   In part (a), suppose that the temperature of the water in the system was (unknown to you) higher when you made the first volume reading (bulb above) than when you made the second reading (bulb below). What effect would this have on your results?

E.8.3   In part (b), something that is not corrected for is the fact that the solubility of air in water decreases somewhat with increasing temperature. What effect would this have on your determination of absolute zero?

E.8.4   Suppose that you used alcohol instead of water in this experiment. What information would you need to be given in order to make your calculations?

E.8.5   Suppose that the thermometer you used gave readings that were always 2°C too low. How would this affect your determination of absolute zero in part (b)?

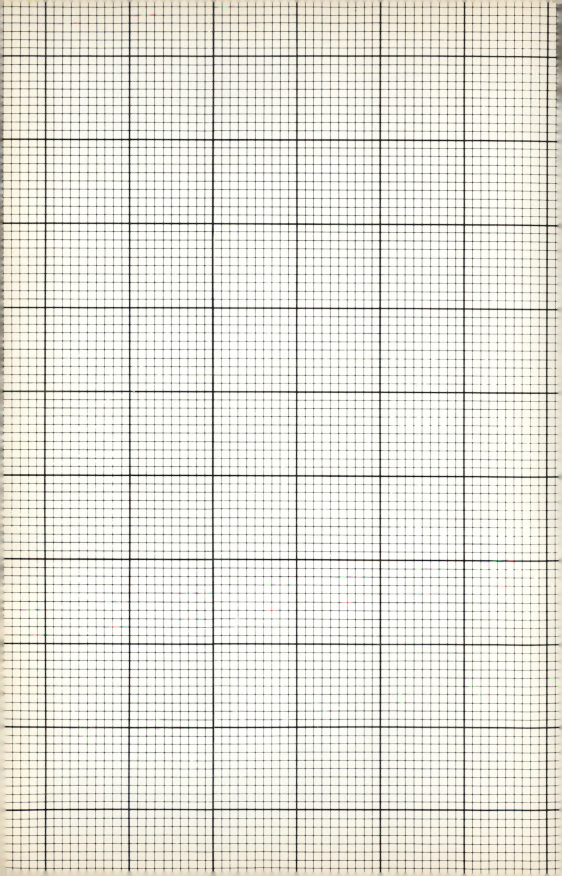

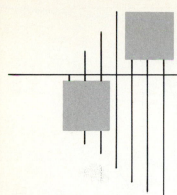

Experiment E.9

Molar Volume of Oxygen

Special Items

15-cm length of Pyrex tubing
(7 mm) sealed at one end
Laboratory barometer
0.1 g KClO₃

Molar volume is the volume occupied by one mole of an ideal gas. For convenience, molar volume is usually given at STP—i.e., at standard temperature (0°C) and standard pressure (1 atmosphere, or 760 mmHg). In this experiment, you will determine the molar volume of oxygen and calculate its value at STP.

When potassium chlorate is heated, the mass loss of the solid phase is due only to evolved oxygen. By determining the volume of oxygen that corresponds to a measured mass loss, you can calculate the volume that corresponds to 32.0 g, or 1 mol, of oxygen. Since the volume of a gas sample changes with temperature and pressure, you must measure them and use them to correct your observed molar volume to STP.

Procedure

■ **WEAR YOUR SAFETY GOGGLES**

Set up the gas-measuring tube and leveling bulb as shown in Figure E.9.1. Attach the rubber connector to the tip of the gas-measuring tube, and set the pinch clamp in an opened position on the lower glass-containing part of the rubber connector. Do not attach the right-angle bend yet, but have it ready. Fill the apparatus with water, and set it aside so that it comes to room temperature. Secure from the stockroom a 15-cm length of Pyrex tubing sealed at one end. Have your instructor place about 0.1 g (not more than 0.15 g) of potassium chlorate in this test tube. Weigh the tube and contents on an analytical balance to the nearest 0.001 g. With a rubber connector, attach the Pyrex tube to a right-angle bend. Raise the leveling bulb until the water level is just at the tip of the gas-measuring tube. Set the pinch clamp so that it closes off the rubber connector. Attach the right-angle bend and tube. Test for leaks by opening the pinch clamp and lowering the leveling bulb. After a short initial drop, the water level in the gas-measuring tube should stay constant. Have your instructor check the apparatus. Now, holding the burner in your left hand

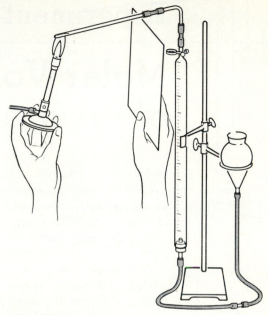

Figure E.9.1

and making sure the pinch clamp is open, gently heat the potassium chlorate. (With your right hand, hold a transite board as shown in Figure E.9.1 to prop up the right-angle bend and to provide a heat shield between the flame and the water.)

■ CAUTION

Do not allow molten KClO$_3$ to come in contact with the rubber connector or any other oxidizable substance.

Continue heating, gradually increasing intensity, until about 40 mL of gas is in the measuring tube. Stop heating, and let come to room temperature. Match up water levels in the measuring tube and the leveling bulb. Record the volume of oxygen evolved.

Record the temperature of the room and the barometric pressure.

Detach the Pyrex tube, and weigh tube and contents.

Data

Initial mass of Pyrex tube and contents   _____

Final mass of Pyrex tube and contents   _____

Volume of oxygen evolved   _____

Temperature   _____

Barometric pressure   _____

Results

Mass of oxygen evolved   _____

Pressure of oxygen (barometric pressure minus
vapor pressure of H_2O)   _____

Molar volume of oxygen under conditions
observed   _____

Molar volume of oxygen at STP   _____

Questions

E.9.1   Potassium chlorate melts at 368°C. What maximum volume
of oxygen, assuming ideal gas behavior, would be produced
by decomposing 0.10 g of $KClO_3$ at 1 atm and 368°C?

E.9.2   At what points in the experiment is the pinch clamp on the
rubber connector left open, and when is it closed?

E.9.3   The accepted value for the molar volume of oxygen at STP is
22.392 liters per mol. Find the relative error in your value in
parts per thousand.

E.9.4   Suppose that you had not heated the $KClO_3$ long enough to
decompose it completely. How would this affect your value
for the molar volume of oxygen?

E.9.5   What would be the value of R, the universal gas constant, if
your value for the molar volume of oxygen were used in place
of the molar volume of an ideal gas at STP (i.e., 22.414 liters
per mol)?

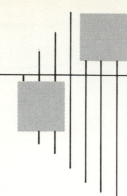

Experiment E.10

Apparent Gram Molecular Weight of Air

Special Items

6 mm glass tubing
Rubber tubing
200-mL graduated cylinder
Laboratory barometer

The apparent gram molecular weight of air is a weighted average of the gram molecular weights of the different gases, mostly nitrogen and oxygen, that make up air. The density and apparent gram molecular weight of air can be determined from measurements of the mass of air in a given volume at two different temperatures, as long as the pressure is held constant. This can be done by first weighing a flask containing air at one temperature, heating the flask (driving out a certain mass of gas), sealing it at the higher temperature, and then reweighing the flask.

The *ideal gas law*, $PV = nRT$, is assumed to be valid; that is, we will assume that air behaves as an ideal gas. The pressure P is the pressure of the gas, which, in this experiment, is equal to the barometric pressure; V is the volume of the container; and R is the *universal gas constant*, 0.08206 L-atm per mol-deg. Temperatures must, of course, be expressed in kelvins.

Recall that n, the number of moles of gas, is simply the mass in grams g divided by the gram molecular weight (GMW), or $n = g/\text{GMW}$. One could then express the ideal gas law as

$$PV = \frac{gRT}{\text{GMW}}$$

The number of grams g of air would then be given by the equation

$$g = \frac{PV(\text{GMW})}{RT} \tag{A}$$

The difference in the mass of gas $(g_1 - g_2)$ at two different temperatures T_1 and T_2, but at constant pressure and volume, may be found from the following equation:

$$g_1 - g_2 = \frac{PV \times \text{GMW}}{R} \left(\frac{1}{T_1} - \frac{1}{T_2} \right)$$

which can be rearranged to give equation B:

$$g_1 - g_2 = \frac{PV \times \text{GMW}}{R} \left(\frac{T_2 - T_1}{T_1 T_2} \right) \qquad \textbf{(B)}$$

In this experiment, you will not measure either g_1 or g_2. You will, however, measure the total mass of gas + container at each of the two temperatures. Since the mass of the container is constant, the difference between the total mass of gas + container at T_1 and the total mass of gas + container at T_2 will be identical to the difference between the mass of gas g_1 at T_1 and the mass of gas g_2 at T_2. In other words,

$$g_1 - g_2 = (\text{total mass})_1 - (\text{total mass})_2$$

By substituting the experimentally determined value of $g_1 - g_2$ into equation B, together with the pressure, volume, and temperatures, you can obtain a value for the apparent gram molecular weight of air.

Procedure

Set us an apparatus consisting of a 125-ml Florence flask, a 600-mL beaker, a No. 0, No. 1, or No. 2 one-holed rubber stopper, a fire-polished 5-cm piece of 6-mm glass tubing, a 5-cm piece of rubber tubing (must fit glass tubing very snugly), a pinch clamp, wire gauze, a ring, and a ring stand clamp (see Figure E.10.1). Choose a stopper (No. 0, No. 1, or No. 2) that will fit snugly in the stop of the flask when the glass tubing is inside the stopper. Moisten the rubber stopper and glass tubing a bit, and, grasping the tubing with a towel, insert the glass tubing into the rubber stopper so that the glass tubing is flush with the bottom of the stopper. Then dry the stopper and glass tubing thoroughly.

It is necessary first to get rid of any residual water adhering to the inside wall of the flask. To do this, set the empty flask on the wire gauze and heat it gently with the flame of your gas burner, starting at the bottom and moving to the top. You should see water condensing and evaporating as you move along. Then allow it to cool.

While you are waiting for the flask to cool, place the

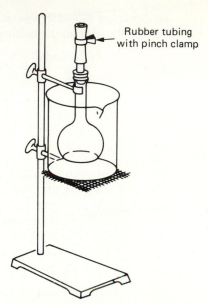

Rubber tubing
with pinch clamp

Figure E.10.1

600-mL beaker on the wire gauze on your ring stand, fill it about half-full of water, and begin heating the water with your gas burner.

As soon as the flask is cool enough to handle, insert the rubber stopper into the flask, and slide the rubber tubing, together with the pinch clamp, onto the glass tubing. Allow the flask to continue cooling with the pinch clamp on loosely (air must pass freely!) until it reaches room temperature.

When the flask has cooled to room temperature, measure and record the temperature to the nearest 0.1°C. Weigh the flask and entire stopper assembly as precisely as your balance will allow, and record the mass.

■ CAUTION

Before heating the flask, ask your instructor to check over your apparatus.

Shut off your gas burner. With the pinch clamp *open*, push the flask down into the hot water in the beaker on the ring stand and clamp the flask to the ring stand in such a way that the flask extends down into the beaker as far as possible. Adjust the water level so that the beaker is filled to within an inch of the top. Heat the water until it boils, and keep it boiling gently for about 10 min.

Shut off the gas burner, and quickly fold the rubber tubing over and pinch it together very tightly with the pinch clamp. Record the temperature of the water to 0.1°C.

Remove the sealed flask from the hot water, and dry it thoroughly with paper towels. Allow it to cool, and obtain the mass on the balance. Wait a few minutes, and weigh it again to make sure air is not leaking into your system. Record the final mass of the system.

If there is time, repeat the entire procedure in order to get a second set of measurements.

Note the position of the stopper; then remove the stopper and fill the flask with tap water. Remove the pinch clamp. Push the stopper into the flask, forcing water up into the rubber tubing, until the stopper is in the same position as before. Carefully remove the stopper, allowing the water in the rubber and glass tubing to pass into the flask. Then pour the water into a dry 200-mL graduated cylinder in order to measure the volume of the system. Record the volume.

Use a barometer to get the barometric pressure, or ask your instructor for the value, and record the pressure.

Data

	Trial 1	Trial 2
Initial temperature (T_1)	————	————
Initial mass (gas + container)	————	————
Final mass (gas + container)	————	————
Mass lost on heating ($g_1 - g_2$)	————	————
Final temperature (T_2)	————	————
Barometric pressure	————	

Results

Show calculation of apparent GMW of air for each trial. Then calculate an average value for the gram molecular weight of air.

Questions

E.10.1   Using your data and equations given in this experiment, calculate the density of air in the flask at T_1 and at T_2.

E.10.2   For a number of reasons, the composition, and therefore the apparent gram molecular weight, of air varies to some extent from place to place and from time to time, depending on relative humidity, pollution, and many other factors. A reasonable estimate of the average value can be made, however, and that value is approximately 28.96 g per mole. If this value is accepted as the "true value," what is the percent error in the value you determined in this experiment?

E.10.3   In the procedural instructions you are directed to heat the flask to remove any water adhering to the inside walls of the flask. If you failed to do this, what effect should not removing the water have on your experimentally determined value for the apparent gram molecular weight of air? Explain.

E.10.4   If the system were not sealed completely at the higher temperature, so that some air got in around the rubber stopper or through the clamped rubber tubing, your calculated value for the apparent gram molecular weight of air would be incorrect. Would your value be too high, or too low? Explain.

E.10.5   Of the following possible errors (assuming the values you obtained were the correct values), which would result in the greatest error in your determination of the apparent gram molecular weight of air, and why?

(a) Suppose you had read T_1 to be 1.0°C lower than you did.

(b) Suppose you had measured the volume to be 1.0 mL more than you did.

(c) Suppose you had measured the first total mass of gas + container to be 1.0 g greater than you did.

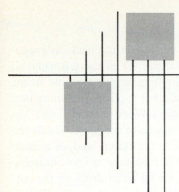

Experiment E.11

Molecular Weight of a Condensable Vapor

Special Items

6-cm-square thin aluminum foil
Laboratory barometer
3 mL of unknown liquid
200-mL graduated cylinder

The molecular weight of a condensable vapor can be determined by adding an excess of the corresponding liquid to a weighed flask, heating the assembly to volatilize all the liquid and to expel the excess vapor, cooling, and then weighing the flask with the condensed liquid. The mass of condensed liquid plus the mass of vapor left uncondensed at room temperature is equal to the mass of vapor that fills the flask at the elevated temperature and the prevailing pressure. If the liquid is relatively nonvolatile at room temperature, then we can neglect the mass left uncondensed at room temperature. Given the mass of vapor that fills the flask at the elevated temperature, we can calculate from the gas laws the volume of the measured mass of vapor at STP. Since at STP 1 mol of ideal gas occupies 22.4 liters, we can calculate the number of moles present and hence the mass per mole, or molecular weight.

Procedure

■ WEAR YOUR SAFETY GOGGLES

Secure a square of aluminum foil (about 6 cm on a side), and fashion a cap for your 125-mL flask by laying the foil on the mouth and folding down the sides. With a pin, poke a *tiny* hole (as small as possible) in the center of the cap. Determine on an analytical balance the mass of the clean, dry flask with cap to the nearest 0.001 g. Get from your instructor about 3 mL of an unknown liquid, and pour it into the flask. Replace the cap securely.

Fill a 600-mL beaker nearly full of water, and, working in a well-ventilated hood, heat to boiling. Clamp the flask at its very top, and suspend it in the beaker as shown in Figure E.11.1. Record the temperature of the boiling water and the barometric pressure.

Watch the liquid in the flask to see when it is all gone.

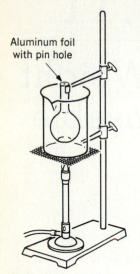

Aluminum foil with pin hole

Figure E.11.1

Do *not* remove the flask from the beaker to check on evaporation, for the vapor may recondense. (You can tell when all the liquid has evaporated by sighting across the hole in the aluminum cap. So long as excess hot vapor is streaming out, there will be a diffraction effect.) When all the liquid has evaporated, including any that may have condensed inside the neck, remove the flask by holding the clamp, and set it aside on the transite board to cool. After the flask is cool, examine the cap to make sure that there are no water drops on the outside surface of the aluminum. Wipe dry if necessary. Weigh the flask, cap, and condensed unknown liquid.

Fill the flask completely full of water, and measure its volume by pouring the water out into a large graduated cylinder provided on the reagent table.

Data

Mass of flask plus cap   _____

Mass of flask plus cap plus liquid   _____

Temperature of boiling water   _____

Barometric pressure   _____

Volume of the flask   _____

Results

Mass of vapor = mass of liquid condensed   _____

Volume of vapor if at STP   _____

Number of moles of vapor   _____

Molecular weight of vapor   _____

Questions

E.11.1   What does the flask contain
 (a) Before adding the volatile liquid?
 (b) At the point when the volatile liquid has completely vaporized?
 (c) At the end of the experiment, at the final weighing?

E.11.2   Why is the temperature of boiling water used for measuring the volume of the vapor instead of the temperature of the liquid after it has cooled to room temperature?

E.11.3   When vaporized, the volatile liquids used in this experiment do not behave exactly as ideal gases. In what direction would you expect the deviations from ideal behavior to lie, and how would this tend to affect your calculated molecular weight?

E.11.4   What restrictions are there on what volatile liquids could be used in this experiment?

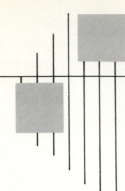

Experiment E.12

Optical Spectrometry

Special Items

Diffraction grating
Cardboard tube
Aluminum foil
Meter stick
Light-bulb setup
Stainless steel spatula
Solid salts (LiCl, NaCl, KCl,
    RbCl, CsCl, CaCl$_2$,
    SrCl$_2$, BaCl$_2$)
Unknown mixtures of solid
    salts
KMnO$_4$ (solid)
3 M H$_2$SO$_4$
H$_2$C$_2$O$_4$ (solid)
MnSO$_4$ (solid)
CuSO$_4 \cdot$ 5H$_2$O (solid)

Optical spectrometry is the analysis of light into its various components. The energy E of a light wave is proportional to its frequency ν:

$$E = h\nu$$

where the proportionality constant h is the Planck constant, having a value of 6.62×10^{-34} joule-sec. The frequency ν, which gives for the light wave the number of wave crests (or wave troughs) passing a given point per second, is inversely proportional to λ, the wavelength of the light wave, where λ gives the distance between successive wave crests or troughs. The product of the frequency and the wavelength equals the velocity of light, generally indicated by c:

$$\nu\lambda = c$$

where c has a value of 3.00×10^{10} cm per sec.

"White light" is a combination of waves of different wavelengths extending from about 400 nm for violet to about 700 nm for red light. The range from 400 to 700 nm is called the *visible spectrum* and is the only range of interest in this experiment, since the detector to be used is the human eye.

The separation of light into various wavelengths can be achieved by passing the light through a grating or prism. The result is a *spectrum* in which each different wavelength is diffracted or bent through a different angle. Diffraction by a prism comes about because short-wavelength light is slowed down more in glass than is long-wavelength light. Diffraction by a grating results because of selective reinforcement of a light wave along a particular direction. As shown in Figure E.12.1, a diffraction grating consists of a series of parallel grooves on a hard, flat surface. Each groove acts as a source of light, sending out a spherical wave in all directions. Reinforcement occurs along a certain angle θ, depending on the wavelength λ and the spacing of the grooves d according to the equation

$$\sin \theta = \frac{BC}{AB} = \frac{\lambda}{d}$$

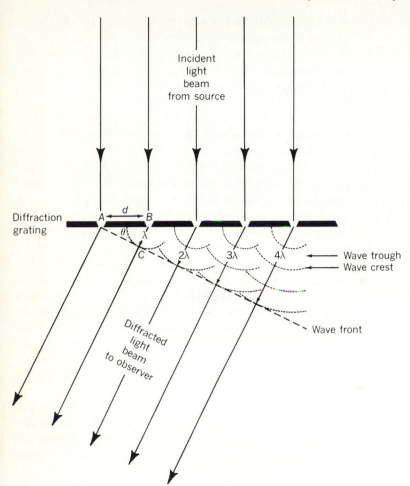

Figure E.12.1

because only in that direction does the maximum of one wave match in step the maximum of another. At other angles there will be destructive interference as crests from one wave superpose on the troughs of another.

Since white light consists of many different wavelengths, it is separated into many angles, each of which corresponds to an image in a different color of the slit through which the light comes from the source. The result is a *continuous spectrum* ranging in color from violet at one end to red at the other. If the source emits light of a single frequency instead of white light, then the result is a *line spectrum* consisting of a single image of the slit at some characteristic color. Several lines at different colors will be observed if the source light is composed of several different frequencies. Another way of classifying spectra is as *emission spectra* or *absorption spectra*. In an emission spectrum, the source, generally an incandescent gas, emits the energies; if only a few characteristic energies are emitted, the result is a *bright-line spectrum*. In an absorption spectrum, the light from a source passes through an absorbing medium, the effect of which is to remove frequencies; the result is a *dark-line spectrum*. If a band of frequencies is absorbed, the result will be a dark area in that part of the spectrum.

Light emission and absorption result from changes in the energy states of atoms of molecules. When an excited atom—e.g., one that is heated in a flame—drops from a high energy state to a lower state, the frequency of the light emitted is characteristic of the energy difference between the two states, and only such frequencies will be emitted as are allowed by the permitted energy states. Similarly, if light of a given energy impinges on an atom, the light can be absorbed if the atom goes from its initial energy state to the permitted higher level. Thus, the lines observed in the emission and absorption spectra of atoms can be understood as resulting from transitions between fixed energy states characteristic of the atom. As a corollary, if the environment of an atom is changed, its energy levels may be affected, in which case the spectrum should show a difference.

This experiment consists of a study of the emission and absorption spectra of some common materials as analyzed by a diffraction grating. Measurement of the grating spacing and the diffraction angles will give the wavelength of the spectral lines. The spectroscope to be used is shown in Figure E.12.2. It consists of a cardboard cylinder of length l, fitted with an aluminum slit at one end and a diffraction grating at the other. A meter stick placed perpendicular to the tube is used to measure the distance a from the slit to the apparent image of the slit resulting from diffraction. Light enters the tube through the slit and is bent by the diffraction grating through the angle θ, thus creating the optical illusion that the image of the slit is on the meter stick. Measurement of a and l defines θ since

$$\sin \theta = \frac{a}{b} = \frac{a}{\sqrt{a^2 + l^2}}$$

Combining this result with the previous relation $\sin \theta = \lambda / d$ leads to

$$\lambda = d\left(\frac{a}{\sqrt{a^2 + l^2}}\right)$$

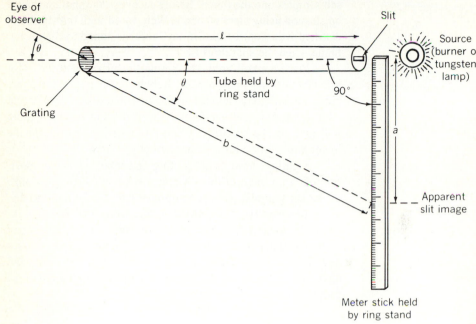

Figure E.12.2

Procedure

Obtain a cardboard cylinder, and attach a diffraction grating to one end, *being careful not to touch the transparent portions of the grating*. Cover the opposite end of the tube with aluminum foil and, with a sharp pin, make a very fine slit about 3 cm in length in the aluminum foil. Align the slit so that it is parallel to the grooves in the grating. Using ring stands and clamps for support, set up the apparatus, using Figure E.12.2 as a guide. Note that the figure shows the apparatus as viewed from the top, i.e., looking down on it. The slit and the grooves in the grating must be vertical.

(a)   **Calibration**   In order to determine the grating spacing d, a constant for your apparatus, you can use a substance that emits light of known wavelength. An excellent standard for this purpose is the so-called sodium D line, for which $\lambda = 589$ nm.

Support a gas burner on a ring stand, clamping it at an angle such that any spillage of salts will end up on the bench top rather than in the burner. Place the gas burner in front of the slit, but *not too close*. Adjust the air control so two distinct cones appear. You will use a stainless steel spatula to insert salt samples into the flame. Before each test, the spatula should be cleaned using steel wool and then rinsed with 6 M HCl and distilled water until little, if any, color is added to the burner flame when the spatula is heated.

Place a small amount (spatula-tip full) of sodium chloride crystals in a small beaker. Replace the cap on the reagent bottle immediately. Dip the stainless steel spatula in water and then in the NaCl. A small amount of the salt on the spatula will give the best results. Place the adhering crystals in the lower part of the flame. Note the color.

Sighting through the grating and tube, adjust the position of the burner so the slit appears bright. A bright image of the slit will also appear to right or left. Determine the distance between the slit and its image, using the meter stick. Have a colleague move a pencil or some other thin object along the meter stick while you signal right or left until marker and slit image are superimposed. Measure the distance and also the tube length. Calculate d and also the number of grooves per centimeter in the grating.

(b)   **Alkali Spectra**   Clean the spatula with 6 M HCl and distilled water, and heat in the flame. Repeat until the distinctive sodium color no longer appears. Test lithium chloride (LiCl) crystals in the same manner as NaCl. Note the color of the flame and the location of the line. Calculate the wavelength, frequency, and energy of the lithium line. Repeat with potassium chloride (KCl) and, if they are available, rubidium chloride (RbCl) and cesium chloride (CsCl). Be sure to replace the caps on the reagent jars immediately after use, because most of the salts are extremely hygroscopic (i.e., pick up water from the air).

(c)   **Alkaline-Earth Spectra**   Repeat part (b), using calcium, strontium, and barium chlorides. Measure line positions in the same manner, and calculate the wavelength, frequency, and energy of the alkaline-earth element spectra. If you have time, you might also try magnesium chloride.

(d)   **Analysis**   Mix sodium chloride and lithium chloride crystals together, and measure the positions of the lines. Compare with your results in (b). Mix lithium chloride and calcium chloride in the same manner. Obtain an unknown from your instructor and analyze it.

(e)   Absorption Spectra

1   Completely dissolve two crystals of potassium permanganate ($KMnO_4$) in approximately 80 mL of water. Note the color of the solution. Point the slit of the spectroscope toward a tungsten light bulb. Describe. Place the potassium permanganate solution between the bulb and the slit. Note what colors are blocked out.

Prepare a solution half as concentrated by pouring out half of what you have and replacing it with water. Examine the spectrum again. Repeat with successive dilutions to one-fourth, one-eighth, etc., of the original concentration until no more change is seen.

If time permits, repeat the above using as light source something other than the tungsten bulb—e.g., sunlight, sodium flame, lithium flame.

2   Make a solution of copper sulfate ($CuSO_4$) by dissolving about 2 g in about 100 mL of water. Observe the absorption spectrum as above. Now add dropwise some dilute ammonia solution, observing the spectrum after each drop until no more change is seen. Then, dropwise again, add dilute hydrochloric acid and note the change in the spectrum.

3   Prepare a solution of potassium permanganate as in step 1. Check the absorption spectrum. Add some dilute sulfuric acid (H_2SO_4). Observe. Then add a crystal of oxalic acid ($H_2C_2O_4$). After several minutes of viewing, add a crystal of manganese sulfate ($MnSO_4$) and immediately observe what happens to the absorption spectrum.

Questions

E.12.1   Why is it important not to spill salt solutions into the burner?

E.12.2   Tell how each of the following errors would affect the calculated value of a measured wavelength:

(*a*)   The distance *a* was measured not from the slit but from the edge of the cardboard cylinder.

(*b*)   The angle between the cardboard cylinder and the meter stick was less than 90°.

(*c*)   The distance *l* was measured not as the length of the cardboard cylinder but as the distance between the grating and the gas burner.

E.12.3   Explain why several apparent slit images, each of a different color, are usually observed when the light from a salt placed in a flame is viewed through a grating.

E.12.4   Suggest a reason why a heated metal, such as tungsten in a light bulb, gives a continuous spectrum (white light at very high temperatures) rather than a line spectrum.

E.12.5   Light coming from the sun gives a spectrum that is nearly continuous, but there are dark lines corresponding mainly to hydrogen and helium energy levels. Explain.

E.12.6   By reference to Figure E.12.1, explain why the diffracted light beam to the observer should appear strong at some angles θ and weak at other values of θ.

E.12.7   If you choose a diffraction grating with more grooves per centimeter, i.e., smaller values of d, what would this do to the distance a measured for the apparent slit image in Figure E.12.2?

E.12.8   What precision can you assign to your experimentally determined value of d in part (a)? How might you modify the setup to improve this precision?

E.12.9   Why might you expect LiCl, NaCl, KCl, RbCl, and CsCl to show different wavelength emission lines?

E.12.10   What elements are probably present if you see emission lines at 670, 420, and 770 nm?

E.12.11   What is the main difference you see between the emission spectrum of, say, NaCl and the absorption spectrum of $KMnO_4$?

E.12.12   How might you use an absorption spectrum, such as that of $KMnO_4$, to follow the kinetics of a reaction? What factors should you make a special effort to control?

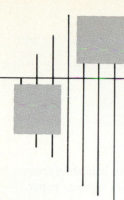

Experiment E.13

Bonding and Molecular Shapes

Special Item

Framework molecular model kit

In this experiment you will be using valence-shell electron-pair repulsion (VSEPR) theory and valence-bond theory to predict and to describe the bonding and shape of a number of molecules and ions. To aid you in visualizing the three-dimensional geometry, you will make use of molecular models.

Molecular Models

The model kits most useful for describing small molecules and ions are framework models and ball-and-stick models. In the case of the latter, the balls represent the atoms; and the sticks, the bonds or lone pairs. The number of holes drilled in a ball usually corresponds to the valence of the atom represented. In the case of the framework kits, atom centers are represented by metal or plastic connectors and bonds by plastic tubes. The number of prongs on a connector corresponds to the valence of an atom. Neither ball-and-stick nor framework models ordinarily give accurate relative atom sizes or bond lengths, but either can be very helpful in visualizing three-dimensional shapes.

Perspective Drawings

It is common practice to use perspective drawings to describe the geometry of molecules. A solid line represents a bond in the plane of the paper, a dashed line indicates a bond directed through the plane away from the observer, and a wedge designates that the bond is coming up out of the plane toward the observer. Perspective drawings of a tetrahedral connector from a framework model kit and of a ball with four sticks coming out in tetrahedral directions are shown in Figure E.13.1

Build models and draw perspective representations showing trigonal bipyramidal (valence = 5) and octahedral (valence = 6) coordination. (*1*) (The numbers in parentheses

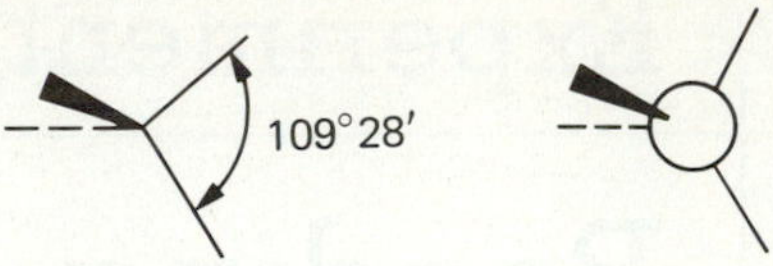

Figure E.13.1

following questions or directions indicate exercises and answers that are to be written out on separate paper.)

Valence-Bond Hybridized Orbitals

According to valence-bond theory, when a central atom in a molecule or ion bonds with other atoms, the valence orbitals of the central atom can be considered to have mixed to give hybridized atomic orbitals of the central atom. The number of atomic orbitals that are mixed is the same as the number of hybridized orbitals that are formed. Mixing one s and one p orbital gives, for example, two sp hybrids, pointing in opposite directions, leading to linear geometry. For each of the following types of hybridization, identify the types (s, p, or d) of atomic orbitals and indicate how many atomic orbitals of each type are mixed to give each of the following sets of hybridized atomic orbitals: sp^2 (triangular), dsp^3 (trigonal bipyramidal), d^2sp^3 (octahedral). Also, tell how many hybridized orbitals are formed in each case. (*2*)

Although the concept of hybridized orbitals is useful in describing bonding, one must know what the geometry is *before* knowing which valence orbitals to mix to give the hybrids. For example, carbon and hydrogen can form compounds in which the hybridization of the valence orbitals of carbon is either sp, sp^2, or sp^3. To decide which is the appropriate type of hybridization for a particular carbon-containing molecule or ion, one must first know if it is linear, triangular, or tetrahedral.

Valence-Shell Electron-Pair Repulsion Theory (VSEPR Theory)

VSEPR theory is a very successful method for predicting shapes of molecules and ions. It can be used to predict the geometry, and valence-bond theory can then be used to describe the hybridization of the valence orbitals of the bonding atoms. The basis for VSEPR theory is the Pauli exclusion principle, which states that for two electrons to occupy the same region of space, they must be of opposite spin; electrons of the same spin tend to stay as far apart as possible in order to minimize repulsions. As a consequence, in molecules or ions, pairs of valence-shell electrons are found to be in geometrical arrangements that provide maximum separation.

The first step in using VSEPR theory is to figure out a Lewis electron dot structure for the molecule or ion, recognizing that when *d* orbitals are available, an atom can have as many as six pairs of valence electrons. The *total* number of valence-shell electron *pairs* (TP) that contribute to shape, that is, the *lone pairs* (LP) plus the sigma (σ) *bonding pairs* (BP), is then counted. Pi (π) bonds are important in determining bond strength and length but are *not* as important in determining shape. Thus, for a double bond, which is made up of one sigma bond and one pi bond, BP = 1, where BP equals the number of sigma bonding pairs. In the structure for NOCl, :Ö::N̈:C̈l:, there is one sigma bond between N and O, one sigma bond between N and Cl, and a lone pair of electrons on the nitrogen; therefore, for NOCl, TP = 3. Note that the lone pairs of electrons on the oxygen atom and on the chlorine atom are *not* included in counting the total pairs, because they are not on the central nitrogen atom. Write out the electron dot structure for NOCl; label the lone pair and the two sigma bonding pairs on the nitrogen atom. (*3*)

The second step in using VSEPR theory is to distribute the sigma bonding pairs and the lone pairs around the central atom so that they will be as far apart as possible (minimum repulsion geometry). For any given number of total pairs of shape-determining electrons, there is one geometrical arrangement that gives minimum repulsions. These are summarized in Table E.13.1. Once the geometry has been established, one can predict the valence-bond hybridization of the valence orbitals of the central atom.

The third step is to make minor adjustments in the geometry to account for the fact that certain electron pairs take up more space and thereby repel neighboring electron pairs to a greater extent than do others. R. J. Gillespie, who is largely responsible for having popularized the use of VSEPR theory, has identified three rules, or assumptions, regarding the relative sizes of different kinds of electron pairs:

1   A lone pair is somewhat larger than a bonding pair of electrons. Consequently, repulsions between neighboring pairs of valence shell electrons decrease in the order LP-LP > LP-BP > BP-BP.

TABLE E.13.1

Total Pairs (TP) (shape-determining)	Predicted Geometry	Predicted Angles	Predicted VB Hybridization	Sketch
2	Linear	180°	sp	
3	Triangular	120°	sp^2	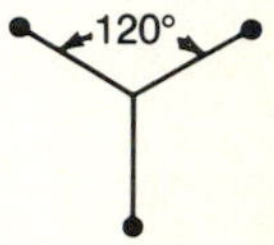
4	Tetrahedral	109.5°	sp^3	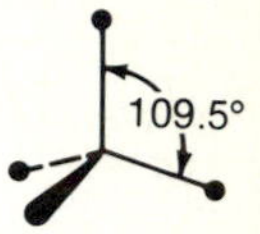
5	Trigonal bipyramidal	90°, 120°	dsp^3	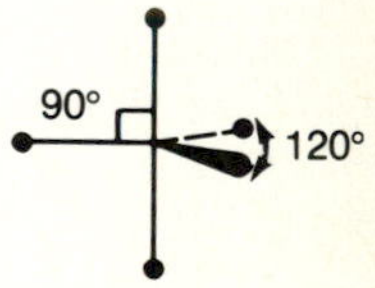
6	Octahedral	90°	d^2sp^3	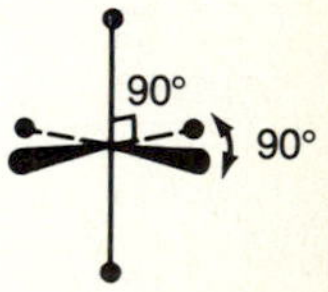

2   The higher the electronegativity of an atom involved in a bond, the smaller the size of the bonding pair. Thus, repulsions between bonds in which bonded atoms have higher electronegativity will be less than between those involving atoms of lower electronegativity.

3   Double and triple bonds are larger than single bonds.

Overall Geometry versus Molecular Geometry

In the ammonia molecule, which has the Lewis electron dot structure

H
:N:H
H

there are three BP and one LP, a total of four pairs (TP = 4) of valence shell electrons surrounding the central nitrogen atom. This suggests tetrahedral *overall geometry* (including the lone pair) and sp^3 hybridization of the nitrogen valence orbitals. Because the lone pair is larger than the three bonding pairs, one would predict that the H—N—H bond angles should be slightly less than the ideal 109.5° tetrahedral angle (actual bond angles are 107°). Note that the *molecular geometry* (atoms only, no lone pairs) is *not* tetrahedral but pyramidal. In general, for what kinds of molecules and ions will the molecular geometry *not* be the same as the overall geometry? Explain. (4)

Procedure

1 For each of the molecules and ions listed below, (a) use VSEPR theory to describe the overall and molecular geometry, (b) use valence-bond theory to describe the hybridization and bonding, and (c) give a perspective drawing, labeled with estimated bond angles. Use molecular models to aid in visualizing the geometry.

$BeCl_2$, BF_3, CCl_4, PF_5, SF_6
CO_2, NO_2, N_2O, $NH_4{}^+$, $NO_3{}^-$
HCN, C_2H_2, C_2H_4, $NO_2{}^+$

2 In molecular ions having trigonal bipyramidal or octahedral overall geometry, one often has more than one choice regarding the placement of lone pairs and bonding pairs. A useful rule of thumb in this situation is that repulsions at angles greater than 90° have little or no effect on bond angles. The arrangement having the smallest number of LP-LP repulsions should be favored. Consider, for example, the ClF_3 molecule, which has trigonal bipyramidal overall geometry. Refer to Figure E.13.2 and show that, for the left structure, at angles of 90° or less, there are no BP-BP repulsions or LP-LP repulsions but there are six LP-BP repulsions. Also count the number of each type of repulsion for the center and right struc-

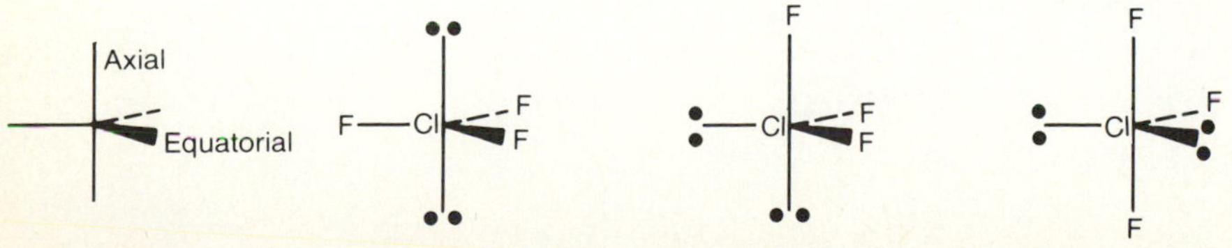

Figure E.13.2

tures. Explain why the observed structure for ClF_3 is the structure on the right, which corresponds to T-shaped molecular geometry.

3   Use VSEPR theory to explain each of the following molecular geometries:

Linear: $AgCl_2^-$, I_3^-
V-shaped: SO_2, NH_2^-, O_3
T-shaped: BrF_3
Square planar: BrF_4^-
Square pyramidal: SbF_5^{2-}

Questions

E.13.1   Explain why in   O   the H$-$C$-$H bond angle is 116°
            $\parallel$
            C
           ╱   ╲
          H   H

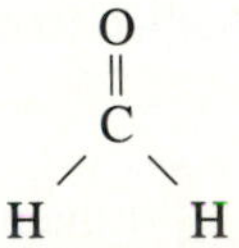

and the O$-$C$-$H bond angle is 122°.

E.13.2   Explain why the H$-$N$-$H bond angle in NH_3 is 107.3°, while the F$-$N$-$F bond angle in NF_3 is 102°.

E.13.3   Explain why in trigonal pyramidal structures, such as PF_5, the axial bond lengths are observed to be longer than the equatorial bond lengths.

E.13.4   Predict the geometry and bonding, and give perspective drawings of PI_3, PCl_3, and PBr_3. Also predict the relative sizes of bond angles I$-$P$-$I, Cl$-$P$-$Cl, and Br$-$P$-$Br in the three molecules. Pauling electronegativity values are: I, 2.5; Cl, 3.0; Br, 2.8.

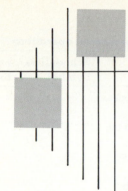

Experiment **E.14**

Molecular Symmetry and Isomers

Special Item

Framework molecular model
   kit

Symmetry

In this experiment, you are asked to construct models of various molecules and examine the spatial relationships of the atoms. Before doing this, you must learn something about the very powerful method of describing these relationships in terms of the symmetry elements that the structure possesses. First we must distinguish between a symmetry element and a symmetry operation. A *symmetry element* is a geometric entity (a point, a line, or a plane) that the structure possesses with reference to which the various superimposable parts of the molecule can be related. A *symmetry operation* is the act of bringing one superimposable part of the molecule into coincidence with another (either physically by manipulating a model, or mentally.) Obviously, symmetry operations are always associated with symmetry elements.

There are two widely used systems for describing symmetry, the Schoenflies (point group) system and the Hermann-Mauguin (H-M), or "international," system. The Schoenflies system is preferred by spectroscopists, the H-M system by crystallographers. In this experiment, the Schoenflies system will be used, but from time to time equivalent H-M symbols will be introduced in order to familiarize you with them.

Five important symmetry elements and their associated operations are shown in Table E.14.1.

A symmetry operation is a movement or combination of movements of a body from one configuration to an *identical* or *equivalent* configuration. Suppose that you had set a molecular model on a table and, with your back turned, someone performed a symmetry operation on it. When you turned around, could you see any difference? Explain. (*1*) (The numbers in parentheses following questions or such directions as

Table E.14.1   Symmetry elements and operations

Symmetry Element	Symmetry Operation	Schoenflies Symbol	Example
Rotation axis	Rotation about the axis	C_n	C_2
Reflection plane	Reflection in the plane	σ	σ
Rotation-reflection axis	Rotation about the axis followed by reflection in a plane perpendicular to the axis	S_n	S_4
Center of inversion	Inversion of all atoms through the center	i	i
Identity	Molecule unchanged	E	E

"Explain" indicate exercises and answers that are to be written out on separate paper.)

Rotation Axis, C_n   A molecule is said to have a *rotation axis* of symmetry if the molecule can be brought into *self-coincidence* by rotating it around an axis by $360°/n$. The value of n, the *order* of the axis, is 1, 2, 3, 4, 5, and 6, respectively, for rotations of 360, 180, 120, 90, 72, and 60°. The corresponding axes are designated C_1, C_2, C_3, C_4, C_5, and C_6. (In H-M notation a C_1 axis would be 1, C_2 would be 2, etc.) A C_1 axis is called a onefold axis; C_2, a twofold axis; C_3, a threefold; etc. As a specific illustration, the H_2O molecule has a C_2 axis; it goes through the O atom and bisects the line between the two H atoms. Rotating an H_2O molecule about its C_2 axis brings one of the H atoms into coincidence with the other.

From a framework molecular model kit, select one tetrahedral (four-pronged) metal connector and two black tubes. Attach the black tubes to the tetrahedral connector to make a representation of the water molecule. Find the C_2 axis, and draw a representation of the H_2O molecule, showing the C_2 axis. (2) Make a second model, except this time use a black tube and a white tube instead of two black tubes. This model represents HOD, in which one of the H's in water has been replaced by a deuterium atom (D). If a C_2 operation is carried out on HOD, is the operation a symmetry operation? Explain. (3)

A onefold axis C_1 is rather special. It really does not imply much symmetry, since rotating any object by 360° of course brings it back to self-coincidence. The rotated object is completely identical with the starting object, and this operation of rotation about C_1 is called the *identity* operation (E). As noted below, the identity operation is important when considering the net effect of several consecutive operations. Thus, for example, two consecutive rotations about C_2, each by 180°, bring a molecule back to its starting orientation and are equivalent to a single identity operation E. This is expressed as $C_2C_2 = E$. The second C_2 means "rotate 180°," the first C_2 means "rotate another 180°," and the "$= E$" means "gives an *identical* configuration." (Note the right-to-left order of operations.) Another way of expressing C_2C_2, is C_2^2, which means, simply, "do the C_2 operation twice." What is meant by $C_3^3 = E$? (4)

If C_2 is a symmetry element of a molecule, then one C_2 operation will take the molecule to an *equivalent* configuration and two C_2 operations will take it to an *identical* configuration.

What is the rotational symmetry of each of the following: circle, square, rectangle, rhombus, hexagon, equilateral triangle, isosceles triangle, regular pentagon? Draw figures showing the rotational symmetry elements in each of the above eight figures. (5) Note that in addition to a principal rotation axis perpendicular to the plane of each figure, it may be possible simultaneously to have other axes of rotation in the plane of the figure. Make sure you find them all.

In a molecule containing a C_n axis, if one atom of a given kind lies off the C_n axis, there must automatically be $n - 1$ more, or a total of n atoms of that kind, each lying in an equivalent configuration, off the C_n axis. Explain. (6)

In molecules with more than one C_n axis, the axis with the highest order (largest n) is called the *principal rotation axis*. In a cartesian coordinate system, the principal rotation axis is the vertical, or z, axis.

Reflection Plane, σ   A molecule is said to have a reflection (mirror) plane of symmetry σ if, when the molecule is divided by the reflection plane, every atom on one side of the plane is opposite a similar atom on the other side of the plane (and at the same distance from the plane). (The H-M symbol for a reflection is m.) All planar molecules have at least one reflection plane of symmetry, the molecular plane. Explain. (7)

Examine your model of the H_2O molecule. Find the two mirror planes (one passes through the O atoms and perpendicularly bisects the line between the H atoms; the other is the molecular plane.) Draw a picture of a water molecule, showing the C_2 axis and the two mirror planes. (8)

One consequence of having a mirror plane in a molecule is that all atoms of a given kind that do not lie in the reflection plane must occur in even numbers. Explain. (9) Also, if there is only a single atom of a particular kind in a molecule, it must lie in every symmetry plane that the molecule has. Explain. (10)

In the case of BF_3 the molecular plane lies perpendicular to the principal rotational axis. Such a plane is called a *horizontal* plane of symmetry and is designated σ_h.

In your model kit there are metal connectors (trigonal bipyramidal) having five prongs, three of which come out from the C_3 axis at 120° angles. Attach three red tubes to the connectors so that they lie in one plane and the two "open" prongs are oriented vertically. This represents a molecule of BF_3.

In addition to the horizontal plane, σ_h, BF_3 has three *vertical* planes, called σ_v', σ_v'', and σ_v'''. Draw a picture of the BF_3 model, indicating the locations of the C_3 axis, the σ_h plane, and the three σ_v planes. (11)

When a molecule has more than one axis of symmetry, a third kind of plane of symmetry can arise. $PtCl_4^{2-}$, which is planar, has two kinds of vertical planes of symmetry. One, designated σ_v, includes the Pt—Cl bonds. The other kind of vertical plane does not include any Pt—Cl bonds but bisects the angle between two successive Pt—Cl bonds. This is called a *dihedral* plane of symmetry and is designated σ_d. From your model kit, select one of the six-pronged metal connectors (octahedral) and four blue tubes. Then construct a model of $PtCl_4^{2-}$. Draw a picture, or several pictures, showing the locations of the rotation axes and the horizontal, vertical, and dihedral mirror planes in $PtCl_4^{2-}$. (12)

Inversion Center, i An *inversion center* is the same as a *center of symmetry*. Its presence in a molecule implies that every point on the molecule can be reflected through the center of the molecule to match an identical point on the opposite side of the molecule. A center of inversion is usually symbolized by i. The H_2O molecule does not possess a center of inversion; a CO_2 (linear) molecule does. Explain. (13) Examine your models of BF_3 and $PtCl_4^{2-}$. Do these molecules have centers of inversion? If so, where? Draw pictures showing the location of the inversion centers, if any. (14)

Rotation-Reflection Axis, S_n If a molecule can be rotated about an axis and then reflected in a horizontal plane to produce an equivalent configuration, the molecule is said to possess a *rotation-reflection axis*, designated as S_n, where n indicates the order of the rotation.

Methane, CH_4, has two kinds of rotation axes—the C_3 axis, which includes C—H bonds, and the C_2 axis, which bisects C—H bonds. Rotation through 180° about a C_2 axis results in a configuration equivalent to the original molecule. However, if the molecule is rotated about the same axis through only 90° and if the resulting orientation is reflected in a plane perpendicular to the axis, the combined operations produce a configuration indistinguishable from the original. Thus, methane has an S_4 rotation-reflection axis that coincides with the C_2 axis.

The important point about the S_4 axis in the methane molecule is that the S_4 operation transforms all hydrogen atoms into each other. In other words, the four hydrogen atoms constitute one set of chemically equivalent atoms. The C_2 axis treats the four hydrogen atoms as two different sets of two hydrogen atoms, one set above the plane and another below the plane. Construct a model of methane using a tetrahedral (four-pronged) connector and a red tube, a yellow tube, a black tube, and a white tube. Each differently colored tube represents one of the H atoms in CH_4. Draw a picture of the model, showing the locations of the colored tubes with respect to the S_4 axis. (*15*)

In general, a molecule has several symmetry elements. They have to be consistent with each other in the sense that an operation about one symmetry element cannot be in contradiction to another symmetry operation about some other symmetry element. How the self-consistency comes about can be seen in Figure E.14.1. It shows a typical possible arrangement of four equivalent points on a molecule. The points could be four identical atoms as, for example, the four Cl atoms in the square-planar complex $PtCl_4{}^{2-}$; they are designated as a_1, a_2, a_3, and a_4 so they can be referred to individually.

Point O (where the Pt would be located) is an inversion center. Point a_1 (coordinates x, y, z) is related by inversion through the center O to a_3 ($-x, -y, -z$); a_2 ($-x, -y, +z$) is related by inversion to a_4 ($+x, +y, -z$). As can be seen, inversion is equivalent to changing all positive coordinates to negative ones, and vice versa.

The line AB is a twofold axis of rotation. Rotation by 180° counterclockwise around the line AB sends point a_1 into a_2, and a_4 into a_3. Simultaneously, a_2 is sent into a_1, and a_3 into a_4. How are the coordinates x, y, z related for two points made equivalent by C_2 rotation? (*16*)

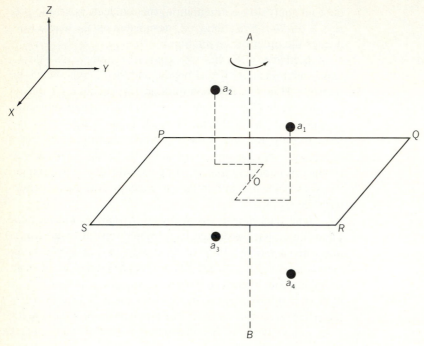

Figure E.14.1

The plane $PQRS$ is a reflection plane. Point a_1 is related by reflection to point a_4; point a_2 is related to a_3. How are the coordinates x, y, z related for two points made equivalent by reflection in a plane? (*17*)

What happens if we carry out successive operations? Suppose, for example, that we first rotate the molecule about C_2 and then reflect it in the mirror plane. The net result would be the same as if we had gone directly through the inversion center. Specifically, rotation of a_2 about the line AB sends a_2 into a_1; subsequent reflection of a_1 through $PQRS$ sends a_1 into a_4. The net result $a_2 \to a_1$ followed by $a_2 \to a_4$ is the same as inverting a_2 directly through O into a_4. This illustrates a general principle: Two successive symmetry operations on a molecule are equivalent to some other symmetry operation characteristic of that molecule. Show how the general coordinates x, y, z change through each of these consecutive operations. (*18*)

The complete set of symmetry operations that characterizes a molecule is said to constitute a *group*. The term "group" is used here in the special sense of mathematics, where there is a branch called *group theory*, which concerns itself with how the elements of a group must be related to each other. The elements of a group may be numbers, mathematical functions, or physical operations. The main requirement is that the combination of two elements of the group (e.g., the product of two numbers or one symmetry operation followed by another) must be equal to some other element of the group. Usually the interrelationships between the elements of a group are represented by what is known as a *multiplication table*. This is simply a matrix where the columns represent one operation, the rows represent another operation, and the intersection of a row and column indicates the combined operation.

Let us consider as a specific example the group composed of the symmetry operations E, C_2, i, and σ_h. This group would be used to describe a molecule containing a twofold rotation axis, an inversion center, and a mirror plane. The multiplication table for the group would be set up as follows:

	C_2	i	σ_h	E
C_2	E	σ_h	i	C_2
i	σ_h	E	C_2	i
σ_h	i	C_2	E	σ_h
E	C_2	i	σ_h	E

The symbols C_2, i, σ_h, and E across the top of the table identify the columns; the symbols C_2, i, σ_h, and E along the left edge identify the rows. Where a column and a row cross, we have put in the entry that indicates the combined result of first carrying out the operation indicated by the column heading and then following it by the operation indicated by the row designation. The letter E stands for the identity operation and indicates that the molecule has been returned to its original orientation.

Isomers

Two substances are *isomers* of each other if they have the same molecular formula but differ in some chemical or physical property. The difference in the property is caused by a difference in molecular structure between the two isomers. Given below are brief descriptions of the various kinds of isomers that are possible.

Conformers   When part of a molecule is rotated around a single bond, a new conformation is produced. The rotation can be by any amount, so an infinite number of conformations are possible, but usually certain specific ones are of interest. Conformations that correspond to a minimum in the potential energy are called *conformers*.

Structural Isomers   When bonds are broken and rearranged so that some of the atoms not bonded to each other in one molecule are joined together in the other, then the molecules are called structural isomers. The number of such structural isomers depends on how many ways the bonds can be rearranged consistent with the valence rules. Structural isomers can be:

1   *Positional isomers*, in which case the relative arrangement of structural units changes but general chemical properties remain the same. For example, an OH group might be moved from one carbon on a chain to another.

2   *Functional group isomers*, in which case atoms are rearranged so as to change the type of chemical reactions the molecule undergoes. For example, an oxygen atom might be moved from an OH group to a position between two carbons, thus changing class properties from those of an alcohol to those of an ether.

Stereoisomers   These always occur in pairs and differ from each other only in the orientation in space of their atoms. Unlike conformations, it is necessary to break and remake bonds to convert one stereoisomer into another, but the remade bonds are always to the same atoms as in the original molecule. Stereoisomers can be:

1   *Geometric isomers*, in which case the molecules are not mirror images but differ from one another in having two particular groups of atoms either *cis* (on the same side of a double bond or ring) or *trans* (on opposite sides of a double bond or ring).

2   *Optical isomers*, in which case one molecule is a nonsuperimposable mirror image of the other.

Optical isomers have the property that one isomer of the pair rotates plane-polarized light in one direction whereas the other isomer rotates it in the opposite direction. Depending on the rotation direction, they are distinguished as *d* (dextrorotatory) or *l* (levorotatory).

Procedure

From your model kit, construct each of the following molecules and study it for its isomer possibilities. Each has been chosen to illustrate a particular topic, so consider it specially from that viewpoint. Draw a picture of each of your structures, and tabulate the symmetry elements.

In drawing pictures of your molecules, particularly of the organic compounds, wherein the carbon always forms four bonds in tetrahedral directions, you may find it advisable to use one of the standard methods of organic chemistry for showing structural relations. Figure E.14.2 shows four different ways of representing a molecule of methane: (*a*) is a projection drawing, most useful for showing which atoms are bonded to which atoms; (*b*) is a line display formula in which the convention is that the atoms following C are bonded to it; (*c*) is a perspective drawing where the wedge is taken to represent a bond coming out of the paper, and the dashed line, going into the paper; (*d*) is another kind of perspective drawing where the central circle represents a carbon atom in the plane of the paper, and lines coming just up to or crossing into the circle are bonds going back into the paper or coming

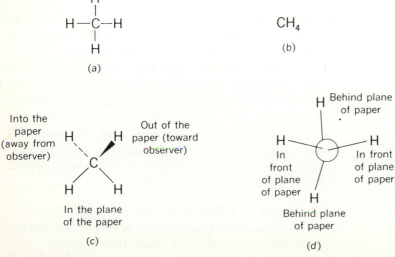

Figure E.14.2

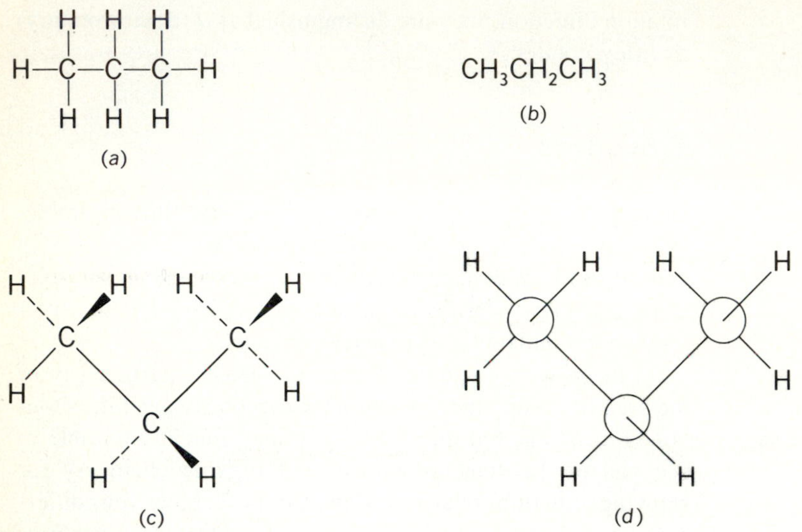

CH$_3$CH$_2$CH$_3$

(b)

(a)

(c)

(d)

Figure E.14.3

out of the paper, respectively. Use whatever method of drawing is most suitable for showing the important features of each molecule. Figure E.14.3 shows four ways of representing another molecule—propane.

(a)   **Ethane (C$_2$H$_6$)**   Look especially for conformers. Which do you consider to be the most important conformations? Why? (*19*) What do you think is the higher-energy conformation of ethane? (*20*)

(b)   **Cyclohexane (C$_6$H$_{12}$)**   Look especially for the conformers of a cyclic compound. A cyclic structure is one in which the atoms are bonded to form a closed figure, or "ring" (e.g., a square, a pentagon, a hexagon). There are a number of isomers of the formula C$_6$H$_{12}$, some of which are cyclic, some not. Work with the cyclic structure having a six-membered ring. Which do you consider to be the most important conformations, and why? (*21*) Why is the ring not planar? (*22*) What do you think is the most stable form, and why? (*23*) Tabulate the symmetry elements for your structures, considering only the carbon skeleton. (*24*)

(c)   **Butyl Alcohol and Dimethyl Ether (C$_4$H$_{10}$O)**   Look especially for structural isomers. How many alcohols of this formula can you make? (*25*) How many ethers? (*26*)

(d)   Butene (C_4H_8)   Look especially for geometrical isomers. (27) What about the possibility of a cyclic structure? (28)

(e)   Hexachloroplatinate ($PtCl_6^{2-}$)   See if you can discover six possible structures. Absorption spectra indicate the following group of symmetry elements: $9C_2, 4C_3, 3C_4, 9\sigma$, and i. Which of your postulated structures appears to be the right one? (29)

Instead of using knowledge of symmetry elements, another way to get structural information is by the method of *isomer number*. This is a chemical technique rather than a physical one and involves substitution of one or more of the atoms of the original molecule to see how many isomers are obtained. As an example of how this works, consider a compound of molecular formula C_3H_6, for which possible structures are

$$CH_2{=}CHCH_3 \qquad\qquad CH_2{-}CH_2$$
$$\text{I} \qquad\qquad\qquad \diagdown\diagup$$
$$CH_2$$
$$\text{II}$$

Suppose it is found by experiment that the chlorine-substitution reaction on C_3H_6 to produce C_3H_5Cl gives a product from which four isomers can be isolated. This immediately tells us the original C_3H_6 must have had structure I. Substitution of one Cl for one H of structure I can lead to

whereas substituting one Cl for H in structure II can give only

Examine the square-planar complex $PtCl_4^{2-}$ and decide how many isomers you would get on monosubstitution (for example, to make $PtCl_3Br^{2-}$) and on disubstitution (for example, $PtCl_2Br_2^{2-}$). Tabulate the symmetry elements for the new products. Do the same thing for $PtCl_6^{2-}$ going to $PtCl_5Br^{2-}$ or $PtCl_4Br_2^{2-}$. (*30*)

(f) **Benzene (C_6H_6)** First construct all the possible structures of the molecular formula C_6H_6. Don't forget straight-chain, branched-chain, doubly bonded, triply bonded, etc., structures. How many structures do you have? (*31*)

Next make monochloro derivatives of each of your structures. How many isomers of the type C_6H_5Cl can you get from each parent C_6H_6? If it is known that benzene gives only one monochlorobenzene, which of your original structures can you eliminate? (*32*)

It is known that the compound dichlorobenzene ($C_6H_4Cl_2$) exists as three distinct isomers. What does this tell you about which of the remaining structures you can eliminate? (*33*)

How many isomers can you make of trichlorobenzene ($C_6H_3Cl_3$)? Show how the number of trichlorobenzenes that you can make from a given dichlorobenzene enables you to decide which isomer of dichlorobenzene you are starting with. (*34*)

Determine and tabulate the symmetry elements for benzene and each of its chlorine derivatives. Consider the complete series C_6H_6, C_6H_5Cl, $C_6H_4Cl_2$, $C_6H_3Cl_3$, $C_6H_2Cl_4$, C_6HCl_5, C_6Cl_6. (*35*)

(g) **Chlorobutane (C_4H_9Cl)** Look especially for optical isomers. Construct all the isomers and classify them as to type.

To clarify the notion of optical isomers, consider the compound CHBrICl (chlorobromoiodomethane). Note that the tetrahedral carbon atom in the center is bonded to four different groups. This is important because it means that there is no plane of symmetry and no inversion center at the carbon. Such a carbon is called an *asymmetric carbon* or an *asymmetric center*; it is generally indicated by a star. In organic compounds, the presence of an asymmetric carbon is a very good indication that there are optical isomers. (Exceptions exist but are quite rare.) The optical isomers of CHBrICl are shown below. It is not possible to tell which is the *d* isomer and which the *l* without measuring the optical rotation.

Make models of these molecules, and be sure you see why they are an isomer pair. When you understand what the situation is, find the asymmetric carbon in each of the chlorobutanes (if there is one) and find the optical isomers.

Although an asymmetric carbon is a handy tool for finding optical isomers, its presence is not absolutely necessary. Helixes, for example, which are important in biological molecules, and some inorganic compounds such as discussed below, represent optically active molecules having no asymmetric carbon atom. The essential requirement is that two structures exist that can be superimposed on each other only after reflection of one of them in a mirror.

(h)   Trisoxalatochromate $[Cr(C_2O_4)_3^{3-}]$ The oxalate ion, $C_2O_4^{2-}$, has a planar structure as follows:

$$\left[\begin{array}{c} O \diagdown \quad \diagup O \\ C-C \\ O \diagup \quad \diagdown O \end{array}\right]^{2-}$$

It is an example of a bidentate ligand, where *ligand* means any group attached to a central metal atom and *bidentate* means that there are two points of attachment. For oxalate, the two points of attachment are through the two singly bonded oxygen atoms.

Chromium in its trivalent state generally forms octahedral complexes with the chromium in the center of an octahedral cage of attached atoms. For a complex such as $CrCl_6^{3-}$, there are six Cr—Cl bonds going out to the corners of an octahedron. For a complex such as $Cr(C_2O_4)_3^{3-}$, there are six Cr—O bonds going out to the octahedral corners, but pairs of octahedral corners are tied together through the rest of an oxalate group.

Construct and classify the isomers of $Cr(C_2O_4)_3^{3-}$. Note what symmetry elements each isomer has. (*36*)

In a similar fashion, study $Cr(C_2O_4)_2Cl_2^{3-}$ and $Cr(C_2O_4)Cl_4^{3-}$. To simplify your drawings, you might find it advisable to use a standard symbolism in which a bidentate ligand is represented by a simple arc between the two attachment positions. (*37*)

Questions

E.14.1   A rhombus is a plane figure with four equal sides in which the angles are not necessarily equal to 90°. Draw a rhombus in the *xy* plane having two 80° internal angles and two 100°

angles. Number the vertices 1 to 4, assigning the number 1 to an 80° corner and going around in a clockwise direction. Starting at point 1, perform the following operations, indicating whether each operation results in a nonequivalent, an equivalent, or an identical configuration. Using the labels 1, 2, 3, and 4, draw representations of the resulting configurations.

(a) C_2^1 about the z axis

(b) i

(c) σ through the xy plane

E.14.2   Draw a hexagon in the xy plane, and number the six vertices consecutively 1 to 6, starting at the top and going around the hexagon clockwise. Each time starting at position 1 (when a choice of position must be made), carry out each of the following operations on the hexagon, and draw a representation of the hexagon in the resulting configuration. Decide whether each resulting configuration is identical, equivalent, or nonequivalent to the starting configuration.

(a) σ through the xz plane

(b) C_6^5 around the z axis

(c) C_3^5 around the z axis

(d) i

(e) σ through the xy plane

E.14.3   Show that D-glucose contains a total of four asymmetric carbon atoms. The structure of D-glucose is as follows:

$$\text{CHO}$$
$$\text{H} \blacktriangleright \text{C} \blacktriangleleft \text{OH}$$
$$\text{HO} \blacktriangleright \text{C} \blacktriangleleft \text{H}$$
$$\text{H} \blacktriangleright \text{C} \blacktriangleleft \text{OH}$$
$$\text{H} \blacktriangleright \text{C} \blacktriangleleft \text{OH}$$
$$\text{CH}_2\text{OH}$$

E.14.4   The following compound has asymmetric carbon atoms, but it is not optically active. Why is it not optically active?

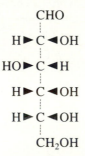

$$\text{CH}_2\text{OH}$$
$$\text{H} \blacktriangleright \text{C} \blacktriangleleft \text{OH}$$
$$\text{H} \blacktriangleright \text{C} \blacktriangleleft \text{OH}$$
$$\text{CH}_2\text{OH}$$

E.14.5   Tabulate the symmetry elements for each of the figures shown in Figure E.14.4. You may find it helpful to make models out of cardboard and adhesive tape.

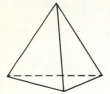

Triangular pyramid
(Only the base is equilateral)

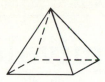

Square pyramid

Tetrahedron (regular)
(All four faces are equilateral)

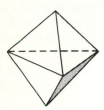

Triangular bipyramid

Octahedron

Dodecahedron

Figure E.14.4

E.14.6 Show that the symmetry element S_2 is equivalent to i. Show also that S_1 is equivalent to σ.

E.14.7 What are the symmetry elements of each of the following?
(a) Snowflake   (f) Hot dog
(b) Bumblebee   (g) Automobile tire
(c) Maple leaf   (h) The Pentagon
(d) Acorn   (i) Your right hand
(e) Watermelon

E.14.8 If an object has symmetry elements E and S_3, what other symmetry elements must it also possess?

E.14.9 Construct the multiplication table for the group containing the symmetry operations E, $C_4{}^1$, $C_4{}^2$, $C_4{}^3$, σ_h, i.

E.14.10 How do the symmetry elements differ in the two structural isomers of trichloroethane? For each of the isomers, draw the probable curve of potential energy vs. angle of rotation around the axis of the C—C bond. Account for the specific shape of each curve.

E.14.11 How would MX_4 (tetrahedral) differ from MX_4 (planar) in the number of isomers that would be obtained on monosubstitution, disubstitution, and trisubstitution of Y for X?

E.14.12 Draw diagrams for the isomers that would be expected for an octahedral complex $M(A—A—A)_2$ where A—A—A is a tridentate ligand. Indicate the symmetry elements of each.

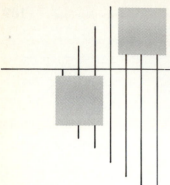

Equivalent Weight of a Metal

Special Items

50-cm string
Laboratory barometer
Two 4-cm strips of metal
   (<0.050 g)
30 mL 3 M HCl

One equivalent of a reducing agent is defined as that mass of the reducing agent that furnishes the Avogadro number of electrons. Similarly, one equivalent of an oxidizing agent is the mass of oxidizing agent that picks up the Avogadro number of electrons. In any oxidation-reduction reaction, the number of equivalents of oxidizing and reducing agent must be equal since the number of electrons gained must equal the number of electrons released.

In this experiment you will study the reaction of a metal on hydrogen ion to form hydrogen gas. The hydrogen gas will be collected and its volume measured at a known pressure and temperature. This will enable you to calculate the number of moles of H_2 formed and hence the number of moles of H^+ (or H_3O^+) used. When H^+ is reduced to hydrogen gas, each H^+ ion picks up one electron; therefore, in this reaction, 1 mol of H^+ is equal to 1 equivalent of H^+. Thus from the moles of H_2 formed, you can calculate the number of equivalents of H^+ (oxidizing agent) used up, which must equal the number of equivalents of metal (reducing agent) used up. Knowing the initial mass of metal, you will calculate the mass of one equivalent of metal in this reaction.

Procedure

■ WEAR YOUR SAFETY GOGGLES

Weigh separately two strips of metal about 4 cm long (maximum mass of each, 0.050 g) to the nearest 0.001 g. For identification, fold the first strip into quarters and the second into thirds. Tie about 25 cm of string to each of these, more or less as shown in Figure E.15.1. Fill a 600-mL beaker about three-fourths full of water. Close off the tip of your gas-measuring tube with a rubber connector and a pinch clamp. Pour about 15 mL of dilute hydrochloric acid into the tube, and carefully fill the rest of it with distilled water. Submerge one of the metal samples to a depth of about 5 cm. Hold your

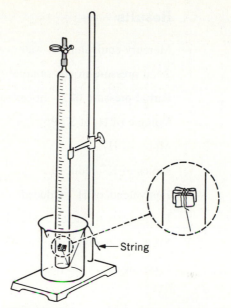

Figure E.15.1

thumb tightly over the mouth of the tube, quickly invert, and set in the beaker of water. Clamp the tube with its mouth just touching the bottom of the beaker as shown in Figure E.15.1. The acid, being denser, diffuses downward, and you should observe evolution of gas when it comes in contact with the metal. When the reaction is complete, measure the difference in water levels inside and outside the measuring tube and the volume of gas in the measuring tube. Insert your thermometer in the beaker so that its bulb is near the mouth of the measuring tube; record the temperatue. Note also the barometric pressure.

Rinse out the beaker and measuring tube, and repeat the experiment with the other sample of metal.

Data

Mass of metal   _____ _____

Volume of hydrogen observed   _____ _____

Difference in water levels   _____ _____

Temperature of water   _____ _____

Barometric pressure   _____ _____

Results

Mercury equivalent of water column   _____ _____

Total pressure in gas sample   _____ _____

Partial pressure of H_2 in gas sample   _____ _____

Volume of H_2 at STP   _____ _____

Moles of H_2 _____ _____

Moles of H^+ reduced   _____ _____

Equivalents of H^+ reduced   _____ _____

Equivalents of metal oxidized   _____ _____

Mass of 1 equivalent of metal   _____ _____

    Average   _____

Questions

E.15.1   Suppose that your sample of metal had partially oxidized before you started the experiment. How would this affect your calculated equivalent weight?

E.15.2   Suppose that you didn't wait long enough before getting the final temperature reading, and the temperature of the gas was higher than the temperature of the water. How would this affect your calculated equivalent weight?

E.15.3   How many electrons were transferred in oxidizing your sample of metal?

E.15.4   How do you determine the pressure of the gas inside the measuring tube before and after the reaction?

E.15.5   In a typical experiment of this kind, it is observed that 0.1362 g of metal liberates 33.6 mL of hydrogen gas at 23°C and 0.982 atm barometric pressure. What is the equivalent weight of the metal?

E.15.6   Suppose that you used dilute phosphoric acid (H_3PO_4) instead of hydrochloric acid (HCl) in this experiment. What difference would it make in your results?

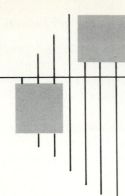

Experiment E.16

Crystal Structure

Special Items

Twenty-seven 2.5-cm and thirteen 5.0-cm balls
60 wooden sticks (toothpicks)

Crystalline solids are characterized by distinctive geometric shapes, which are thought to arise from definite orderly patterns of the constituent atoms. The order of arrangement can be represented by the unit cell. The unit cell is defined as the smallest portion of the space lattice that, when moved a distance equal to its own dimensions in various directions, generates the whole space lattice. In this experiment, you will first construct models of various unit cells and, assuming that atoms can be represented as rigid spheres, calculate the fraction of the volume that is empty space. Then you will construct models of the unit cells of various compounds and show the relation to the simplest formula. Finally, you will investigate the close-packing of spheres. (If you have access to a microscope, you might want to look at some crystals under the microscope while you are doing this experiment.)

Procedure

■ WEAR YOUR SAFETY GOGGLES

Obtain a supply of Styrofoam balls and wooden sticks. Construct each of the models indicated below. When using the sticks to join two balls, insert one end into one ball and then push the second ball on the point sticking out. When finished, remove all sticks by pulling them straight out and return them with the balls to the reagent table.

(a) Construct a single unit cell of the simple cubic lattice (Figure E.16.1). The dots in the figure represent centers of atoms; in your model, spheres should be in contact. By using only four more balls, extend your model in the c direction so that you end up with two unit cells in contact. Note that four of the balls are shared by the two unit cells. In similar fashion, extend the model in the a and b directions. Continue building until you have used a total of 27 balls and have formed a large cube containing eight unit cells.

Focus your attention on one unit cell. Note that because the spheres are in contact along the cube edge, the edge length of the unit cell (distance from sphere center to sphere center)

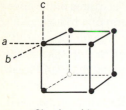

Simple cubic   Body-centered cubic   Face-centered cubic

Figure E.16.1

is only twice the radius of one sphere. Record the edge length in terms of sphere radius on the data sheet.

Now focus your attention on the sphere hidden in the center of your model. What fraction of this sphere lies in each unit cell? Record. Note that any sphere lying at the corner of a cubic unit cell contributes this same fraction of a sphere to the unit cell. The equivalent of how many spheres are there within a simple cubic unit cell? Record.

(b) Construct a single unit cell of the body-centered cubic lattice (Figure E.16.1). Start with the center sphere, and work outward. Note that the spheres are in contact along the cube diagonals but not along the cube edges. Record the cube-diagonal length in terms of sphere radius. Record also the equivalent number of spheres within the unit cell.

(c) Construct a single unit cell of the face-centered cubic lattice (Figure E.16.1). Note that the spheres are in contact along the face diagonals; so first construct a face. Record the face-diagonal length in terms of sphere radius and the equivalent number of spheres within the unit cell.

(d) Construct a single unit cell of the NaCl lattice. Use 5-cm balls for the chloride ions and 2.5-cm balls for the sodium ions. Start with a chloride ion, and attach four sodium ions at right angles all in the same plane. Insert a chloride between each pair of sodiums in the same plane. With the plane horizontal, add sodiums above and below each chloride and chlorides above and below each of the four sodiums. Your final model should contain 27 balls (13 chlorides and 14 sodiums). The equivalent of how many sodium ions is within the unit cell? Of how many chloride ions? Compare with the simple formula.

(e) Construct a unit cell (face-centered cubic) in which each face center is occupied by a small ball *A* and each corner by a large ball *B*. Determine the equivalent number of *A* and *B* spheres within the unit cell.

(f) Construct a tetrahedral hole as follows: Place three of the large spheres on your desk in contact in the form of a triangle. Join them with sticks. Place a fourth large sphere on top over the pocket formed by the other three. Take off the top sphere, put a small sphere in its place, and then try to put the fourth large sphere back on top. The small sphere is in the tetrahedral hole formed by four spheres in contact. Decide whether the small sphere fits loosely in the hole or is too big for it.

Construct an octahedral hole as follows: Place three large spheres on your desk in contact in the form of a triangle. Join them with sticks. Make a second identical unit. Set the second unit flat on top of the first unit so that the triangular planes are parallel to each other. Rotate the top unit around a vertical axis so the top spheres nestle into the pockets between the bottom spheres but the hole in the upper triangle is just above the hole in the lower triangle. Note that the six spheres form an octahedron about a central cavity. Determine whether you can fit a small sphere in the octahedral hole.

Make a portion of a close-packed layer as follows: Place a large sphere on your desk. Cluster around it in the same plane as many other large spheres as can be accommodated next to the original sphere. Note the number of near neighbors and the geometric figure formed by their centers. Join the unit together with sticks. Construct another identical unit. Stack the second unit over the first so the holes lie on the same vertical axes. Determine the distance between layers. Slide the top layer so its spheres nestle over holes in the bottom layer. Determine the distance between layers and compare with sphere diameter. Note the formation of tetrahedral and octahedral holes between layers.

Make a third portion of a close-packed layer. Stack any three units over each other and determine how many kinds of stacking are possible.

Data

(a) Edge length in terms of sphere radius r _____

Fraction of interior sphere in each unit cell   _____

Number of spheres per unit cell   _____

(b) Cube-diagonal length in terms of r _____

Number of spheres per unit cell   _____

(c) Face-diagonal length in terms of r _____

Number of spheres per unit cell   _____

(d)   Sodium ions per unit cell _____

Chloride ions per unit cell _____

(e) A spheres per unit cell _____

B spheres per unit cell _____

(f)   Observation on tetrahedral hole:

Observation on octahedral hole:

Observations on close-packed layers:

Results

(a)   Volume of unit cell in terms of r _____

Volume of spheres within unit cell of terms of r
(volume of sphere $= 4\pi r^3/3$) _____

Fraction of unit cell that is empty space _____

(b)   Edge length of unit cell in terms of r _____

Volume of unit cell in terms of r _____

Volume of spheres within unit cell in terms of r _____

Fraction of unit cell that is empty space _____

(c)   Edge length of unit cell in terms of r _____

Volume of unit cell in terms of r _____

Volume of spheres within unit cell in terms of r _____

Fraction of unit cell that is empty space _____

(d)   Simplest formula _____

(e)   Simplest formula _____

(f)   Size of Tetrahedral Hole
Edge length of tetrahedron in terms of r _____

Distance from vertex to center in terms of r _____

Distance from vertex occupied by large sphere   _____

Possible radius for small sphere in hole   _____

Size of Octahedral Hole
Edge length of octahedron in terms of r _____

Distance from vertex to center in terms of r _____

Distance from vertex occupied by large sphere   _____

Possible radius for small sphere in hole   _____

Distance between Close-Packed Layers
(Consider a tetrahedron formed by three spheres in one layer and a fourth sphere in the adjacent layer.)

Edge length of tetrahedron in terms of r _____

Perpendicular distance from vertex to opposite face (= altitude of the tetrahedron) in terms of r _____

Questions

E.16.1   For space lattices in which spheres are at each lattice point and the spheres are in contact, find the unit cell density if the spheres are 2.00 g per cm^3 and if the unit cell is
(a) Simple cubic
(b) Body-centered cubic
(c) Face-centered cubic

E.16.2   In a close-packed three-dimensional structure composed of identical spheres, what is the ratio of the number of spheres to the number of
(a) Octahedral holes?
(b) Tetrahedral holes?

E.16.3   Copper metal crystallizes in a cubic closest-packed structure. The atomic radius of copper is 0.128 nm. Find the density of copper.

E.16.4   Tungsten metal crystallizes in a structure having a body-centered cubic unit cell that is 0.316 nm on a side. The density is 19.35 g per cm^3. Find the Avogadro number from these data.

E.16.5   A certain metallic oxide M_xO_y crystallizes with the oxygen atoms located at every lattice site in a cubic closest-packed array and the metal atoms in each of the tetrahedral holes. What is the empirical formula of the oxide?

E.16.6   How large a sphere can you place inside the hole located at the center of a face-centered cubic cell? Express your answer in terms of r, the radius of the equivalent spheres of which the cubic array is constructed.

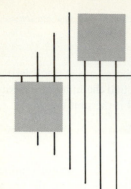

Experiment E.17

Temperature and Vapor Pressure of Water

In this experiment you will investigate the relationship between the temperature and vapor pressure of water. Through mathematical manipulation and graphical treatment of the data, you will obtain an estimate of the molar heat of vaporization of water, $\Delta\overline{H}_{vap}$ (the bar above H means "per mole"), the amount of heat that must be added to convert a mole of liquid water to water vapor at the boiling point and 1 atm of external pressure. Something interesting is that in the experiment you will *not* carry out any measurements at the boiling point. Rather, you will make a series of measurements of vapor pressure at different temperatures, ranging from 50 to 80°C. Although molar heat of vaporization changes slightly with temperature, the fact that the change is small makes the procedure quite valid.

Method

A sample of air will be trapped in an inverted test tube immersed in a water bath. The water bath will be heated to a temperature exceeding 80°C and then will be allowed to cool. As it cools, temperature and gas volume readings will be recorded. The number of moles of water vapor in the gas phase changes with temperature, but the amount of "waterless" air remains constant (assuming negligible solubility variation with temperature).

If the number of moles of "waterless" air in the gas sample were known, then the partial pressure of air could be calculated at each temperature and the vapor pressure of water obtained by the difference from the barometric pressure,

P_{atmos}, according to

$$P_{air} = P_{atmos} - P_{H_2O\,vapor}$$

The number of moles of air can be found from measurement of the volume, temperature, and pressure of the trapped gas at a temperature near 0°C, where the water vapor content is less than 1 percent and can be neglected.

The estimate of the molar heat of vaporization will be made by use of the Clapeyron equation,

$$\log \frac{p}{p_0} = \frac{-\Delta \overline{H}_{vap}}{2.303\,RT} + C$$

where p is the equilibrium vapor pressure, p_0 is a reference pressure (defined as 1 atm), T is the temperature in kelvins, R is the universal gas constant, and C is a constant that depends on the particular liquid. If $\Delta \overline{H}_{vap}$ is expressed in joules per mole, $R = 8.31$ joules per mol-deg, and the expression reduces to

$$\log \frac{p}{p_0} = \frac{-\Delta \overline{H}_{vap}}{19.15\,T} + C$$

The vapor pressure will be determined at a series of temperatures. Then the logarithm of p/p_0 will be plotted versus the inverse of the absolute temperature readings. The plot of $\log (p/p_0)$ versus $1/T$ should give a straight line of slope = $-\Delta \overline{H}_{vap}/19.15$.

Procedure

1   Cut out a slip of paper 8.5 × 1 cm in size and, using your ruler, carefully draw neat lines (a scale) on the paper 1 mm apart. Label the paper, and attach it to the outside of a 90-mm test tube facing inward, as shown in Figure E.17.1, using waterproof transparent tape. The left end of the scale should

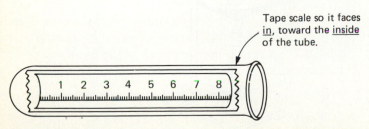

Tape scale so it faces in, toward the inside of the tube.

Figure E.17.1

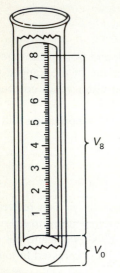

Figure E.17.2

be 1 cm from the bottom of the test tube. The scale should be read by looking *through* the test tube.

Using a burette, fill the test tube until the bottom of the meniscus just reaches the bottom of your scale, the "zero mark." Record the volume used as V_0. Next, use the burette to add enough water to reach the 8.0 mark with the bottom of the meniscus. Again, record the volume (i.e., from the zero mark to the 8.0 mark) in your laboratory journal, this time as V_8 (see Figure E.17.2).

2   Fasten the test tube next to the bottom end of the thermometer, as shown in Figure E.17.3.

3   Place about 200 cm^3 of crushed ice in your 600-mL beaker, and then nearly fill the beaker with distilled water. Stir the slush until the temperature has dropped to $\approx 0°C$, and then remove nearly all the excess ice.

4   Fill the test tube you prepared earlier with distilled water until it is about 2 cm from the top. Cover the top with your finger. Invert the test tube, submerge it in the ice-water in your 600-mL beaker, and remove your finger. An air sample of approximately 1.5 mL should be trapped. Attach the thermometer to a clamp on a ring stand (using a split stopper) such that the test tube is kept submerged (Figure E.17.3). Add more water to the beaker if necessary to make sure that the air trapped in the test tube is surrounded by water.

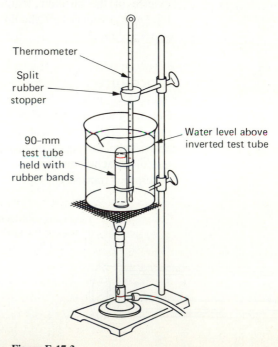

Figure E.17.3

5   With the ice-water mixture surrounding the test tube at $\approx 0°C$, take a reading of the "length" of the air column in your test tube (to the bottom of the meniscus), as measured by the scale taped to the tube, and record the reading in your laboratory notebook. This reading later will be converted to a volume.

6   Remove any crushed ice remaining, and, if necessary, add enough distilled water to the beaker to make certain the inverted test tube is completely submerged.

7   Heat the beaker with a gas burner to approximately 85°C, or until the gas expands beyond the 8.0 mark on your scale.

8   Remove the burner, and allow the beaker to cool, stirring the water bath frequently to avoid thermal gradients. About every 4° between 80 and 50°C, record the temperature of the water to the nearest 0.1°C. At the same time, take readings of the "length" of the air column, recording the data in your laboratory notebook.

9   Obtain a reading of the barometric pressure, p_{atmos}.

Data

Record volumes V_0 and V_8 in your laboratory journal. Also set up two data columns, labeled "Temperature, °C" and "Air column readings." Enter the measurements as you carry out the experiment. Finally, record the barometric pressure, p_{atmos}.

Results

1   Set up six columns of information in your laboratory journal with the following headings:

T, °C	Gas volume, mL	p_{air}, atm	p_{H_2O}, atm	$\log \dfrac{p_{H_2O}}{p_0}$	$\dfrac{1}{T}$, K^{-1}

2   Using the air column readings from the scale taped to your test tube, together with V_0 and V_8, convert the readings to volumes in mL. Before entering the values into the "Gas volume" column, subtract 0.2 mL from each value to correct for the inverted meniscus.

3   Calculate the number of moles of "waterless" air in the sample, using the data obtained near 0°C.

4   For each temperature reading, calculate a value for the partial pressure of air in the gas sample, entering data in the "p_{air}" column.

$$p_{air} = \frac{n_{air} RT}{V}$$

5   Calculate the vapor pressure of water at each temperature, and enter the values in the column labeled '' p_{H_2O}, atm.''

6   Divide each p_{H_2O} value by p_0, where p_0 is defined to be exactly 1 atm. This is done in order to be able to take logarithms of pure numbers rather than numbers with units. Then take the $\log_{10}$ of each p/p_0 value. Also calculate the inverse of each temperature value, i.e., $1/T$, K^{-1}, and record the information in the appropriate columns.

7   Plot $\log(p_{H_2O}/p_0)$ on the y axis versus $1/T$ on the x axis of the graph paper provided, and draw the best straight line through the set of points. Determine the slope $\Delta \log(p_{H_2O}/p_0)/\Delta(1/T)$ and, using the Clapeyron equation, calculate $\Delta \overline{H}_{vap}$ for water.

8   Read the value of the vapor pressure of water at 65°C from the graph.

Questions

E.17.1   Explain why a plot of $\log(p_{H_2O}/p_0)$ versus $1/T$ gives a straight line of slope $-\Delta \overline{H}_{vap}/19.15$.

E.17.2   Explain how you will obtain the vapor pressure readings from the experimental setup.

E.17.3   If $V_0 = 1.80$ mL and $V_8 = 5.30$ mL, write an equation that could be used to convert scale readings to volumes.

E.17.4   In part 2 of the Results, you are directed to correct the gas volume values "for the inverted meniscus." Explain, using a sketch.

E.17.5   Suppose the test tube were not completely immersed in the water bath. How would this affect your experimental value of $\Delta \overline{H}_{vap}$?

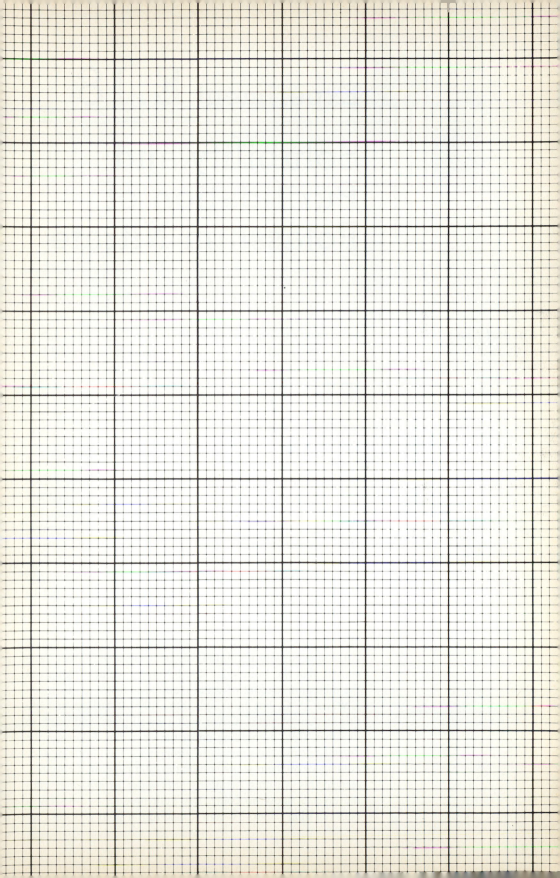

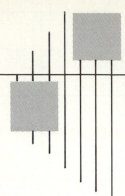

Experiment E.18

Solubility and Purification of Potassium Nitrate

Special Items

50 mL distilled water
40 g KNO$_3$
10-g mixture of KNO$_3$ and Cu(NO$_3$)$_2$

The change of solubility with temperature varies from salt to salt. In this experiment, you will determine how the solubility of potassium nitrate changes with temperature and use this property to purify a sample of potassium nitrate that is contaminated with a colored salt.

Procedure

■ **WEAR YOUR SAFETY GOGGLES**

(a) Measure out 10 mL of water into your large test tube. Weigh out on the platform balance 4 g of potassium nitrate, and add to the test tube. Heat gently until all the solid disappears, but do not boil the water. Insert your thermometer in the tube, and allow to cool. Stir gently. Record the temperature at which solid begins to crystallize.

Repeat the entire experiment using 10 mL of water and 8 g of potassium nitrate. Repeat successively with 12 and 16 g of potassium nitrate.

(b) Obtain from your instructor a 10-g sample of potassium nitrate contaminated with a colored salt such as copper nitrate. Assuming that the sample is 75% KNO$_3$ by weight, figure out the minimum amount of water required to dissolve this much KNO$_3$ at 75°C. Put the sample into a 100-mL beaker, and add this amount of water. Cover with a watch glass, and warm to dissolve. Cool by holding under the water tap. Filter out the crystals. Rinse the crystals by dropping a few drops of very cold water on them while still on the filter paper in the funnel. Repeat this washing, *using as little water as possible*, until the crystals show no color. Dry your product by pressing it between filter papers. Weigh it, and turn it in to your instructor. He or she will tell you the original composition of the mixture so that you can calculate the percentage recovery.

Data

(a)   Temperature at which crystals appear:

With 4 g KNO_3 per 10 g H_2O _____

With 8 g KNO_3 per 10 g H_2O _____

With 12 g KNO_3 per 10 g H_2O _____

With 16 g KNO_3 per 10 g H_2O _____

(b)   Mass of product recovered   _____

Results

(a)   Plot a solubility curve for potassium nitrate, showing molality of the saturated solution vs. temperature (horizontal axis).

(b)   Grams of KNO_3 in original mixture   _____

Percentage recovery   _____

Questions

E.18.1   In part (a), the solubility behavior of pure KNO_3 is observed; in part (b), you use information from part (a) to figure out how much water to add at 75°C. In part (b), the presence of $Cu(NO_3)_2$ is expected to depress the solubility of KNO_3 somewhat, owing to the common-ion effect. Does this mean that you should add more or less water than the amount determined from part (a)? Explain.

E.18.2   Suppose that the KNO_3 solutions tended to become supersaturated before crystallization was observed. How would this affect your data?

E.18.3   From the procedure given, you can deduce that the solubility of KNO_3 increases with temperature. Does this correspond to an endothermic or an exothermic heat of solution? Explain.

E.18.4   The solubility of $Cu(NO_3)_2$ is considerably higher than that of KNO_3 at T_1 and about the same as KNO_3 at T_2. Which temperature must be higher, T_1 or T_2, for the separation to be effective using the procedure given in this experiment? Explain.

E.18.5   The solubility product constant K_{sp} of KNO_3 equals the molality of K^+ times the molality of NO_3^- in a saturated solution, that is, $K_{sp} = [K^+][NO_3^-]$. Find the K_{sp} of KNO_3 at 20°C and at 80°C. The heat of solution may be obtained from the relation

$$\log \frac{K_{sp1}}{K_{sp2}} = \frac{\Delta H}{19.15} \left(\frac{1}{T_1} - \frac{1}{T_2} \right)$$

where ΔH is the heat of solution in joules per mole and T is temperature in kelvins. Find the heat of solution of KNO_3 in this temperature range.

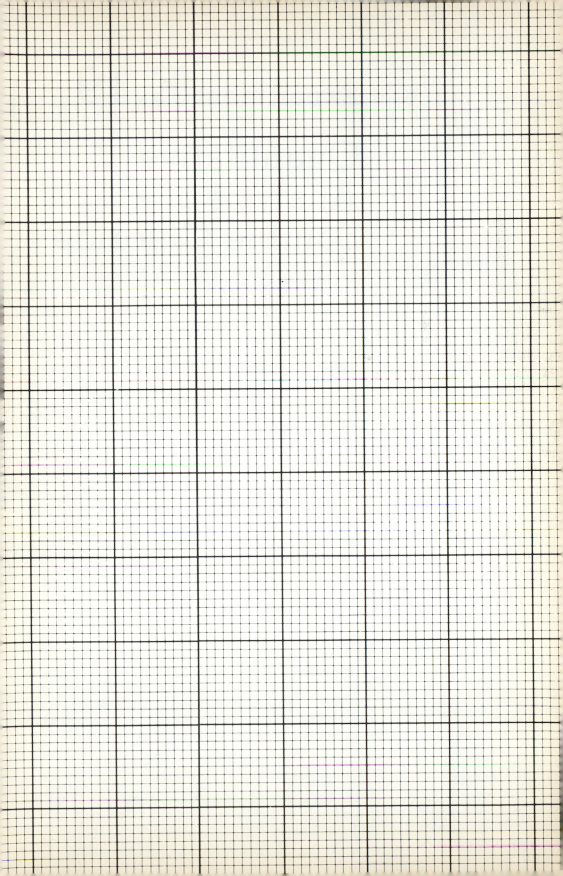

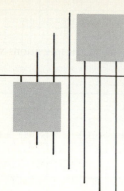

Quantitative Analysis of a Soluble Chloride

Special Items

Diphenylcarbazone and bromophenol blue indicator

50-mL burette

Three 250-mL Erlenmeyer flasks

One 100-mL volumetric flask, one 10-mL graduated cylinder, one 25-mL pipette

Standardized sodium chloride solution (0.02 M)

Mercuric nitrate solution (ca. 0.0125 M)

0.050 M nitric acid

Crude rock salt

Many years ago, a large part of upstate New York was under water. As this water evaporated, huge deposits of sodium chloride accumulated. They were subsequently covered by limestone, as skeletons of marine animals accumulated and were compacted. Today, as a result, we have 2000 ft below Cayuga's waters a large bed of almost pure sodium chloride. It is not pure enough for human consumption, but the Cargill Salt Company mines it and sells it for industrial use. In this experiment the problem is to determine quantitatively the chloride content of Cayuga rock salt. The method to be used is a titration method depending on the fact that chloride ion cannot exist free in solution simultaneously with mercuric ion in any appreciable concentration.

Mercuric ion (Hg^{2+}) reacts with chloride ion (Cl^-) to form $HgCl_2$, which although quite soluble does not dissociate appreciably. Dropwise addition of mercuric nitrate solution (Hg^{2+}, NO_3^-) to a sodium chloride solution (Na^+, Cl^-) leaves little mercuric ion free as long as there is sufficient chloride ion to form the $HgCl_2$ species for the equilibrium

$$Hg^2 + 2Cl^- \rightleftharpoons HgCl_2$$

Appearance of excess mercuric ion can occur only when all the chloride has been used up by addition of an equivalent amount of Hg^{2+}. Excess mercuric ion can be detected by use of diphenylcarbazone (DPC), which is an organic molecule that combines with Hg^{2+} to form a blue-violet complex:

$$Hg^{2+} + 2DPC + 2H_2O \rightleftharpoons Hg(DPC)_2 + 2H_3O^+$$

Colorless   Colorless   Blue-violet

The key to accurate determination of chloride is to use the appearance of the blue-violet color to tell the exact point at which free Hg^{2+} starts to accumulate. Complications will occur if the solution is basic, since Hg^{2+} reacts with OH^- to form mercuric oxide:

$$Hg^{2+} + 2OH^- \rightarrow HgO + H_2O$$

The effect of such a reaction would be to steal away the Hg^{2+} and prevent it from forming the colored complex. To ensure against such a complication, you can maintain the solution on the acid side at a pH of 3 by using nitric acid. The pH can be monitored by use of a second indicator, bromophenol blue, which is yellow when the pH is 3 or lower and blue when the pH is 3.6 or higher. In practice, the two indicators bromophenol blue and diphenylcarbazone are combined into a single indicating solution. Assuming that the solution stays at pH 3, the color change at the end point will be from yellow (characteristic of bromophenol blue) to blue-violet (characteristic of the mercury-DPC complex). As the end point of the titration is approached, the blue-violet color becomes more persistent until one drop is able to color it permanently. Be careful not to add the mercuric nitrate too rapidly near the end point; otherwise, you may overshoot.

Procedure

■ **WEAR YOUR SAFETY GOGGLES**

Obtain from the stockroom one 50-mL burette, three 250-mL Erlenmeyer flasks, one 100-mL volumetric flask, one 10-mL graduated cylinder, and one 25-mL pipette. Examine the pipette and the burette for leaks before using in the experiment by testing with water solutions. In order to read the level of liquid in the burette properly, hold a piece of paper with a dark band drawn on it behind the burette on the level of the bottom of the meniscus. During the titration take special care in washing the burette tip with squirts of distilled water from your wash bottle so as to ensure addition of all the solution. Often a fraction of a drop will cling to the tip of the burette after an addition. Finally, rest the flask containing an unknown solution on a white background such as a paper towel in order to see the color changes at the end point more clearly.

Use your pipette to measure out 25 mL of standardized sodium chloride solution into each of the three Erlenmeyer flasks. Add about 1 mL of indicator solution to each. If a blue-violet or red develops, add 0.05 M nitric acid dropwise until the color changes to yellow, and then add 15 drops in excess. Using your burette, titrate with $Hg(NO_3)_2$ to the first persistent violet. Record the volumes. Calculate the exact concentration of the $Hg(NO_3)_2$ solution.

Now obtain some of the crude rock salt and grind it up in a mortar. Accurately weigh about 0.15 g of the mixture, carefully transfer it to the volumetric flask, and fill to the 100-mL mark with distilled water.

Now repeat the titration procedure using the rock-salt solution in place of the standard NaCl. Be sure to rinse your pipette with the new solution to be analyzed. Record your results. Calculate the percentage by weight of chloride in Cayuga rock salt.

Data and Results

Questions

E.19.1   In order to get the DPC complex color to show up, you must add a small amount of $Hg(NO_3)_2$ solution in excess after reaching the equivalence point in the titrations. How does the procedure you use cancel out any possible error that this might cause in your determination of the percentage chloride in your sample of Cayuga rock salt?

E.19.2   What is the percentage of NaCl in your sample of Cayuga rock salt, assuming all the chloride came from NaCl?

E.19.3   What error, if any, might be caused if, in pipetting out the three 25-mL samples of standardized NaCl solutions, (a) the pipette contained some drops of distilled water, (b) the Erlenmeyer flasks contained some distilled water when the NaCl was added?

E.19.4   In excess chloride ion concentrations, Hg^{2+} ion forms the complex ion $HgCl_4^{2-}$. In the titration of chloride ion by Hg^{2+}, why is this not a problem?

E.19.5   What effect would each of the following have on a determination of the percentage of NaCl in a sample of Cayuga rock salt?
(a) The sample contained some sand.
(b) The sample contained some KCl.
(c) The pH of the titration solution was not kept low.
(d) The sample contained some Br^- ion (K_{sp} of $HgBr_2 \approx 10^{-22}$).

E.19.6   Obtain from your classmates at least nine values of the percentage of chloride in Cayuga rock salt. Calculate an average, an average deviation, and a standard deviation. Express your result in terms of the uncertainty.

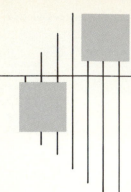

Experiment E.20

Alloys of Tin and Lead

Special Items

Ruler
Scissors
Brass or copper sheet 5 × 10 cm
4 g tinfoil
4 g lead foil
Five temperature standards

An alloy can be defined as a combination of two or more elements that has metallic properties. Many common alloys are solid solutions of one metal in another. When one metal is dissolved in another, we might expect the properties of the product to be different from that of either pure component. In this experiment, you will prepare various alloys of tin and lead (solder) and will determine what happens to the melting point as the composition is changed. To find the melting points of the alloys, you will compare their fusibility with that of standards of known melting point.

Procedure

■ **WEAR YOUR SAFETY GOGGLES**

Prepare 2 g of an alloy that is 20 percent tin and 80 percent lead by weight as follows: On the reagent shelf, you will find tinfoil and lead foil, each labeled by mass per unit area. With ruler and scissors, cut the desired amount of each. Press the two metals into a compact ball, and place in a clean crucible. Heat to fusion. Stir well, and skim off the oxide layer. Using crucible tongs, pour the molten alloy on a cold surface, such as the desk top. Press the alloy *immediately* with the desk padlock to flatten it out. In similar fashion, prepare 2 g of each of the following: 40% Sn–60% Pb, 60% Sn–40% Pb, 80% Sn–20% Pb. Try to make the final products of about equal thickness.

Determine the approximate melting points of your alloys as follows: Lay your transite board on the ring as shown in Figure E.20.1. Place a strip of brass, about 5 cm wide and 10 cm long, so that one end sticks out over the edge of the board about 2 cm. Cut a thin strip (about 5 × 2 mm) of each alloy and of each standard available on the reagent shelf. On the end of the brass strip arrange the standards in order of increasing melting point followed by the unknowns in order of increasing tin content. Press the samples flat against the brass strip. Carefully observing the nine strips of metal, heat

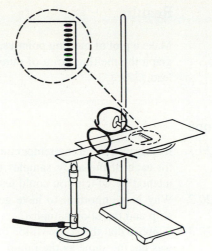

Figure E.20.1

with a low flame the free end of the brass strip. Note where each alloy melts relative to the standards. Poke the samples occasionally with a file tip so as not to overlook any melting.

Data

Results

Make a plot of melting point vs. composition. Include on your graph the melting point of pure tin (232°C) and that of pure lead (327°C).

Questions

E.20.1   It is assumed that the temperature of the brass strip is uniform for each of the alloy samples being heated on it. Propose a method by which you could test this assumption.

E.20.2   Why is it important to have good contact between the brass strip and the alloy samples?

E.20.3   Suppose that the composition of the oxide layers skimmed off was not the same as the liquid alloys. If the ratio of tin oxide to lead oxide was consistently higher than the ratio of tin to lead in the alloys, how would your results be affected?

E.20.4   Obtain from at least nine of your classmates values for the melting point of their 60% Pb–40% Sn alloys. Calculate an average value and an average deviation. What range of uncertainty can you place on your value?

E.20.5   Use your graph to predict the melting point of a 20% Sn–80% Pb alloy.

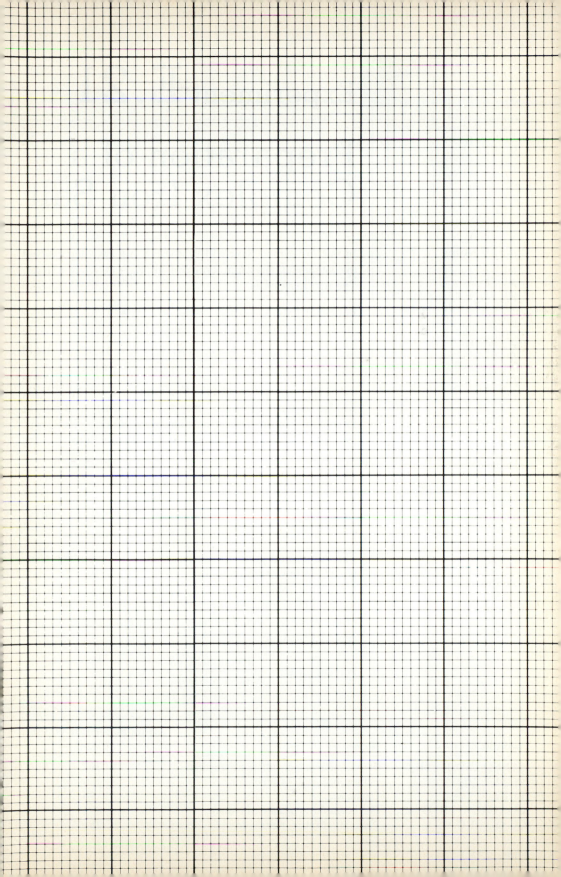

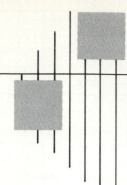

Experiment **E.21**

Preparation of a Complex Iron Salt

Special Items

30 mL distilled water
25 mL 1 M $H_2C_2O_4$
10 mL saturated $K_2C_2O_4$ soln
  (300 g $K_2C_2O_4 \cdot H_2O$ per
  liter)
20 mL 3% H_2O_2
10 mL 95% ethyl alcohol
5 drops 3 M H_2SO_4
5 g $Fe(NH_4)_2(SO_4)_2 \cdot 6H_2O$

In this experiment you will synthesize in a two-step process the compound potassium ferric oxalate, $K_3Fe(C_2O_4)_3 \cdot 3H_2O$. The first reaction is

$$Fe(NH_4)_2(SO_4)_2 \cdot 6H_2O + H_2C_2O_4 \xrightarrow{\Delta}$$

White solid (heat) Solution

$$FeC_2O_4 \cdot 2H_2O + (NH_4)_2SO_4 + H_2SO_4 + 4H_2O$$

Yellow solid   Soluble   Soluble

The main product of the first reaction, $FeC_2O_4 \cdot 2H_2O$, will be separated from the other products, which will remain in solution. In the second reaction, the $FeC_2O_4 \cdot 2H_2O$ will be converted to $K_3Fe(C_2O_4)_3 \cdot 3H_2O$, according to

$$2[FeC_2O_4 \cdot 2H_2O] + H_2C_2O_4 + H_2O_2 + 3K_2C_2O_4 \rightarrow$$

Oxalic   Hydrogen
acid   peroxide

$$2[K_3Fe(C_2O_4)_3 \cdot 3H_2O]$$

Green crystals

The names and gram molecular weights of important compounds in this experiment are listed in the table.

Your objective is to prepare as great a yield of $K_3Fe(C_2O_4)_3 \cdot 3H_2O$ as possible. To measure the efficiency of the procedure you will calculate something called the percent yield, where

$$\text{Percent yield} = \frac{\text{actual yield}}{\text{theoretical yield}} \times 100\%$$

Formula	Name	Gram Molecular Weight, g/mol
$Fe(NH_4)_2(SO_4)_2 \cdot 6H_2O$	Ferrous ammonium sulfate hexahydrate	392
$FeC_2O_4 \cdot 2H_2O$	Ferrous oxalate dihydrate	180
$H_2C_2O_4$	Oxalic acid	90
$K_3Fe(C_2O_4)_3 \cdot 3H_2O$	Potassium ferric oxalate trihydrate	491
$K_2C_2O_4 \cdot H_2O$	Potassium oxalate monohydrate	184

The *theoretical yield* is the calculated maximum amount of product that might be obtained under ideal conditions from the starting materials. In an experiment, the theoretical yield is seldom, if ever, reached. In this experiment, the theoretical yield will be the maximum number of grams of $K_3Fe(C_2O_4)_3 \cdot 3H_2O$ that might be obtained from the specified amounts of starting materials (reactants). In determining the percent yield, it will be necessary to calculate the moles of each reactant and then find the *limiting reagent*. The limiting reagent determines the theoretical yield of $K_3Fe(C_2O_4)_3 \cdot 3H_2O$.

Procedure

■**WEAR YOUR SAFETY GOGGLES**

In your 200-mL beaker, dissolve approximately 5 g (measure on analytical balance) of $Fe(NH_4)_2(SO_4)_2 \cdot 6H_2O$ in 15 mL of distilled water to which 5 drops of 3 M H_2SO_4 have been added. To this solution add 25 mL (use your graduated cylinder) of 1 M $H_2C_2O_4$ (oxalic acid) solution. Heat the mixture to boiling, stirring continuously to prevent spattering. This solution is very susceptible to boilover, so you must not leave it unattended. If it should start to boil over, immediately remove the flame. Allow the yellow precipitate of $FeC_2O_4 \cdot 2H_2O$ to settle. Using large beaker tongs, carefully decant the hot supernatant liquids into the sink, retaining the solid in the beaker. Add 20 mL of distilled water to wash the precipitate, warm, stir, allow the solid to settle, and again decant the liquid into the sink, retaining the solid.

To the solid ($FeC_2O_4 \cdot 2H_2O$) in the beaker, add 10 mL of saturated (300 g $K_2C_2O_4 \cdot H_2O$ per liter) potassium oxalate ($K_2C_2O_4 \cdot H_2O$) solution. Support your thermometer using a ring stand, a one-holed stopper, and a clamp, and place the thermometer in the solution. Carefully heat the solution to 40°C. Obtain 20 mL of 3% H_2O_2 in your 25-mL graduate, and add the 20 mL of H_2O_2, *very slowly*, a few drops at a time, stirring continuously and keeping the temperature near 40°C. Do not be too concerned if some $Fe(OH)_3$ (slimy red precipitate) appears. After adding all the H_2O_2, heat to boiling. To the boiling solution, add 8 mL of 1.0 M $H_2C_2O_4$—the first 5 mL all at once and the last 3 mL very slowly—keeping the solution boiling. Set up a funnel with filter paper on a ring stand (see Figures A.12 and A.13), and filter the boiling solution into a clean 100-mL beaker. Allow to cool to room temperature, and then add 10 mL of ethyl alcohol to the beaker. Label the beaker with your name, cover it with your largest beaker, and place it in your laboratory desk until the next laboratory period.

The reason that the beaker should be placed in your desk is that the product, $K_3Fe(C_2O_4)_3 \cdot 3H_2O$, decomposes in the presence of light. The products of the decomposition are carbon dioxide (from the oxidation of $C_2O_4{}^{2-}$) and a complex of ferrous, Fe(II), ion [from the reduction of Fe(III)]. The reason for covering the beaker is to prevent complete evaporation of the water, which would make it impossible to separate the pure $K_3Fe(C_2O_4)_3 \cdot 3H_2O$ crystals from the other salts, which should have remained in solution. Also, slower evaporation encourages the development of larger, purer crystals.

At the beginning of the next laboratory session, assemble a regular filtration setup, as described in Part A of this manual. Using your stirring rod equipped with a rubber policeman, transfer the crystals of $K_3Fe(C_2O_4)_3 \cdot 3H_2O$ from the beaker to the filter paper in the funnel. When the crystals are free of solution, remove the filter paper and spread the crystals out so they can dry. Obtain the mass of a stoppered glass vial or test tube on the analytical balance. When the crystals are dry, scrape them from the filter paper into the preweighed container. To obtain the mass of the $K_3Fe(C_2O_4)_3 \cdot 3H_2O$, simply subtract the mass of the empty container from the mass of the full container. Submit your container of crystals to your instructor.

Data and Results

Record all qualitative observations, and calculate your percent yield based on the starting mass of $Fe(NH_4)_2(SO_4)_2 \cdot 6H_2O$, assuming that all other reagents are available in excess.

Questions

E.21.1   Suppose that you carried out only the first step of the synthesis described in this experiment, that is, the reaction between $Fe(NH_4)_2(SO_4)_2 \cdot 6H_2O$ and $H_2C_2O_4$ to give the main product $FeC_2O_4 \cdot 2H_2O(s)$ and several by-products. Given: 4.88 g of $Fe(NH_4)_2(SO_4)_2 \cdot 6H_2O$ and 25.5 mL of 1.133 M $H_2C_2O_4$,
(a) Find the *limiting reagent*.
(b) Calculate the *theoretical yield* of $FeC_2O_4 \cdot 2H_2O$.

E.21.2   Given that saturated $K_2C_2O_4$ solution contains 299.5 g of $K_2C_2O_4 \cdot H_2O$ per liter of solution at room temperature, calculate the number of moles of $K_2C_2O_4$ contained in 10.00 mL of saturated $K_2C_2O_4$ solution.

E.21.3   Calculate the number of grams of $H_2C_2O_4$ contained in 8.031 mL of 1.002 M $H_2C_2O_4$ solution.

E.21.4   Calculate the number of moles of H_2O_2 contained in 19.5 mL of a solution that contains 3.00% H_2O_2 by weight. Assume the density of the solution is 1.01 g/mL.

E.21.5   Why might it make more sense to calculate the percent yield for each step of the synthesis independently, rather than to calculate only the percent yield of the final product?

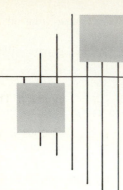

Experiment E.22

Analysis of a Complex Iron Salt

Special Items

1 g $K_2Fe(C_2O_4)_3 \cdot 3H_2O$
0.5 g $Na_2C_2O_4$ (primary standard)
100 g Drierite™, indicating
300 mL 0.014 M KMnO$_4$ (2.21 g KMnO$_4$/liter)
50 mL 3 M H$_2$SO$_4$ (167 mL 18 M H$_2$SO$_4$/liter)
1 g zinc dust (reagent grade)
1 Weighing bottle
1 Mortar and pestle
1 Desiccator
1 Oven, capable of maintaining 120°C temperature

In chemistry research it is not acceptable practice to prepare a compound and publish the synthesis without first having confirmed the identity of the compount by analysis. The analysis may be done instrumentally (x-ray diffraction, nmr, ir, etc.) or chemically or by a combination of instrumental and chemical methods. In Experiment E.21 your aim was to synthesize the compound potassium ferric oxalate [that is, K tris-oxalatoferrate(III)], $K_3Fe(C_2O_4)_3 \cdot 3H_2O$. In this experiment you will analyze the compound you prepared by titration with permanganate. The analysis will have the following stages:

(a) Standardization of a permanganate solution
(b) Analysis for oxalate in $K_3Fe(C_2O_4)_3 \cdot 3H_2O$
(c) Analysis for iron in $K_3Fe(C_2O_4)_3 \cdot 3H_2O$

In acidic solution, permanganate, which is purple, reacts with reducing agents to give nearly colorless Mn^{2+}. The permanganate will serve as its own indicator; i.e., when all the reducing agent has been oxidized by the MnO_4^-, a drop or two of MnO_4^- added in excess will result in a definite pink coloration of the solution. One problem with permanganate solutions, especially when they are acidic, is that they are quite unstable if not handled properly. A freshly prepared solution tends to react with any dust or organic matter present in the water to form reduced species such as $MnO_2(s)$, which act to catalyze further decomposition. Fortunately, it is rather simple to remove the $MnO_2(s)$, which settles to the bottom of the bottle, by carefully siphoning off the KMnO$_4$ solution after it has been allowed to react for a day or two. The resulting KMnO$_4$ solution is stable for weeks.

A permanganate solution having a concentration of approximately 0.014 M will be standardized by titration with primary standard sodium oxalate, $Na_2C_2O_4$:

$$5C_2O_4^{2-} + 2MnO_4^- + 16H^+ \rightarrow$$

$$10CO_2(g) + 2Mn^{2+} + 8H_2O$$

A portion of the standardized permanganate will be used to titrate the oxalate ($C_2O_4^{2-}$) portion of your $K_3Fe(C_2O_4)_3 \cdot 3H_2O$ complex. The reaction is essentially the same as the reaction between sodium oxalate and permanganate. The main products (in acidic solution) are $CO_2(g)$, Fe^{3+}, Mn^{2+}, and H_2O. From the mass of $K_3Fe(C_2O_4)_3 \cdot 3H_2O$ used, you can get the moles of complex. From the volume and molarity of the $KMnO_4$ used, you can obtain the moles of $KMnO_4$ and thereby the moles of $C_2O_4^{2-}$ per mole of complex, which we hope will be about 3.

As it was in the original complex, the iron at this point is in the +3 oxidation state. To analyze for iron using permanganate, it is necessary first to reduce the iron quantitatively from the +3 state (ferric) to the +2 state (ferrous). This is accomplished by adding an excess of zinc dust and boiling the solution until all the Fe^{3+} has been reduced:

$$2Fe^{3+} + Zn(s) \rightarrow 2Fe^{2+} + Zn^{2+}$$

The zinc ion produced is stable and will not interfere with subsequent portions of the analysis. The excess zinc metal is removed by simple filtration.

The final step is to titrate the Fe^{2+} by oxidizing it with standardized acidic permanganate solution. The main reaction products are Fe^{3+}, Mn^{2+}, and H_2O. From the volume and concentration of the MnO_4^- required, you can obtain the moles of Fe^{2+} and thereby the moles of iron per mole of the original complex.

Procedure

(a)

■WEAR YOUR SAFETY GOGGLES

Standardization of a Permanganate Solution Obtain a weighing bottle and a desiccator from your laboratory instructor. Clean and dry the weighing bottle thoroughly in the oven in your laboratory, which should be held at a temperature of about 120°C. Get a bottle of anhydrous $CaSO_4$ (Drierite), and put enough of the desiccant in the bottom of your desiccator to give a depth of about 1.0 cm. *Immediately* cover the bottle of Drierite and the desiccator. Weigh the empty bottle on the analytical balance and record the mass. Then, using the triple-beam balance, add approximately 0.5 g of sodium oxalate, $Na_2C_2O_4$, primary standard to the weighing bottle. Dry the sodium oxalate for about an hour, seal it, and place it in your desiccator. Weigh the bottle containing the sodium oxalate on the analytical balance. Be sure to enter all data in your laboratory notebook.

Thoroughly clean, rinse, and dry a 300-mL Florence flask, which you will be using for storage of your $KMnO_4$ solution. The stockroom has prepared a $KMnO_4$ solution that is approximately 0.014 M, allowed it to form some $MnO_2(s)$, and siphoned it into the large siphon bottle in your laboratory. Obtain approximately 300 mL of this solution in your clean, dry flask, cover the top with a 50-mL beaker (inverted), and store it in a dark place, such as your desk. Be sure to label the flask with your name. Each student in the class will perform one or two titrations, and the class will then share the results in order to get a better value for the molarity of the $KMnO_4$ solution.

Clean a 250-mL titration flask thoroughly, and then, using the triple beam balance, weigh a 0.12- to 0.15-g sample of sodium oxalate from the previously weighed weighing bottle into the flask. Immediately seal the weighing bottle, and weigh it on the analytical balance to get the exact mass of sodium oxalate removed. To estimate how much sodium oxalate there is in 0.12 to 0.15 g, refer to the total mass of sodium oxalate in the bottle and take the appropriate fraction of it; for example, if there is a total of 0.50 g, take a little less than a third.

Set up a clean burette on a ring stand, and rinse it with 4 to 5 mL of your $KMnO_4$ solution. Also, set up a ring stand with a ring, wire gauze, and gas burner. Obtain about 25 mL of 3 M H_2SO_4 solution in your clean 25-mL graduated cylinder. Add 40 to 50 mL of distilled water to the flask containing the $Na_2C_2O_4$; also pour in 10 mL of the 3 M H_2SO_4 (which gives acid solution for the titration). Fill the burette with $KMnO_4$ solution and then put the $KMnO_4$ solution back in your desk with an inverted beaker sitting on the top. Record the initial burette reading, and add about 8 to 10 mL of the $KMnO_4$ to the flask. The reaction between $KMnO_4$ and oxalate is very slow at first, but it is catalyzed by the Mn^{2+} that forms in the reduction of the $KMnO_4$. To get the reaction started, place the flask on the ring stand and heat the solution with a gas burner until the violet color of the $KMnO_4$ disappears. Place the flask under the burette, and titrate the warm solution until the solution retains a pink color for 20 to 30 sec. Record the final burette reading. You now have sufficient data to calculate the molarity of the $KMnO_4$ solution. Repeat this standardization procedure if there is time remaining at the end of the next stage of the procedure; if not, perform only one standardization titration. Share your value for the molarity of the $KMnO_4$ with the others in your section.

(b) **Analysis for Oxalate in $K_3Fe(C_2O_4)_3 \cdot 3H_2O$** Clean and dry a small beaker, and then weigh the beaker on the analytical balance. Clean a mortar and pestle (see typical chemistry laboratory equipment pictures in the front of your laboratory manual), and dry them thoroughly. Using the triple-beam balance, weigh out about a gram of the $K_3Fe(C_2O_4)_3 \cdot 3H_2O$ crystals you prepared in Experiment E.21. Grind the crystals to powder, and transfer them to the preweighed beaker. Measure the mass of the beaker plus the crystals. Clean and dry a titration flask, and place 0.15 to 0.20 g of the $K_3Fe(C_2O_4)_3 \cdot 3H_2O$ crystals in the flask. Add about 30 mL of distilled water, and swirl the contents. Then add about 10 mL of 3 M H_2SO_4. Swirl the contents until the sample has dissolved. If you have trouble with a brownish color in the solution, add a bit more 3 M H_2SO_4, which should relieve the problem.

Fill your burette with standardized $KMnO_4$, cover the $KMnO_4$ flask, and return it to its storage place. Record the initial burette reading, and add about 10 mL of the $KMnO_4$ to the solution in the flask. As you did in the standardization procedure, heat the solution to initiate the reaction, and then titrate the warm solution until the pink color of excess $KMnO_4$ persists 20 to 30 sec. Record the final burette reading. *Retain the solution in the flask for the iron analysis.* If there is time, perform a second titration and again retain the final solution. If there is any time after that, repeat the standardization of $KMnO_4$ procedure.

(c) **Analysis for Iron in $K_3Fe(C_2O_4)_3 \cdot 3H_2O$** Set up a ring stand, equip it with a ring and wire gauze, and set the flask containing Fe^{3+} and Mn^{2+} on the wire gauze. [At this point you may observe a trace of brown precipitate (possibly MnO_2) in the flask. This should disappear after heating with the zinc dust.] Heat the solution until it almost boils, and add a small amount of analytical reagent grade zinc dust (volume equal to two pencil erasers) to the hot solution. Cover the beaker, and wait until the yellow color disappears. At this point the Fe^{3+} will have been converted to Fe^{2+} (Fe^{3+} is yellow; Fe^{2+} is colorless). While you are waiting, get ready to filter the hot solution. It is necessary to filter it quickly because Fe^{2+} is reoxidized by O_2 and much of the acid gets used up in reacting with zinc.

Refer to Part A in the front of your laboratory manual for information about filtration. Set up a funnel and equip it with regular qualitative filter papaer. Clean a second titration flask, and set it under the funnel. Grasping the hot flask with a towel, decant the hot solution through the filter paper. Then rinse the flask with water into the filter paper. Finally, rinse the funnel with about 5 additional milliliters of distilled water. The titration flask should now contain all the iron originally in the sample of $K_3Fe(C_2O_4)_3 \cdot 3H_2O$. Rinse a cleaned burette with about 5 mL of standardized $KMnO_4$ solution, fill the burette, and record the initial reading. Then titrate the Fe^{2+} solution until the solution turns pink and remains pink for 20 to 30 sec. Record the final burette reading. If you were able earlier to complete more than one titration of the oxalate in the complex and you have time remaining in this laboratory period, carry out a second iron determination. You now have sufficient data to determine the moles of Fe per mole of what you think is $K_3Fe(C_2O_4)_3 \cdot 3H_2O$.

Questions

Prelaboratory

E.22.1   Prepare a detailed flowchart showing reagents and species present at each stage of the procedure you will use to analyze the complex iron salt.

E.22.2   Given that in the titration of the complex compound the reactants are the iron complex, $KMnO_4$, and H_2SO_4 and the products are $CO_2(g)$, $Fe_2(SO_4)_3$, $MnSO_4$, K_2SO_4, and H_2O, write a balanced equation for the reaction, including both reacting species and "spectator ions."

E.22.3   Explain in detail how you will be able to obtain a value for the number of moles of oxalate per mole of complex compound from the data obtained in this experiment.

E.22.4   Explain how you will be able to determine the number of moles of iron per mole of complex compound from the data obtained in this experiment.

Postlaboratory

E.22.5   How would each of the following tend to affect your calculated number of moles of oxalate per mole of complex?
  (a) The complex was contaminated with some $FeC_2O_4 \cdot 2H_2O$.
  (b) After standardization, some of your MnO_4^- decomposed to $MnO_2(s)$ before you titrated the complex.

E.22.6   Calculate the percent error in the value you obtained for the number of moles of iron per mole of complex, assuming that the number should be exactly 1.0000.

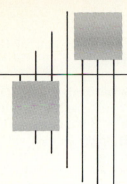

Experiment E.23

Hydrogen Peroxide

Special Items

50 mL distilled water
21 mL 3% H_2O_2
5 mL 0.001 M KMnO$_4$
5 mL 0.10 M KI
6 mL 1 M KBr
3 mL 3 M NaOH
3 mL 0.10 M FeCl$_3$
3 mL bromine water
5 mL 3 M H$_2$SO$_4$
3 g barium peroxide

In this experiment, you will prepare hydrogen peroxide (H_2O_2) by the reaction of barium peroxide (BaO_2) with sulfuric acid. You will then investigate its oxidizing and reducing ability so as to be able to rank the two half-reactions

$$O_2(g) + 2H_3O^+ + 2e^- \rightarrow H_2O_2 + 2H_2O$$

$$H_2O_2 + 2H_3O^+ + 2e^- \rightarrow 4H_2O$$

relative to the half-reaction

$$Br_2 + 2e^- \rightarrow 2Br^-$$

By observing whether H_2O_2 oxidizes Br^- to Br_2 and whether H_2O_2 reduces Br_2 to Br^-, you will determine the order of the three reduction potentials, and hence you will be able to decide whether H_2O_2 is stable to disproportionation into O_2 and H_2O.

If H_2O_2 is not stable to disproportionation, then it must exist only because the disproportionation is slow. In such a case, it would make sense to search for a catalyst to speed the disporportionation.

Procedure

(a) Mix 3 g barium peroxide with 50 mL of water in a small flask, and cool under the water tap. Add dropwise about 2 mL of dilute sulfuric acid while still cooling under the running water. Filter off the barium sulfate precipitate. Add a few milliliters of the filtrate to an acidified solution of potassium iodide. Also, try the reaction of a few milliliters of the filtrate with an acidified solution of 0.001 M KMnO$_4$. Test for possible evolution of oxygen (e.g., with a glowing splint).

■ **WEAR YOUR SAFETY GOGGLES**

(b) Using the 3% H_2O_2 solution provided on the reagent shelf, add about 3 mL to each of two test tubes. Add several drops of dilute H_2SO_4 to each. To test tube 1, add about 3 mL of bromine water; to test tube 2, add about 3 mL of 1 M KBr. Stir, and observe for about 5 min.

(c) To each of five test tubes, add about 3 mL of 3% H_2O_2. Label the test tubes 0, H_3O^+, OH^-, Fe^{3+}, and Br^-. To the first, add 3 mL of water; to the second, 3 mL of dilute H_2SO_4; to the third, 3 mL of dilute NaOH; to the fourth, 3 mL of acidified $FeCl_3$ solution; and to the last, 3 mL of 1 M KBr and a few drops of dilute H_2SO_4. Place the five test tubes in a 600-mL beaker one-third full of water. Set on a wire gauze, and heat. Note the order in which reaction occurs in the various test tubes as the temperature increases.

Data

(a) Record the observations made.

(b) Record all observations.

(c) Record observations and the order in which reaction occurs.

Results

(a)   What do you conclude about the reaction of BaO_2 with sulfuric acid?

(b)   Write the three half-reactions in order of decreasing reduction potential. What do you conclude about the stability of H_2O_2 to disproportionation?

(c)   Rank the substances tried in order of decreasing effectiveness as catalysts for this reaction.

Questions

E.23.1   What is the general meaning of *disproportionation reaction*? If H_2O_2 were to disproportionate into H_2O and O_2, what oxidation state changes would occur?

E.23.2   What is the maximum number of moles of H_2O_2 you could produce in part (*a*)?

E.23.3   Hydrogen peroxide sometimes behaves as an oxidizing agent, and at other times as a reducing agent. What conditions, acidic or basic, would favor having H_2O_2 behave as an oxidizing agent? Explain.

E.23.4   Try to deduce from the procedure in part (*a*) whether the dissolution of BaO_2 in water and the reaction of BaO_2 with sulfuric acid are exothermic or endothermic reactions.

E.23.5   Offer an explanation for the observed effect on the disporportionation rate found for OH^- in part (*c*) of the procedure.

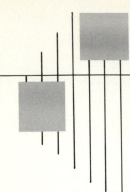

Experiment E.24

Analysis of Hydrogen Peroxide Solution

Special Items

Burette (if not in desk)
100 mL distilled water
100 mL standardized $KMnO_4$
  (ca. 0.1 M)
20 mL unknown (about 3% H_2O_2)
30 mL 3 M H_2SO_4

Most commercial solutions of hydrogen peroxide are about 3% H_2O_2, that is, 3 g of H_2O_2 per 100 g of solution. In this experiment, you will analyze an unknown solution by titrating it with potassium permanganate solution. In acid solution, MnO_4^- oxidizes H_2O_2 to form O_2 and colorless Mn^{2+}. Thus, when a solution of $KMnO_4$ is added dropwise to an acidified solution of H_2O_2, each drop is decolorized until all the H_2O_2 is used up. The next drop added remains colored. By knowing the concentration of the $KMnO_4$ solution and the volume needed to react with all the H_2O_2, you will be able to calculate the number of moles of H_2O_2 oxidized. If the weight of the original H_2O_2 solution is known, its concentration as percent by weight can be calculated.

Procedure

■ **WEAR YOUR SAFETY GOGGLES**

Read Part A to learn how to set up and use a burette. Construct a burette as shown in Figure A.9. Using a clean, dry beaker and a funnel, fill it with the standard solution of $KMnO_4$ (approximately 0.1 M; its exact concentration is given on the bottle).

Weigh your clean 300-mL flask on the platform balance to the nearest 0.1 g. Add to it about 10 mL of the peroxide solution provided, and immediately reweigh. Add about 50 mL of distilled water and 15 mL of dilute sulfuric acid. Swirl to mix.

Record the initial reading on the burette to the nearest 0.1 mL. Titrate by running a few milliliters of the $KMnO_4$ solution into the flask and swirling. Continue successive additions of $KMnO_4$ until the color begins to persist. Then add the $KMnO_4$ dropwise, swirling after each addition, until one drop produces a pink color that persists for at least 1 min. The color is more readily seen if there is a piece of white paper under the flask. Record the burette reading.

Repeat the titration on a duplicate sample.

Data

Mass of flask   _____ _____

Mass of flask plus H_2O_2 solution   _____ _____

Molarity of $KMnO_4$ solution   _____ _____

Initial burette reading   _____ _____

Final burette reading   _____ _____

Results

Write a balanced equation for the reduction of MnO_4^- by H_2O_2 in acid solution, and use it to calculate the number of moles of H_2O_2 in the sample. Calculate the weight percentage H_2O_2 in the sample.

Questions

E.24.1 What are the molarity and the normality of a 3.00% H_2O_2 solution if the density of the solution is 1.03 g/mL?

E.24.2 Suppose that during the titration some of the H_2O_2 reacts with itself by disproportionation. How would this affect your calculated weight percentage of H_2O_2 in the sample?

E.24.3 A color-blind student cannot perceive the pink titration end point until the color is quite intense. How is the calculated value of the weight percentage of H_2O_2 in the sample going to be affected?

E.24.4 What is incorrect about the following equation?

$$6H_3O^+ + 2MnO_4^- + 11H_2O_2 \rightarrow 2Mn^{2+} + 8O_2 + 20H_2O$$

E.24.5 In a *neutral* solution, permanganate reacts with peroxide to give $MnO_2(s)$. Write a balanced equation for this reaction. Calculate the number of moles of $KMnO_4$ required to react with the amount of H_2O_2 in your sample if the reaction is carried out in such a neutral solution.

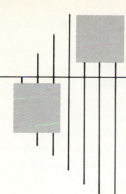

Experiment E.25

Titration of Antacids

When your stomach hurts, if you have reason to believe the cause is not appendicitis, the chances are quite good that you will decide to take one of the following tablets or powders:

Alka-Seltzer (aspirin, monocalcium phosphate, sodium bicarbonate, and citric acid)

Bromo Seltzer (acetaminophen, phenacetin, potassium bromide, caffeine, sodium bicarbonate, and citric acid)

Digel (magnesium carbonate, aluminum hydroxide, magnesium hydroxide, and activated methylpolysiloxane)

Milk of Magnesia (magnesium hydroxide)

Rolaids (aluminum sodium carbonate hydroxide)

Tums (calcium carbonate, magnesium carbonate, magnesium trisilicate)

In this experiment, you will have an opportunity to test the antacid capacity of some of these. You will do this by the method of back-titration, which will be described later. It should be pointed out, however, that the simple ability to neutralize acid is not the only criterion for choosing one preparation over another, since there may be side effects that could vary with the individual. For example, sodium bicarbonate, $NaHCO_3$, is a water-soluble carbonate that is capable of neutralizing too much acid in the stomach and can actually make the stomach become neutral. Since the natural state of the stomach is acidic, the stomach may then overcompensate and try to produce a large amount of acid in a phenomenon called *rebound acidity*. This rebound acidity repeated often over a long period of time may help to produce a peptic ulcer. Also, because $NaHCO_3$ can be absorbed into the blood readily, large amounts can upset the acid-base balance of the body. At the least, it is likely to overwork the kidneys, which try to maintain the balance.

Magnesium carbonate, $MgCO_3$, has an advantage over sodium bicarbonate in that the magnesium carbonate is insoluble in basic or neutral solutions and therefore cannot dissolve further once the acid in the stomach has been neutralized.

However, in the alimentary tract it is converted to the soluble bicarbonate and may act as a mild laxative in large doses.

Aluminum hydroxide, $Al(OH)_3$, also becomes insoluble when the acid in the stomach has been neutralized, and therefore it too cannot cause excess alkalinity. Further, the aluminum ion has an astringent effect (constricts vessels, preventing discharges), which tends to produce constipation. Evidently, proper formulation of an antacid requires careful attention to a balancing of the physiological effects so as to minimize undesirable side effects.

Method

The normal pH of the stomach is low. Consequently, in testing these preparations, we will not wish to measure the ability of these tablets to bring the pH to a neutral 7. Rather, we will want to find out their ability to bring the pH to just above 3. One method that might be considered would be to dissolve the tablets in water and then add acid until the pH goes below 3. This will not work with good speed, as many of the tablets contain compounds not soluble in water. A better method is that of *back-titration*. An amount of acid that is more than enough to bring the pH below 3 is added to the sample. The excess acid is then titrated with a strong base, and a simple subtraction permits determination of the amount of acid neutralized by the sample.

Procedure

■ WEAR YOUR SAFETY GOGGLES

From Table E.25.1, find the amount of tablet you are to use, and break your tablet into such pieces. Fold a piece of paper over the sample, and pulverize it carefully with something hard. Weigh paper and sample (and later weigh the paper after removing the sample).

Transfer the pulverized sample to your largest clean beaker. With a 25-mL pipette, add 50 mL of standardized 0.1 M HCl to the beaker. With a stirring rod, assist the reaction in going to completion. Do not expect the entire tablet to dissolve, as it may contain insoluble binder and filler. Support the beaker on a ring stand, being sure to place a wire gauze on the ring. Heat the solution until it almost boils to remove CO_2. Cool the beaker by placing some cold water in a pan and setting the beaker in it with your crucible tongs. Add a few drops of methyl-orange indicator. It should be red in the

Table E.25.1

Product	Recommended Dose	Fraction of Tablet To Be Used for Sample	Approximate Number of Grams in That Size of Sample
Alka-Seltzer	1 or 2 tablets	$\frac{1}{4}$	0.9
Bromo Seltzer	7 g	Powder	2
Digel	2 tablets	$\frac{1}{2}$	0.35
Milk of Magnesia	2–4 tablets	$\frac{1}{2}$ or less	0.25
Rolaids	1–2 tablets	$\frac{1}{2}$	0.75
Tums	1–2 tablets	$\frac{1}{3}$ or less	0.3

acid solution. If it is not, confer with your teaching assistant. Titrate the solution with standardized 0.1 M NaOH solution until the color changes from red to yellow. The color change (end point) will occur at approximately pH = 4.

From the weight of sample and the volume of base used, calculate the equivalents neutralized per gram of sample.

Questions

Prelaboratory

E.25.1   Clearly explain, using numbered steps, how you will analyze your sample of antacid in this experiment.

E.25.2   What error, if any, in the calculated number of equivalents per gram of sample might be caused by each of the following?
(a) Your sample was not thoroughly pulverized before adding the HCl.
(b) Some of the sample adhered to the weighing paper and was not transferred to the beaker.
(c) Not all the CO_2 was boiled out of the solution before back-titration.

Postlaboratory

E.25.3   The formulation of one commercial product includes both aluminum hydroxide and magnesium hydroxide. Suggest a possible reason for this particular combination.

E.25.4   What governs the choice of indicator in this experiment? Suppose that you used phenolphthalein; how would it affect your results?

E.25.5   What data would you need to evaluate the claim that a certain product neutralizes 47 times its weight in excess stomach acidity?

E.25.6   Why is it desirable in the method of back-titration to have a considerable amount (more than 10 mL) of acid added in excess?

E.25.7   Obtain from each of the other students who analyzed the same antacid that you studied their values for the number of equivalents neutralized per gram of sample. Calculate an average (mean) value, and determine an average deviation.

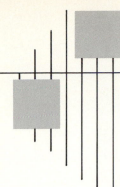

Experiment **E.26**

Chemical Kinetics— Gas Evolution

Special Items

40 mL 0.10 *M* KI
20 mL 3% H_2O_2
30 mL distilled water

NOTE: Students work in pairs.

The rate of a chemical reaction can be determined by noting the rate at which one of the products appears or the rate at which one of the reactants is used up. The rate depends on the concentration of reactants and must be determined experimentally since it cannot be deduced from the balanced equation for the overall reaction. In this experiment, you will study the rate of decomposition of hydrogen peroxide to form oxygen according to the net equation

$$2H_2O_2 \rightarrow 2H_2O + O_2(g)$$

By measuring the rate at which oxygen is evolved, you will investigate how the rate changes with varying initial concentrations of H_2O_2 and iodide catalyst. Although I^- does not appear in the overall equation, it affects the rate; and you will study the effect of changing its concentration on the rate of oxygen evolution.

At the end of the experiment, you will summarize your results by attempting to write a rate law for the reaction, showing the dependence on the concentrations of H_2O_2 and I^-. Your rate law will have the form

Rate of oxygen evolution $= k[H_2O_2]^n [I^-]^m$

and your problem is to determine the numerical values of n and m.

Procedure

■ WEAR YOUR SAFETY GOGGLES

(Select a laboratory partner who will work with you on this experiment.) Assemble the apparatus shown in Fig. E.26.1. Fill the trough with water at room temperature. (It may be

155

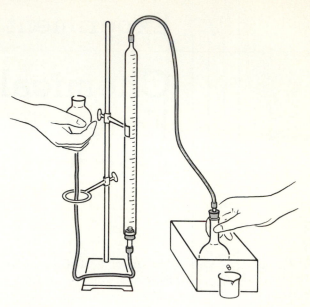

Figure E.26.1

necessary to add hot water from a beaker.) Add room-temperature water to the gas-measuring assembly until the measuring tube is filled when the leveling bulb is at the same level.
Check for leaks by lowering the bulb.

(a) Add 10 mL of 0.10 M KI and 15 mL of distilled water to the
flask. Swirl the flasks so that the solution comes to bath temperature. Add 5 mL of the 3% H_2O_2, and quickly stopper the
flask. One student should keep swirling the flask in the bath
as vigorously as possible throughout the experiment. The other
can observe the volume of oxygen evolved at various intervals. The first reading of volume should be taken when approximately 2 mL have evolved. One student matches up the
water levels by manipulation of the leveling bulb and reads
the volume while the other (still swirling the flask) records the
time at the instant the volume is read off. Readings should be
taken at 2-mL intervals until 14 mL of oxygen have accumulated.

(b) Rinse the flask well with distilled water, and drain thoroughly.
Repeat the experiment, this time first adding 10 mL of 0.10
M KI and 10 mL of H_2O, swirling, and then adding 10 mL of
H_2O_2. Quickly stopper, and take readings as before.

(c) Rinse the flask and repeat. First add 20 mL of 0.10 *M* KI and 5 mL of H_2O, swirl, and then add 5 mL of H_2O_2.

Data

(a)		(b)		(c)	
Vol.	Time	Vol.	Time	Vol.	Time
___	___	___	___	___	___
___	___	___	___	___	___
___	___	___	___	___	___
___	___	___	___	___	___
___	___	___	___	___	___
___	___	___	___	___	___
___	___	___	___	___	___

Results

Plot your data for each run, showing volume of oxygen evolved (ordinate) vs. elapsed time (abscissa). Zero time for each run corresponds to the time of the first volume reading (2 mL). Measure the slope of each of your three curves for the beginning of the experiment. This slope gives you an idea of the rate of oxygen evolution at the start of each run.

Decide how the initial rate of oxygen evolution is affected by doubling the concentration of H_2O_2 and of I^-. Write a rate law for the reaction.

Questions

E.26.1 Why are you directed to swirl the flask vigorously during each reaction? How might your results be affected if you neglected to swirl the flask?

E.26.2 Propose a mechanism for the reaction consistent with the rate law you determined.

E.26.3 Does the rate of oxygen evolution increase or decrease with time? Explain.

E.26.4 Use your rate-law equation to estimate the time required to produce 5 mL of oxygen after mixing 20 mL of 0.20 M KI, 5 mL of H_2O, and 5 mL of 3% H_2O_2.

E.26.5 Suppose that the bottle of H_2O_2 solution you used in this experiment were mislabeled; the concentration was actually 6% but the label said 3% H_2O_2. How would the use of this solution affect your data, and how would the rate-law expression be affected?

E.26.6 Oxygen gas is slightly soluble in water. What effect should this have on your data and on your rate equation?

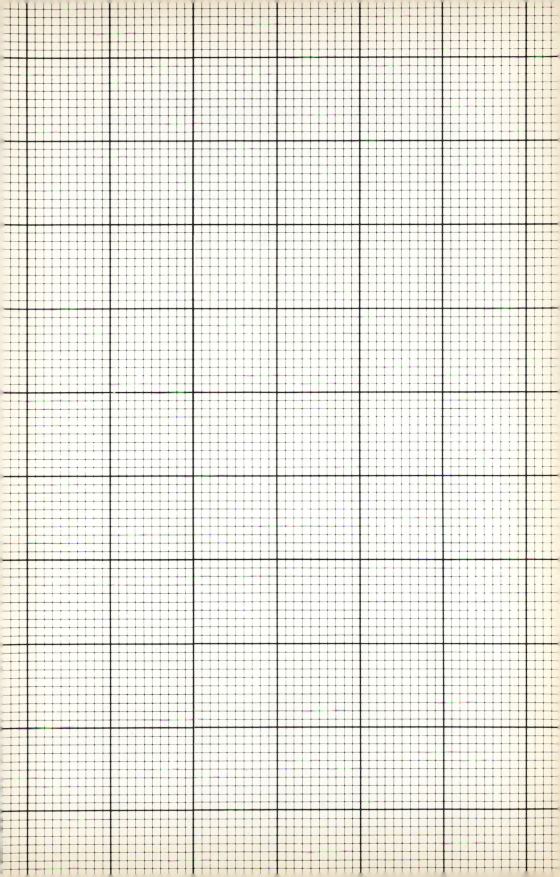

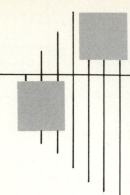

Experiment E.27

Chemical Kinetics—Iodine Clock Reaction

Special Items
Clock with second hand
0.16 M KI solution
0.12 M $(NH_4)_2S_2O_8$ solution
0.10 M EDTA solution
Solid starch indicator

Peroxydisulfate ion reacts with iodide ion in aqueous solution to give iodine and sulfate ion:

$$2I^- + S_2O_8^{2-} \rightarrow I_2 + 2SO_4^{2-}$$

Although this is a correct balanced equation for the *net* reaction, it says nothing about the actual *mechanism* of the reaction, which describes in detail how the reacting species come together in each of a series of *elementary reaction steps*. The sum of all of the elementary steps must give the balanced net equation. Consider, for example, the following hypothetical two-step mechanism:

Step 1	$A + 2B \rightleftharpoons AB_2$	(slow)
Step 2	$AB_2 + A \rightleftharpoons A_2 + B_2$	(fast)
Net reaction	$2A + 2B \rightleftharpoons A_2 + B_2$	

Substances A and B are called the *reactants*, because they appear on the left side of the equation for the net reaction, and A_2 and B_2 are considered to be the *products*, because they are on the right side in the equation for the net reaction. Note that AB_2 does not appear in the equation for the net reaction; it is produced in step 1 and consumed in step 2. A substance such as AB_2 is called an *intermediate*.

A rate expression may be written for each of these elementary reaction steps. In the rate expression for a reaction step, the concentration of each reactant and each product is written inside brackets, and each concentration is raised to a power. The power (exponent) to which each concentration is raised is equal to the number of atoms or molecules of each substance that combine in that step. The total number of reactant particles that come together in a reaction step is called the *molecularity* of the reaction step; for example, since three particles combine in reaction step 1, its molecularity is 3, and it is said to be *termolecular*. (Step 2 is *bimolecular*, because two particles are involved.) The rate of the first step of the above reaction would be expressed as

$$\text{Rate}_1 = k_1[A][B]^2 - k_{-1}[AB_2]$$

The expression $k_1[A][B]^2$ is the rate of the forward reaction for step 1; and $k_{-1}[AB_2]$ is the rate of the reverse reaction for step 1. The "rate" of step 1, then, is the forward rate minus the reverse rate. The constant k_1 is called the *specific rate constant* of the forward reaction, and k_{-1} is the specific rate constant of the reverse reaction for the first step. Specific rate constants depend on temperature.

We can almost never know the detailed mechanism of a reaction with absolute certainty, because we have no way of observing the individual particles as they react. We can, however, come up with a very good idea of what the mechanism must be by carrying out well-chosen experiments. What usually is done is to postulate a number of reasonable mechanisms for a reaction and then carry out experiments designed to narrow the list of possibilities. If the list is narrowed to a single mechanism that stands up to repeated testing, it may be accepted as the actual mechanism. For many reactions there is one elementary step that is much slower than any of the other steps and is thereby rate-determining for the overall reaction. The rate expression for the rate-determining step will then be the rate expression for the entire reaction.

In the hypothetical mechanism described above, the first step is much slower than the second, so the first step is rate-determining. As any reaction progresses with time, the concentrations of reactants diminish and product concentrations increase; this nearly always results in a slowing down of the rate. Analysis of a reaction after it has been going on for some time is very complex, because one must keep track of concentrations of both reactants and products. At the very beginning of a reaction, however, only the reactants are present; since product concentrations are zero or negligible, "back-reactions" can be ignored.

The *initial rate* of a reaction, then, is determined solely by the rate expression for the forward rate of the rate-determining step. This initial rate expression is called the *rate law* for the reaction. For the hypothetical reaction described above, then, the rate law would be

$$\text{Rate} = k[A][B]^2$$

where k, the observed specific rate constant for the net reaction, is equal to k_1, the specific rate constant of the forward reaction in the slow step.

Sometimes the rate-determining step involves *intermediates* or *catalysts*, substances that do not appear in the net equation. Recall that an intermediate is something *not* put in at the beginning but that is generated in an elementary step. A catalyst is something put in at the beginning that is consumed in one step but regenerated in another step; thus the catalyst does not disappear in a net sense. Since neither catalysts nor intermediates show up in the net balanced equation, these situations lead to rate laws that are somewhat complicated. In the reaction of iodide ion with peroxydisulfate ion, for example, the rate-determining step for the reaction when no catalyst is present involves only iodide ion and peroxydisulfate ion; in the presence of a catalyst, however, the mechanism is changed because the catalyst appears in the rate-determining step.

In this experiment, you will determine the orders of iodide ion and peroxydisulfate ion only for the reaction *without* a catalyst. Thus the rate expression for the slowest step, and therefore the *rate law* for the overall reaction, may be expressed as

$$\text{Rate} = k[\text{I}^-]^x[\text{S}_2\text{O}_8{}^{2-}]^y$$

where x and y are the orders of iodide ion and peroxydisulfate ion, respectively. The order for the reaction as a whole is the sum of $x + y$; in other words, if the order of iodide were 2 and the order of peroxydisulfate were 1, the *reaction order* would be 3, and it would be called a "third-order reaction."

Note that the reaction order gives the molecularity of the rate-determining step. To determine the order of iodide ion, you will carry out one set of experiments in which the iodide ion concentration is varied, while the peroxydisulfate ion concentration is held constant; this will allow you to determine x. To find a value for y, you will perform a second set of experiments in which the peroxydisulfate ion concentration is varied while the iodide ion concentration is held constant.

The initial rate of the reaction will be determined by measuring the time it takes to generate a certain amount of iodine, the *same* amount in every trial, according to reaction A, which is, of course, the reaction you are investigating. The faster the initial rate, the shorter the time that will be required to produce the same amount of iodine.

$$2I^- + S_2O_8^{2-} \rightarrow I_2 + 2SO_4^{2-} \qquad \text{(A)}$$

The experiment has been designed to give you a clear signal when the correct amount of iodine has been produced. In each trial, you will add a certain amount of thiosulfate ion, $S_2O_3^{2-}$, the identical amount every time. Thiosulfate reacts with iodine as fast as iodine is produced, converting it back to iodide ion, according to the reaction

$$2S_2O_3^{2-} + I_2 \rightarrow S_4O_6^{2-} + 2I^- \qquad \text{(B)}$$

Reaction B is very fast. Thus, iodine will not have a chance to build up in the solution until all of the thiosulfate has been consumed. Any buildup of iodine in the solution indicates that the thiosulfate has been used up, and that, of course, means that the constant amount of iodine has been produced.

To detect the presence of excess iodine in the solution, you will add some starch to one of the solutions before mixing the iodide and peroxydisulfate. As soon as iodine begins to build up in the solution, it will react immediately to form a dark-blue complex with the starch according to reaction C:

$$I_2 + \text{starch} \rightarrow I_2\text{-starch complex} \qquad \text{(C)}$$

The time period as measured from the time of mixing the peroxydisulfate with the iodide in the presence of starch and thiosulfate to the appearance of the dark-blue color, then, is the time it takes to produce a certain amount of iodine. This amount will be equal in moles to half the number of moles of thiosulfate added plus whatever amount of excess iodine is required to turn the starch blue. The average reaction rate during this initial time period, the *initial rate*, is inversely proportional to the time period; the faster the rate, the shorter the time period.

You will carry out a series of experiments to find the orders of iodide ion and peroxydisulfate ion. In each experiment you will double the concentration of either iodide ion or peroxydisulfate ion. If you find that doubling the concentration of one reactant while holding the concentration of the other reactant constant (*a*) has no effect on the reaction rate, the order of that reactant is zero; (*b*) doubles the initial rate, the order of that reactant is 1; or (*c*) quadruples the initial rate, the order of that reactant is 2. By analyzing your data, you should be able to determine the order of each reactant as well as the overall order of the reaction. This knowledge should allow you to begin to develop a reasonable mechanism for the reaction.

There are a few important guidelines regarding mechanisms. One is, of course, that the sum of the elementary steps must give the balanced net equation. Another regards the elementary steps themselves. An elementary step describes a collision between one set of particles that results in the production of another set of particles. It is not reasonable to suggest that more than three particles could come together simultaneously in one step. The probability of even three particles colliding is very minute. Almost all collisions are *bimolecular*, involving *two* particles. Unimolecular (one-particle) elementary processes are extremely rare in chemistry, with the exception of certain molecular configurational changes and nuclear decay processes.

Procedure

Students should work in pairs, one student mixing the reagents and the other timing the reaction. Before beginning the experiment, each pair should obtain the following quantities of reagents in *very* clean beakers:

0.16 M KI solution, 300 mL
0.12 M $(NH_4)_2S_2O_8$ solution, 300 mL
0.0055 M $Na_2S_2O_3$ solution, 200 mL
0.10 M EDTA solution, 5 mL
Distilled water, 600 mL
Solid soluble starch indicator, 5 g

You will be adding a bit of EDTA (ethylenediaminetetraacetate) to each reaction mixture to remove impurities that otherwise might catalyze the reaction. Because these reactions are very sensitive to temperature changes and to the presence of impurities, it is important to maintain constant temperature ($\approx$ 22°C), and it is *essential* to use scrupulously clean glassware.

(a) Determination of the Orders of Iodide and Peroxydisulfate Ions in the Rate Law

Trial 1

■ WEAR YOUR
SAFETY
GOGGLES

Label two clean graduated cylinders, one "$S_2O_3{}^{2-}$ and KI," the other "$S_2O_8{}^{2-}$." Throughout the experiment, use the graduated cylinders for only the solutions as labeled. Measure 40 mL of 0.16 M KI, 20 mL of 0.0055 M $Na_2S_2O_3$, a drop of 0.1 M EDTA, and 0.2 g (about half as much as you could pile on a dime) of soluble starch into a clean Erlenmeyer flask. Into a *second* Erlenmeyer flask, measure 35 mL of distilled water and 5 mL of 0.12 M $(NH_4)_2S_2O_8$ solution. Noting the starting time (use a timer or clock with a sweep second hand), quickly pour the contents of the second flask (containing the peroxydisulfate) into the first flask. Pour the mixture back and forth between the two flasks several times. Then gently swirl the flask containing the reaction mixture for as long as it takes for the solution to turn from colorless to deep blue. Record the reaction time in seconds. (*Note:* The reaction time for this trial should be considerably longer than for any of the others; it might require more than 10 min.)

Trial 2   Measure 40 mL of 0.16 M KI, 20 mL of 0.0055 M Na$_2$S$_2$O$_3$, a drop of EDTA, and 0.2 g of soluble starch into one clean Erlenmeyer flask. Into a second Erlenmeyer flask, place 10 mL of 0.12 M (NH$_4$)$_2$S$_2$O$_8$ and 30 mL of distilled water. Repeat, as in trial 1, recording the reaction time in seconds.

Trial 3   Measure 40 mL of 0.16 M KI, 20 mL of 0.0055 M Na$_2$S$_2$O$_3$, a drop of EDTA, and 0.2 g of soluble starch into one clean Erlenmeyer flask. Into a second Erlenmeyer flask, place 20 mL of 0.12 M (NH$_4$)$_2$S$_2$O$_8$ and 20 mL of distilled water. Repeat, as before, recording the reaction time in seconds.

Trial 4   Measure 5 mL of 0.16 M KI, 20 mL of 0.0055 M Na$_2$S$_2$O$_3$, a drop of EDTA, and 0.2 g of soluble starch into one clean Erlenmeyer flask. Into a second Erlenmeyer flask, place 40 mL of 0.12 M (NH$_4$)$_2$S$_2$O$_8$ and 35 mL of distilled water. Repeat, as before, recording the reaction time in seconds.

Trial 5   Measure 10 mL of 0.16 M KI, 20 mL of 0.0055 M Na$_2$S$_2$O$_3$, a drop of EDTA, and 0.2 g of soluble starch into one clean Erlenmeyer flask. Into a second Erlenmeyer flask, place 40 mL of 0.12 M (NH$_4$)$_2$S$_2$O$_8$ and 30 mL of distilled water. Repeat, as before, recording the reaction time in seconds.

Trial 6   Measure 20 mL of 0.16 M KI, 20 mL of 0.0055 M Na$_2$S$_2$O$_3$, a drop of EDTA, and 0.2 g of soluble starch into one clean Erlenmeyer flask. Into a second Erlenmeyer flask, place 40 mL of 0.12 M (NH$_4$)$_2$S$_2$O$_8$ and 20 mL of distilled water. Repeat, as before, recording the reaction time in seconds.

Trial 7   Measure 40 mL of 0.16 M KI, 20 mL of 0.0055 M Na$_2$S$_2$O$_3$, a drop of EDTA, and 0.2 g of soluble starch into one clean Erlenmeyer flask. Into a second Erlenmeyer flask, place 40 mL of 0.12 M (NH$_4$)$_2$S$_2$O$_8$. Repeat, as before, recording the reaction time in seconds.

(b)   **The Effect of Increasing the Temperature on Initial Rate**

Trial 8   As in trial 1 in part (*a*), measure 40 mL of 0.16 M KI, 20 mL of 0.0055 M Na$_2$S$_2$O$_3$, a drop of EDTA, and 0.2 g of soluble starch into one clean Erlenmeyer flask; into a second Erlenmeyer flask, place 5 mL of 0.12 M (NH_4)$_2S_2O_8$ and 35 mL of distilled water. Using separate clean thermometers, note the temperature of each solution, and heat each flask enough to raise the temperature of the contents by 10°C. Then combine the contents, and note the time of reaction as in part (a).

Data and Results

(a)   The average rate of appearance of iodine in the initial time period for each reaction is the "initial rate." The initial rate should be expressed in terms of moles per liter (molarity) of iodine per second. The number of moles of iodine produced in every trial is the same and is just equal to half the number of moles of thiosulfate, $S_2O_3{}^{2-}$, consumed. Calculate the number of moles of iodine produced in each trial (up to the time that the color changes).

In the column "Initial concentrations," give the concentration of each reactant *after* mixing (total volume = 100 mL) but *before* reacting. Don't forget that when solutions are mixed the concentrations of all of the materials are lowered. You can obtain final concentrations by using the formula $M_iV_i = M_fV_f$, where M_i and V_i are the initial molarity and volume and M_f and V_f are the final molarity and volume.

Trial	Initial Concentrations, M (After Mixing, Before Reacting)		Time, sec	Initial Rate, mol/liter-sec
	I^-	$S_2O_8^{2-}$		
1	_____	_____	_____	_____
2	_____	_____	_____	_____
3	_____	_____	_____	_____
4	_____	_____	_____	_____
5	_____	_____	_____	_____
6	_____	_____	_____	_____
7	_____	_____	_____	_____

Determination of x and y in the rate law, Rate $= k[I^-]^x[S_2O_8^{2-}]^y$:

$x =$ _____ $y =$ _____

Overall reaction order $(x + y)$ _____

(b)   Calculate a value for the initial rate when the temperature has been raised by 10°C (Trial 8).

Average temperature in the two flasks before heating: _____

Final average temperature in the flasks after heating: _____

Calculated initial rate at this temperature. _____

Questions

E.27.1   Suggest a reasonable mechanism for the reaction of iodide ion with peroxydisulfate ion.

E.27.2   Calculate an average value for k, the specific rate constant, for trials 1 through 7. Be sure to give the correct units.

E.27.3   Calculate a value for k for trial 8. Why is it different from the average value of k for trials 1 through 7?

E.27.4   In questions 2 and 3 you determined two specific rate constants, k at one temperature (call it k_1 at T_1) and k at a higher temperature (k_2 at T_2). From this information, together with the following equation, you can obtain a value for the activation energy E_a for the reaction:

$$\ln \frac{k_2}{k_1} = \frac{E_a}{R}\left(\frac{1}{T_1} - \frac{1}{T_2}\right)$$

Given that $R = 8.31$ joules/mol · deg, calculate a value for the activation energy.

E.27.5   You have calculated initial rates in terms of the rate of appearance of iodine. Compare the rate of appearance of iodine with the rate of:

(a) Disappearance of I^-

(b) Appearance of SO_4^{2-}

E.27.6   What percentage of the total number of moles of $S_2O_8^{2-}$ ion present at the beginning is consumed in trial 1?

E.27.7   According to the design of the experiment, the amount of iodide ion present in the solution remains effectively constant during the entire time period from the beginning of the reaction until the appearance of the iodine-starch complex. Explain.

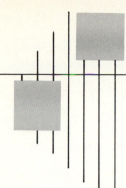

Experiment **E.28**

Chemical Kinetics— Spectrophotometric Monitoring

Special Items

(Amount per student; no wastage allowed for)
Spectronic 20 or 21
Six spectrophotometer cuvettes
40 mL 10^{-4} M methyl orange in 3 M $HClO_4$
150 mL 3 M $HClO_4$
10 mL 0.25 M $SnCl_2 \cdot 2H_2O$ in 3 M $HClO_4$
20 mL 2 M HCl
Stopwatch or timer
Seven 10-mL graduated cylinders
Seven disposable Pasteur pipettes
1-mL pipette with 0.01 graduations
2-mL volumetric pipette

Study Part A, Section 2h, at the beginning of this manual to learn about the proper use of the spectrophotometer. In this experiment you will use the spectrophotometer in order to measure the concentration of a chemical species, methyl orange, as it changes with time in a reacting mixture. The reaction to be studied is the reduction of methyl orange by stannous chloride. It is a complicated reaction that can be written for stoichiometry as

$$CH_3 \diagdown N \text{---} \bigcirc \text{---} N\text{=}N\text{---} \bigcirc \text{---} SO_3H + 2Sn^{2+}$$

$$+ 4H_3O^+ \rightarrow \quad CH_3 \diagdown N \text{---} \bigcirc \text{---} NH_2$$

$$+ H_2N\text{---} \bigcirc \text{---} SO_3H + 2Sn^{4+} + 4H_2O$$

The mechanism is more involved than this because, for example, chloride ion is observed to be a good catalyst for the reaction. For the catalyzed reaction the rate law can be written

$$\text{Rate} = -\frac{d[\text{MeO}]}{dt} = k[\text{MeO}][\text{Sn}^{2+}]^x[\text{Cl}^-]^y \qquad (1)$$

where [MeO] is the methyl-orange concentration, d/dt is its derivative with respect to time, and x and y are two exponents to be determined. When the concentrations of Sn^{2+} and Cl^- are much larger than the concentration of MeO, they do not

169

change much with time, so they can be considered constants that can be lumped together with the rate constant k to give

$$\text{Rate} = -\frac{d[\text{MeO}]}{dt} = k'[\text{MeO}] \tag{2}$$

Under such conditions the observed rate is proportional to the first power of the methyl-orange concentration; that is, the reaction is first order with respect to MeO. It is actually a *pseudo*-first-order reaction, and k' is called the pseudo-first-order rate constant.

The rate constant k' can be determined experimentally under various conditions. Equation (2) can be rearranged to give

$$\frac{d[\text{MeO}]}{dt} = -k'\,dt \tag{3}$$

Taking the integral of both sides for time from $t = 0$ to $t = t$, and for methyl-orange concentration from initial concentration ($[\text{MeO}]_0$) to concentration at time $= t$ ($[\text{MeO}]_t$) yields

$$\ln[\text{MeO}]_t = -k't + \ln[\text{MeO}]_0 \tag{4}$$

The term $\ln[\text{MeO}]_0$ is a constant, and a plot of $\ln[\text{MeO}]_t$ versus time will yield a straight line with a slope of $-k'$.

The exponents x and y may be also determined experimentally. From equations (2) and (3),

$$k' = k[\text{Sn}^{2+}]^x[\text{Cl}^-]^y \tag{5}$$

Taking the logarithm of both sides gives

$$\ln k' = \ln\left(k[\text{Sn}^{2+}]^x[\text{Cl}^-]^y\right) \tag{6}$$

which can be expressed as

$$\ln k' = x\ln[\text{Sn}^{2+}] + \ln\left(k\,[\text{Cl}^-]^y\right) \tag{7}$$

If $[\text{Cl}]^-$ is kept constant and $[\text{Sn}^{2+}]$ is varied, $\ln(k[\text{Cl}^-]^y)$ is a constant, and a plot of $\ln k'$ versus $\ln[\text{Sn}^{2+}]$ will yield a straight line with a slope of x. Similarly, when $[\text{Cl}^-]$ is varied and $[\text{Sn}^{2+}]$ is kept constant, the equation may be written as

$$\ln k' = y\ln[\text{Cl}^-] + \ln\left(k[\text{Sn}^{2+}]^x\right) \tag{8}$$

where $\ln(k[\text{Sn}^{2+}]^x)$ is a constant. In this case, a plot of $\ln k'$ versus $\ln[\text{Cl}^-]$ yields a straight line with a slope of y.

In this experiment you will determine the values of the exponents x and y. First you will obtain a Beer's law plot for methyl orange [part (b)]. Then you will find k' for six different concentrations of Sn^{2+} by monitoring the reaction (using the absorbance of methyl orange) for solutions where all variables are kept constant except the concentration of Sn^{2+} [part (b)]. You will then find x by plotting $\ln k'$ versus $\ln [Sn^{2+}]$, according to equation (7). Finally, in part (c) you will find k' for five different concentrations of Cl^- by holding all variables constant except $[Cl^-]$. A plot of $\ln k'$ versus $\ln [Cl^-]$ will then yield y.

The ultimate goal of a kinetics study is to understand the mechanism of the reaction. In our case, although we have written Sn^{2+} as the formula of the reducing agent, the reaction mechanism probably involves a collision between a methyl-orange molecule and a complex containing Sn^{2+}. The observed fact that MeO does not react very fast with Sn^{2+} in the absence of Cl^- suggests that the reaction path does not simply involve MeO and Sn^{2+} but goes via a Sn^{2+}-Cl^- complex. The nature of this complex might be suggested by a knowledge of the exponents x and y.

Procedure

(a)   Find out if solutions of methyl orange obey Beer's law. To do this, measure the absorbance of MeO as a function of concentration, diluting the stock solution (10^{-4} M methyl orange in 3 M $HClO_4$) with successively larger volumes of 3 M $HClO_4$. Use about six dilutions, in the range 10^{-4} to 10^{-7} M. For example, in seven different 10-mL graduated cylinders, prepare solutions using the following amounts of MeO solution and diluting to 10 mL with 3 M $HClO_4$:

1   5 mL of 10^{-4} M MeO stock, giving 5×10^{-5} M MeO
2   4 mL of 10^{-4} M MeO stock, giving 4×10^{-5} M MeO
3   3 mL of 10^{-4} M MeO stock, giving 3×10^{-5} M MeO
4   2 mL of 10^{-4} M MeO stock, giving 2×10^{-5} M MeO
5   1 mL of 10^{-4} M MeO stock, giving 1×10^{-5} M MeO
6   1 mL of solution 1, giving 5×10^{-6} M MeO
7   1 mL of solution 6, giving 5×10^{-7} M MeO

■ CAUTION   **Perchloric acid, $HClO_4$, is a powerful oxidizing agent that must not be mixed with reducing agents, such as paper or other organic compounds. Handle and dispose of it only as directed by your instructor.**

Measure the absorbance of each solution at 515 nm, the absorption maximum of methyl orange, with 3 M $HClO_4$ in the reference cell.

Make a Beer's-law plot of absorbance vs. methyl-orange concentration. If the points do not fall on a straight line, draw a smooth curve through them. This will serve as your calibration curve. If Beer's law does apply, absorbance is directly proportional to concentration, and a plot of ln absorbance vs. time gives k' directly. Otherwise, you must go from absorbance to concentration via the calibration curve and then plot ln concentration vs. time.

(b) Examine the dependence of the rate on Sn^{2+} concentration. Get 10 mL of 0.25 M $SnCl_2 \cdot 2H_2O$ solution (do not expose to air too long or it will oxidize) and prepare reaction mixtures containing Sn^{2+} concentrations ranging from 0.005 to 0.010 M at constant chloride concentration (say, 0.3 M), constant hydronium-ion concentration (say, 2.8 M), and constant methyl-orange concentration (say, 2×10^{-5} M). This may be done by combining 2 mL of stock solution (10^{-4} M methyl orange in 3 M $HClO_4$), 1.5 mL of 2 M HCl, plus 6.5 mL of 3 M $HClO_4$ in a 10-mL graduated cylinder and adding different amounts of 0.25 M $SnCl_2$ for successive runs as follows:

1 0.20 mL, giving 0.005 M Sn^{2+}
2 0.24 mL, giving 0.006 M Sn^{2+}
3 0.28 mL, giving 0.007 M Sn^{2+}
4 0.32 mL, giving 0.008 M Sn^{2+}
5 0.36 mL, giving 0.009 M Sn^{2+}
6 0.40 mL, giving 0.010 M Sn^{2+}

At the start of any particular run ($t = 0$), using a 1-mL pipette with 0.01-mL graduations, pipette the proper amount of stannous chloride solution into the cylinder and stir. Using a Pasteur (disposable) pipette, quickly rinse the spectrophotometer cell (cuvette) with a small portion of the solution. Discard the rinse, and immediately place a portion of the solution to be examined in the cuvette. Read the absorbance at 30-sec intervals until the reaction flattens out. Readings may be taken more frequently, for example, every 15 sec, at the start of the reaction when absorbance is changing quickly. Be sure to check the instrument adjustments before each run.

For each of the above runs, determine k' from a plot of ln [MeO] versus time as shown in equation (4). Then determine the exponent x from a plot of ln k' versus ln [Sn^{2+}].

(c) Examine the dependence of the rate on chloride concentration. Proceed as in (b), except prepare the reaction mixtures so that they contain chloride at concentrations ranging from 0.1 to

0.5 M, the H_3O^+ being held constant at 2.8 M and the methyl orange at 2×10^{-5} M. To do this, pipette into a 10-mL graduated cylinder 2 mL of the stock solution (10^{-4} M methyl orange in 3 M $HClO_4$) and on successive runs the following amounts of 2 M HCl:

1   0.5 mL, giving 0.1 M Cl^-
2   1.0 mL, giving 0.2 M Cl^-
3   1.5 mL, giving 0.3 M Cl^-
4   2.0 mL, giving 0.4 M Cl^-
5   2.5 mL, giving 0.5 M Cl^-

Dilute to the 10-mL mark with 3 M $HClO_4$, and at zero time pipette in 0.2 mL of 0.25 M $SnCl_2$. Introduce each solution as quickly as possible into the rinsed cell, and read the absorbance at 30-sec intervals until the reaction mixture stops changing appreciably.

For each of the above runs, determine k' from a plot of ln [MeO] versus time and then get the exponent y from a plot of ln k' versus ln [Cl^-].

Questions

Prelaboratory

E.28.1   In part (*b*) you are directed to make six different Sn^{2+} solutions, each of a different molarity. Calculate the volume of 0.25 M $SnCl_2$ that must be added to the other reagents to give each of the solutions.

E.28.2   If your Beer's-law plot of absorbance vs. methyl-orange concentration gives points that do not fall on a straight line, you must prepare a calibration curve. Explain what a calibration curve is and how it can be used.

Postlaboratory

E.28.3   What assumptions are made in the design of this experiment regarding each of the following?
(*a*) Possible absorption of light at 515 nm by reactants or products other than methyl orange
(*b*) Possible formation of a chloride complex with Sn^{4+}, one of the reaction products
(*c*) Possible effects on the reaction rate by temperature changes

E.28.4   If $-d[\text{MeO}]/dt = k'[\text{MeO}]$, show that the slope of a plot of ln [MeO] versus time is $-k'$.

E.28.5   Give the rationale behind the strategy used to obtain the exponents x and y.

E.28.6 Why in part (*b*) do you have to run so many different Sn^{2+} concentrations?

E.28.7 Why can you not look at Sn^{2+} or Cl^- concentrations by use of the spectrophotometer to monitor the reaction rate?

E.28.8 Suggest a mechanism for the reaction of MeO with Sn^{2+} on the basis of your observations in this experiment.

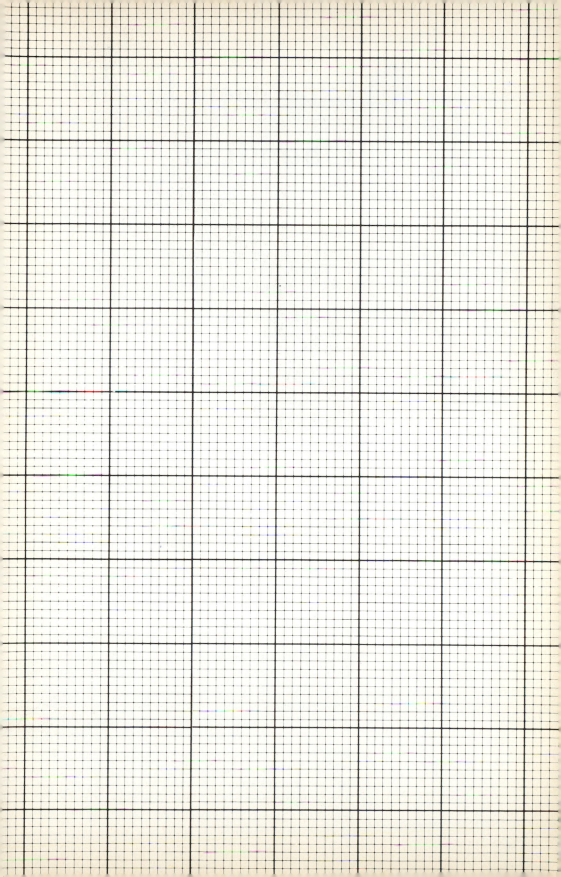

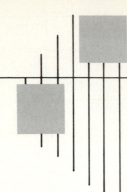

Chemical Equilibrium

Special Items

Without spectrophotometer:
Ruler
100 mL distilled water
30 mL 0.0020 M NaNCS
15 mL 0.20 M Fe(NO$_3$)$_3$
(must be colorless at start)
Using Spectronic 20 or 21:
30 mL 2.0×10^{-4} M NaNCS
15 mL 0.20 M Fe(NO$_3$)$_3$
6 cuvettes for spectrophotometer

Chemical reactions occur so as to approach a state of chemical equilibrium. The equilibrium state can be characterized by specifying its equilibrium constant, i.e., by indicating the numerical value of the mass-action expression. In this experiment you will determine the value of the equilibrium constant for the reaction between ferric ion (Fe^{3+}) and isothiocyanate ion (NCS$^-$),

$$\text{Fe}^{3+} + \text{NCS}^- \rightleftharpoons \text{FeNCS}^{2+}$$

for which the equilibrium condition is

$$\frac{[\text{FeNCS}^{2+}]}{[\text{Fe}^{3+}][\text{NCS}^-]} = K$$

To find the value of K, it is necessary to determine the concentration of each of the species Fe^{3+}, NCS$^-$, and FeNCS^{2+} in the system at equilibrium. This will be done colorimetrically, taking advantage of the fact that FeNCS^{2+} is the only highly colored species in the solution.

The color intensity of a solution depends on the concentration of the colored species and on the depth of solution viewed. Thus, 2 cm of a 0.1 M solution of a colored species appears to have the same color intensity as 1 cm of a 0.2 M solution. Consequently, if the depths of two solutions of unequal concentrations are chosen so that the solutions appear equally colored, then the ratio of concentrations is simply the inverse of the ratio of the two depths. It should be noted that this procedure permits only a comparison between concentrations. It does not give an absolute value of either one of the concentrations. To know absolute values, it is necessary to compare with a standard of known concentration.

For color determination of FeNCS^{2+} concentration, you must have a standard solution in which the concentration of FeNCS^{2+} is known. Such a solution can be prepared by starting with a small known concentration of NCS$^-$ and adding such a large excess of Fe^{3+} that essentially all the NCS$^-$ is converted to FeNCS^{2+}. Under these conditions, you can assume that the final concentration of $FeNCS^{2+}$ is equal to the initial concentration of NCS^-.

Procedure

(a)   **Without Spectrophotometer**   Thoroughly clean six 15-cm test tubes, rinse with distilled water, and let drain. To each of these test tubes, add 5 mL of 0.0020 M NaNCS. To the first test tube, add 5 mL of 0.20 M Fe$(NO_3)_3$. This tube will serve as standard. For the other test tubes, proceed as follows: Add 10 mL of 0.20 M Fe$(NO_3)_3$ to your graduated cylinder, fill to 25 mL with distilled water, and stir thoroughly to mix. Pour 5 mL of the resulting diluted solution (0.080 M Fe^{3+}) into test tube 2. Discard all but 10 mL of the diluted solution that is in the graduated cylinder, refill with distilled water to 25 mL, and stir thoroughly. Add 5 mL of the resulting solution (0.032 M Fe^{3+}) to test tube 3. Discard all but 10 mL of the solution in the cylinder, and again dilute to 25 mL. Continue this procedure until you have added to each successive test tube 5 mL of progressively more dilute Fe^{3+} solution.

Now the problem is to determine the concentration of $FeNCS^{2+}$ in each test tube relative to the standard in test tube 1. Compare the color intensity in test tube 1 with that in each of the other test tubes. Take the two tubes to be compared, hold them side by side, and wrap a strip of paper around both. Look down through the solutions toward a white paper on your desk, as shown in Figure E.29.1, or toward a strong light source. If color intensities appear identical, record this fact. If not, take test tube 1 and pour out some of the standard into a clean, dry beaker (you may need to pour some back!) until the color intensities appear identical. Measure the heights of solutions in the two tubes being compared. Do this comparison for all five tubes.

(b)   **With Spectrophotometer**   (Refer to Part A, Section 2h, for instructions on use of your spectrophotometer.) Determine the equilibrium constant $K = [FeNCS^{2+}]/[Fe^{3+}][NCS^-]$ for the equilibrium

$$Fe^{3+} + NCS^- \rightleftharpoons FeNCS^{2+}$$

by using the spectrophotometer to measure the concentration of $FeNCS^{2+}$ in various equilibrium mixtures. To do this proceed as follows:

Thoroughly clean six 15-cm test tubes and six 10-cm spectrophotometer cuvettes or test tubes. Rinse thoroughly

■ **WEAR YOUR SAFETY GOGGLES**

Figure E.29.1

with distilled water, and flame-dry. To each of the large test tubes, add 5.0 mL of 2.0×10^{-4} M NaNCS. To the first test tube, add 5.0 mL of 0.20 M Fe$(NO_3)_3$, and stir. The resulting solution is your standard, in which you can assume all the NCS$^-$ has been converted to FeNCS^{2+}. For the other test tubes, do the following: Measure 10.0 mL of 0.20 M Fe$(NO_3)_3$ in a 25-mL graduated cylinder, fill to 25.0 mL with distilled water, and stir thoroughly to mix. Measure 5.0 mL of the resulting 0.080 M Fe^{3+} into test tube 2. Discard all but 10.0 mL of the 0.080 M Fe^{3+}, refill the graduated cylinder to 25.0 mL with distilled water, and stir thoroughly. Add 5.0 mL of the resulting 0.032 M Fe^{3+} to test tube 3. Again discard all but 10.0 mL of the contents of the graduated cylinder, refill to 25.0 mL with distilled water, stir, and add 5.0 mL to test tube 4. Repeat with test tubes 5 and 6.

Rinse out each of the small test tubes (or cuvettes) with a few milliliters of the corresponding solutions from test tubes 1 through 6, and then transfer enough of the solution to fill about half full. Take the six small test tubes, and measure the absorbance of the solutions in the spectrophotometer. All you need do is insert each tube successively in the sample holder, push it all the way down, close the lid, and read the scale. Make all measurements at the same wavelength, that at which the solution in test tube 1 shows maximum absorbance.

Assuming Beer's law is valid, you can state that the absorbance of any two of your solutions is in the same ratio as their concentrations of FeNCS^{2+}. Knowing the concentration in test tube 1, you can calculate it in all the other tubes. Do this, and use your results to get values for the equilibrium concentrations of Fe^{3+} and NCS$^-$ left uncombined. Calculate K for each of your test tubes.

If you have time, repeat the series of absorbance measurements at several different wavelengths (e.g., 450, 550, 650 nm). What effect does changing the wavelength of the measurement have on your determined value of K?

Data

(a)   Without Spectrophotometer

Test Tube	Height of Liquid	Comparison Height of Standard
2	_____	_____
3	_____	_____

4   _____   _____

5   _____   _____

6   _____   _____

(b)   **With Spectrophotometer**
Wavelength of maximum absorbance   _____

Test Tube	Absorbance
2	_____
3	_____
4	_____
5	_____
6	_____

Results

Test Tube	Initial Conc.		Equilibrium Conc.			
	$[Fe^{3+}]$	$[NCS^-]$	$[FeNCS^{2+}]$	$[Fe^{3+}]$	$[NCS^-]$	K
1	____	____	____	____		
2	____	____	____	____	____	____
3	____	____	____	____	____	____
4	____	____	____	____	____	____
5	____	____	____	____	____	____
6	____	____	____	____	____	____

In calculating initial concentrations, assume that $Fe(NO_3)_3$ and NaNCS are each completely dissociated. Remember also that mixing two solutions dilutes both of them.

In calculating equilibrium concentrations, assume that in test tube 1 all the initial NCS^- has been converted to $FeNCS^{2+}$. For the other test tubes, calculate $FeNCS^{2+}$ from the ratio of heights in the color comparison or the absorbance values. Equilibrium concentrations of Fe^{3+} and NCS^- are obtained by subtracting $FeNCS^{2+}$ formed from initial Fe^{3+} and NCS^-.

For each of test tubes 2 to 6, calculate the value of K. Decide which of these values is most reliable.

Questions

E.29.1   In this experiment you examine the equilibrium $Fe^{3+} + NCS^-$ $\rightleftharpoons FeNCS^{2+}$. Competing somewhat with this equilibrium are equilibria such as $FeNCS^{2+} + NCS^- \rightleftharpoons Fe(NCS)_2^+$ and $Fe(NCS)_2^+ + NCS^- \rightleftharpoons Fe(NCS)_3$.

(a) To allow you to ignore these other equilibria, how must their equilibrium constants compare with that of the equilibrium being studied?

(b) Given that all ions are colorless except for $FeNCS^{2+}$, what effect should these competing equilibria have on the value of K determined in this experiment? Explain.

E.29.2   How reasonable was your assumption that all the NCS^- in test tube 1 was converted to $FeNCS^{2+}$? (Calculate the percentage of NCS^- converted to $FeNCS^{2+}$ using your best value of K.)

E.29.3   Why are the values of K determined for test tubes 3, 4, and 5 probably more reliable than those determined for tubes 2 or 6?

E.29.4   In your own words, give the rationale behind the procedure and methodology in this experiment.

E.29.5   In part (b) you are directed to use the wavelength that gives maximum absorbance in test tube 1 for measurement of absorbances in the other sample tubes. Explain.

E.29.6 What is the purpose of a monochromator in a spectrophotometer? How might your results be affected if it were not included in the design of the instrument?

E.29.7 How would the operation of a spectrophotometer be affected if the light-source tungsten lamp were replaced by a gas burner flame to which NaCl has been added?

E.29.8 Show how Beer's law leads to the well-known observation that at equal color intensities the concentration of a colored species is inversely proportional to the depth of solution needed to give equal color.

E.29.9 How would you redesign the spectrophotometer so that you could replace the grating by a prism to select out desired wavelengths? Draw a sketch. Recall that blue light is refracted more than red light.

E.29.10 Why should it make such a big difference if you do not use closely matched containers for your samples in the spectrophotometer?

E.29.11 In part (b), why can you assume that in the first tube all the NCS^- is converted to $FeNCS^{2+}$? How do you go about calculating the concentration of $FeNCS^{2+}$ in this solution?

E.29.12 Why in part (b) is it important to rinse out the small test tubes with the various solutions before proceeding to fill them for the measuring of the absorption spectrum?

Solubility Product of Cupric Iodate

Special Items

Without Spectrophotometer:
Ruler
Burette (if not in desk)
Distilled water
35 mL 0.150 M $CuSO_4$
35 mL 0.320 M HIO_3
2 mL concentrated NH_3
Using Spectronic 20 or 21:
50 mL 0.150 M $CuSO_4$
25 mL 0.320 M HIO_3
Six cuvettes (or scratch-free test tubes)

When a saturated solution of a strong electrolyte is in equilibrium with excess solid, the equilibrium can be described by use of the solubility product. In the case of cupric iodate, the equilibrium of interest is

$$Cu(IO_3)_2(s) \rightleftharpoons Cu^{2+} + 2IO_3^-$$

and the equilibrium condition is

$$[Cu^{2+}][IO_3^-]^2 = K_{sp}$$

This means that in any saturated solution of cupric iodate, the concentration of cupric ion multiplied by the square of the concentration of iodate ion is equal to a characteristic number. The individual concentrations of Cu^{2+} and IO_3^- may be different in going from one saturated solution to another, but the ion product $[Cu^{2+}][IO_3^-]^2$ always stays constant. In other words, K_{sp} always has the same numerical value (assuming constant temperature) for any saturated solution of cupric iodate. The main purpose of this experiment is to determine this numerical value.

To find out the value of K_{sp} for $Cu(IO_3)_2$, it is necessary to know the concentration of Cu^{2+} and the concentration of IO_3^- at equilibrium in a saturated solution. The concentration of Cu^{2+} in a solution can be rather easily determined by comparing the color of the solution with the color of a solution of known concentration (see Experiment E.29). However, in this experiment the concentration of Cu^{2+} is too low to produce an easily observable color; therefore, it will be advisable to intensify the color by adding aqueous ammonia to convert Cu^{2+} to intensely colored $Cu(NH_3)_4^{2+}$. To determine the concentration of IO_3^- in the solution, it will be necessary to proceed indirectly. You will mix known solutions of Cu^{2+} and IO_3^- to precipitate $Cu(IO_3)_2$. From the initial and final concentrations of Cu^{2+}, you will know how much Cu^{2+} is precipitated. This will enable you to calculate how much IO_3^- is precipitated. (Note that for each Cu^{2+} ion precipitated, two

IO_3^- ions must go along with it.) Knowing how much IO_3^- there was in the initial solution and how much of it precipitated, you will be able to calculate the concentration of IO_3^- left in solution.

Procedure

(a)   Without Spectrophotometer   (As a check, you will carry out three determinations of K_{sp}, starting with solutions of different concentration.) Set up a burette as shown in Figure A.8. Rinse the burette once with distilled water and then twice with 5-mL portions of 0.150 M $CuSO_4$. Fill the burette to about the 25-mL mark with 0.150 M $CuSO_4$. Clean and dry three test tubes and your graduated cylinder. Label the test tubes 1, 2, and 3. Into each test tube, run out exactly 5.00 mL of 0.150 M $CuSO_4$; into the graduated cylinder, 1.00 mL.

■ **WEAR YOUR SAFETY GOGGLES**

Rinse out the burette thoroughly with distilled water and then twice with 5-mL portions of 0.320 M HIO_3 (iodic acid, a strong electrolyte). Fill the burette to about the 25-mL mark with 0.320 M HIO_3. Run into test tube 1 exactly 4.00 mL; into test tube 2, exactly 4.50 mL; into test tube 3, exactly 5.00 mL. To bring all volumes to 10 mL, add 20 drops (1 mL) of distilled water to test tube 1 and 10 drops to test tube 2.

With a stirring rod, stir the solution in test tube 1 until a precipitate forms. It may be necessary to scratch the side of the test tube with the rod to start precipitation. Remove the rod, rinse and dry it, and stir test tube 2. Repeat for test tube 3. During the next 10 min, shake each test tube frequently.

Filter the solution from test tube 1 through a double thickness of filter paper into a clean, dry test tube. Discard the solid precipitate. To the liquid filtrate, add a few drops of the concentrated ammonia solution in the hood. If a precipitate remains, add a few more drops of concentrated NH_3 until the solution clears. For color comparison, prepare a standard by adding distilled water to the 1.00 mL of 0.150 M $CuSO_4$ in the graduated cylinder to bring the volume to 20 mL. Add about 1 mL of the concentrated ammonia solution. Using the technique described in Experiment E.29 (see especially Figure E.29.1), determine quantitatively the depth of standard solution in a test tube required to match the color intensity of solution 1. Record the two heights of solutions compared. Recall from Experiment E.29 that the concentrations are inversely proportional to the heights that give equal color intensity.

In similar fashion, filter solutions 2 and 3, using a double thickness of filter paper each time, and measure the height of color standard required to give the same intensity as each.

(b)   **With Spectronic 20 or 21**   (Refer to Part A, Section 2h for instructions on use of your spectrophotometer.) Determine the solubility product $K_{sp} = [Cu^{2+}][IO_3^-]^2$ for the equilibrium

$$Cu(IO_3)_2(s) \rightleftharpoons Cu^{2+} + 2IO_3^-$$

by using the spectrophotometer to measure the concentration of Cu^{2+} in equilibrium with excess solid cupric iodate. To do this, proceed as follows.

Clean and flame-dry five 15-cm and five cuvettes or scratch-free 10-cm test tubes. Clean but do not dry a 25-mL graduated cylinder. From a clean burette, rinsed and filled with 0.150 M $CuSO_4$, run 5.40 mL of the solution into one of the 15-cm test tubes. Label this test tube 1. In like manner, draw down 5.70 mL into tube 2, 6.00 mL into tube 3, and 6.30 mL into tube 4. Into the 25-mL cylinder dispense an accurately measured portion of about 2.00 mL of the solution, and fill the cylinder to the 25.00-mL mark. Make sure this diluted solution is thoroughly stirred, and then pour about 10 mL into test tube 5. Stopper it and put it aside.

Empty the burette and refill with 0.320 M HIO_3 solution. Rinse twice with distilled water and twice with a few milliliters of HIO_3 solution before filling. Measure 6.60 mL of this solution into tube 1, 6.30 mL into tube 2, 6.00 mL into tube 3, and 5.70 mL into tube 4.

With a stirring rod, stir the solution in test tube 1 until a precipitate forms. It may be necessary to scratch the side of the test tube with the rod to start precipitation. Remove the rod, rinse and dry it, and stir test tube 2. Repeat with tubes 3 and 4. When crystallization has been initiated in all four of the test tubes, stopper them and put them aside for a day, if possible.

Filter, through a double thickness of filter paper, a portion of each of the supernatant solutions from tubes 1 to 4 into separate 10-cm test tubes. These tubes may be labeled at the very top but not so low that it could interfere with the light beam in the spectrophotometer. Keep all the levels in the several tubes at about the same height, i.e., about halfway. Get an equal volume of the standard from test tube 5. Add 5 drops of concentrated ammonia to each of the five test tubes and stir well. Measure the absorbance of each solution in the spectrophotometer. Insert each tube successively in the sample holder, push it all the way down, close the lid, and read the scale. Make all measurements at the same wavelength, that at which the solution in test tube 5 shows maximum absorbance.

Assuming that Beer's law is valid, the absorbance of your solutions will be proportional to the concentration of $Cu(NH_3)_4^{2+}$. You can assume that in test tube 5 the concentration of $Cu(NH_3)_4^{2+}$ is determined by complete conversion of the Cu^{2+} by the NH_3. Since you know how much you diluted the stock $0.150\ M$ $CuSO_4$ to make up test tube 5, you know what its Cu^{2+} concentration was before NH_3 addition and what the $Cu(NH_3)_4^{2+}$ concentration was after addition. By comparing absorbances you can now calculate the $Cu(NH_3)_4^{2+}$ in each of the other test tubes. Set this equal to the Cu^{2+} concentration that was there before NH_3 addition. These Cu^{2+} concentrations represent final equilibrium concentrations after $Cu(IO_3)_2$ has precipitated. Knowing the initial $[Cu^{2+}]$ before precipitation and the final $[Cu^{2+}]$ after precipitation, you can calculate how much Cu^{2+} precipitated, how much IO_3^- must have precipitated as well, and how much IO_3^- is left in each solution.

Calculate for each of your test tubes the final equilibrium ion product $[Cu^{2+}][IO_3^-]^2$.

Data and Results

(a) Without Spectrophotometer

	1	2	3
Initial burette reading (CuSO₄ solution)	———	———	———
Final burette reading (CuSO₄ solution)	———	———	———
Initial burette reading (HIO₃ solution)	———	———	———
Final burette reading (HIO₃ solution)	———	———	———
Height of filtered solution in tube	———	———	———
Comparison height of color standard	———	———	———

	1	2	3	4
Concentration of Cu^{2+} in 10-mL mixture:				
Before precipitation (calculated)		___	___	___
After precipitation (from color comparison)		___	___	___
Decrease of Cu^{2+} concentration		___	___	___
Decrease of IO_3^- concentration		___	___	___
Concentration of IO_3^- in 10-mL mixture:				
Before precipitation		___	___	___
After precipitation		___	___	___
Solubility product $= [Cu^{2+}][IO_3^-]^2$		___	___	___

(b)   With Spectrophotometer

	1	2	3	4
Volume of 0.150 M $CuSO_4$	___	___	___	___

Precise volume of 0.150 M $CuSO_4$ placed in 25-mL graduated cylinder	___
Concentration of $CuSO_4$ in test tube 5	___

	1	2	3	4
Volume of 0.320 M HIO_3	___	___	___	___

Wavelength of maximum absorbance in test tube 5 (standard)	___

	5	1	2	3	4
Absorbance values for $Cu(NH_3)_4^{2+}$ solutions	___	___	___	___	___

	5	**1**	**2**	**3**	**4**
Concen- tration of $Cu(NH_3)_4^{2+}$	____	____	____	____	____

	1	**2**	**3**	**4**
Concentration of Cu^{2+} before precipitation	____	____	____	____
Concentration of Cu^{2+} after precipitation	____	____	____	____
Decrease of Cu^{2+} concentration	____	____	____	____
Decrease of IO_3^{2-} concentration	____	____	____	____
Concentration of IO_3^{2-} after precipitation	____	____	____	____

Calculation of equilibrium solubility products $[Cu^{2+}][IO_3^-]^2$

Test Tube
    1

    2

    3

    4

Questions

E.30.1   Why do you rinse the burette with 5-mL portions of solution before filling?

E.30.2   Owing to the reaction $NH_3 + H_2O \rightleftharpoons NH_4^+ + OH^-$, aqueous ammonia solutions are basic. The K_{sp} of $Cu(OH)_2(s)$ is 1.6×10^{-19} at 25°C. What does the procedure given in this experiment imply about the relative concentration of Cu^{2+} in equilibrium with $Cu(OH)_2(s)$ and $Cu(NH_3)_4^{2+}$? What problem might occur if NH_3 is not added to Cu^{2+} solutions in excess?

E.30.3   Suppose that in part (*a*) the stirring rod you used in order to initiate precipitation of $Cu(IO_3)_2(s)$ in test tube 2 was contaminated with a small amount of pure $Cu(IO_3)_2(s)$. Would this extra $Cu(IO_3)_2(s)$ have any effect on your results for test tube 2? How about if you had used the contaminated stirring rod to stir the filtrate (Cu^{2+} solution) from test tube 2 after adding the NH_3?

E.30.4   In part (*b*) you are directed to use the wavelength that gives maximum absorbance in test tube 5 for measurement of absorbances in the other sample tubes. Explain.

E.30.5   In part (*b*) it is suggested that if possible you allow your test tubes to stand for a day after crystallization has been initiated. Explain. How are your results likely to be affected if you cannot wait before proceeding?

E.30.6   Suppose in part (*b*) that some solid $Cu(IO_3)_2$ got through your filter paper. How would it affect subsequent results?

pH Indicators and Titration

Special Items

Wide range of indicators,
   such as methyl violet,
   thymol blue, erythrosine,
   bromophenol blue, methyl
   orange, methyl red,
   p-nitrophenol,
   bromthymol blue, cresol
   red, phenolphthalein,
   thymolphthalein, and
   alizarin yellow, to cover
   pH range from 1 to 13
50 mL of each of about 10
   buffers to cover pH range
   from 1 to 13
25 mL 0.010 *M* unknown
   weak acid
25 mL 0.050 *M* unknown
   weak base
pH meter (optional)

An indicator, such as litmus or phenolphthalein, changes color when the pH (the negative logarithm of the hydronium-ion concentration) of the environment changes. In principle, any molecule could serve as an indicator, provided its color were a function of pH, but, in fact, to be useful the color change has to be reasonably sharp (within two pH units) and the amount of material that produces the color must be very small. Otherwise, the indicator becomes an important species modifying other equilibria, and that would not be desirable.

In the simplest case, an indicator, which is generally a complicated organic molecule, can be considered to be a weak acid. As such we can write it as HIn, where In stands for all the complex of rings and double bonds that produces color. The H is shown separately because its removal is what changes the color, In^- having a different absorption spectrum from HIn. If we write the dissociation of HIn in aqueous solution as a typical weak-acid dissociation, we have

$$HIn + H_2O \rightleftharpoons H_3O^+ + In^-$$

and at equilibrium

$$K = \frac{[H_3O^+][In^-]}{[HIn]}$$

where K is a specific numerical constant characteristic of the indicator. Rewriting the above equation, we have

$$\frac{[HIn]}{[In^-]} = \frac{1}{K}[H_3O^+]$$

which shows that the relative concentration of HIn and In^- is directly proportional to the hydronium-ion concentration. In other words, the greater the hydronium-ion concentration (i.e., the lower the pH), the more of the indicator in the HIn form. Stated another way, in highly acid solution most of the indicator is present as HIn, whereas in highly basic solution most of the indicator is present as In^-. If now the HIn molecule has a color different from that of the In^- ion, the actual color in the solution becomes strongly dependent on pH, since it is the color of the species in larger concentration that is generally "seen." Setting a concentration ratio of 10 as optimal for "seeing" the color of the dominant species, we can see that the "acid" color dominates when $[H_3O^+] = 10K$ and the "base" color dominates when $[H_3O^+] = 0.1K$. This means that the changeover from acid color to base color will take place at a hydronium-ion concentration that depends on K of the indicator, and the hydronium-ion concentration must change by a factor of 100 (i.e., the pH has to change by two units) to ensure a clear change in color.

In this experiment you will study about 10 different indicators to find at what pH their colors change and thus find out the K value of each indicator. The mode of operation will be to add a bit of the indicator to solutions of various pH, these solutions being made by taking various buffers for which the pH is known. After you have investigated the various indicators, you will use this information to determine the dissociation constants of a weak acid and a weak base. Then you will attempt to make a mixture of several of these indicators (a so-called universal indicator) that will provide known color changes stepwise over a wide pH range. Using this universal indicator, you are then asked to perform two titrations, (1) NaOH versus HCl and (2) NaOH versus $HC_2H_3O_2$, to find out what is similar and what is different between two such systems.

Procedure

■ WEAR YOUR SAFETY GOGGLES

(a) Get about 50 mL of a buffer solution of known pH, and place 5 mL in each of 10 clean, dry test tubes. Add 2 drops of each indicator to different test tubes, and record the colors observed. Repeat the entire procedure with each of the buffer solutions provided. Construct a pH color chart summarizing your results.

(b) Using your clean, dry graduated cylinder, secure from the reagent table 25 mL of the unknown acid solution, which contains 0.010 mol HX per liter. Place 5 mL of the acid solution in each of five clean, dry test tubes. Add 2 drops of a different indicator to each test tube. Note the colors obtained.

Proceeding in similar fashion, note the color produced by adding each indicator to the 0.050 M solution of unknown base, which is provided on the reagent table.

(c) Using the data from (a), concoct a recipe for a universal indicator, using not more than seven indicators. Check your predictions with tests on the various buffer solutions.

(d)   Using your universal indicator, perform two titrations:

1   0.1 M HCl versus 0.1 M NaOH
2   0.1 M $HC_2H_3O_2$ versus 0.1 M NaOH

recording in each case the pH of the solution as a function of each milliliter of base added to 10 mL of acid. Carry out the addition to 10 mL beyond the equivalence point. Plot your data, account for the shape of each curve, and explain any differences observed.

(e)   If you have time, repeat the above by adding stepwise 0.1 M NH_3 to 10 mL of 0.10 M HCl and then by adding stepwise 0.1 M NH_3 to 10 mL of 0.10 M $HC_2H_3O_2$.

(f)   If you have a pH meter available, check any one of the curves you have obtained (preferably the NH_3 versus $HC_2H_3O_2$). Add the base in smaller increments (0.1 mL), and read to 0.1 pH unit. See Experiment E.38 for instructions on how to operate a pH meter.

Data and Results

(a)   Colors of Indicators versus pH
Enter the pH values for each buffer, the identities of the indicators tested, and the color of each indicator in each buffer in the table below.

pH of Buffers

Indicator

(b)   Color of Indicators with Unknown Acid and Unknown Base

Indicator	Color with Unknown Acid	Color with Unknown Base
_____	_____	_____
_____	_____	_____
_____	_____	_____
_____	_____	_____
_____	_____	_____
_____	_____	_____
_____	_____	_____
_____	_____	_____
_____	_____	_____
_____	_____	_____

Assuming that your acid is monoprotic HX, calculate the dissociation constant for $HX + H_2O \rightleftharpoons H_3O^+ + X^-$. Note that the concentration of X^- equals the concentration of H_3O^+ and that the concentration of HX is the initial concentration of HX minus that which dissociated.

Calculate the constant of $MOH \rightleftharpoons M^+ + OH^-$ of your unknown base, noting that $[OH^-] = 1.0 \times 10^{-14}/[H_3O^+]$.

(c)   Recipe for Universal Indicator

(d)   Titration Data, Results, and Discussion

  1   0.1 M HCl versus 0.1 M NaOH

  2   0.1 M HC$_2$H$_3$O$_2$ versus 0.1 M NaOH

(e)   Titration Data, Results, and Discussion

  1   0.1 M NH$_3$ versus 0.1 M HCl

  2   0.1 M NH$_3$ versus 0.1 M HC$_2$H$_3$O$_2$

(f)   Titration Data Using pH Meter
Choice of acid and base for titration

Volume and concentration of acid used

mL Base Added	pH	mL Base Added	pH	mL Base Added	pH	mL Base Added	pH
———	———	———	———	———	———	———	———
———	———	———	———	———	———	———	———
———	———	———	———	———	———	———	———
———	———	———	———	———	———	———	———
———	———	———	———	———	———	———	———
———	———	———	———	———	———	———	———

Questions

E.31.1   The dissociation constant for methyl red is approximately 10^{-5}, and that of thymol blue is approximately 10^{-9}. Which indicator would be preferred, and why, for a titration of
(a) 0.1 M CH_3NH_2 ($K_{diss} = 5 \times 10^{-4}$) with 0.1 M HCl?
(b) 0.1 M HNO_2 ($K_{diss} = 5 \times 10^{-4}$) with 0.1 M NaOH?

E.31.2   Litmus is red below pH 4.5 and blue above pH 8.3; bromocresol purple is yellow below pH 5.2 and purple above pH 6.8. Which indicator would be preferred for a titration in which the equivalence point is pH 6? Explain.

E.31.3   The molecular (HIn) form of the indicator paranitrophenol is colorless, and the ionic form (In^-) is yellow. Using Le Châtelier's principle, predict the color of paranitrophenol in basic solution and in acidic solution.

E.31.4   One problem with single indicators is that they change color over a pH range instead of precisely at a given pH. Explain how you could obtain a sharper end point with a mixture of two indicators than with a single indicator.

E.31.5   If phenolphthalein changes from colorless to red at pH = 9, what can you conclude about its dissociation constant?

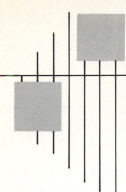

Hydrolysis and Amphoterism

Special Items

150 mL distilled water
1.5 g NaCl
Ca. 2 g $NaC_2H_3O_2$
0.6 g $Cu(NO_3)_2 \cdot 3H_2O$
Ca. 1.5 g $KAl(SO_4)_2 \cdot 12H_2O$
10 g Dry Ice
20 mL 6 M NaOH
20 mL 6 M HCl
10 mL 1.0 M $CaCl_2$
10 mL 0.5 M $ZnCl_2$
5 mL 0.0010 M $KAl(SO_4)_2$
5 mL 0.00010 M Na_2S
Indicators as in Experiment
E.31

Some ions react with water to form a weak electrolyte plus either H_3O^+ or OH^-. Such a reaction is called "hydrolysis." The extent of hydrolysis for a particular ion can be described by a hydrolysis constant, $K_h = [HX][OH^-]/[X^-]$ or $K_h = [MOH][H_3O^+]/[M^+]$, the value of which can be determined by measuring the pH of a solution containing the ion in question. In this experiment, you will determine the extent of hydrolysis in various solutions, using indicators to measure pH as done in Experiment E.31. From the measured pH and the solution concentration, you will be able to calculate hydrolysis constants.

Cations that hydrolyze to a moderate extent (not too much and not too little) form amphoteric hydroxides or oxides; that is, they neutralize both acids and bases. In the second part of this experiment, you will investigate several substances to see if they are amphoteric. As a simple test for amphoterism, you will determine if particular hydroxides and oxides dissolve in both acidic and basic solutions. If you find that a given oxide, for example, dissolves only in basic solution, then it is not amphoteric but instead may be classified as an acid oxide; if it dissolves only in acidic solution, it is a basic oxide. Only if it dissolves in both acidic and basic solutions is it amphoteric.

Procedure

(a)

WEAR YOUR SAFETY GOGGLES

Heat 150 mL of distilled water in a clean flask to boiling (to expel dissolved CO_2). Cool, with a small beaker inverted over the mouth. Pour 5 mL of the water into each of five clean, dry test tubes. To the first, add 2 drops of methyl-orange indicator; to the second, methyl red; to the third, bromthymol blue; to the fourth, phenolphthalein; and to the fifth, alizarin yellow. Compare the colors observed with those in the color chart of Experiment E.31, part (a) and note the pH.

Prepare 25 mL of 1.0 M NaCl by weighing out 1.5 g of

solid NaCl on the platform balance, placing it in your clean graduated cylinder, and adding with stirring enough of the boiled distilled water to make 25 mL of solution. Determine the indicator colors as before.

In similar fashion, prepare 25 mL of 1.0 M $NaC_2H_3O_2$ (sodium acetate) after first calculating the mass of solid needed (if in doubt, check your calculation with your instructor). Measure the pH as before.

Prepare a Cu^{2+} solution as follows: Weigh out on the platform balance 0.6 g of $Cu(NO_3)_2 \cdot 3H_2O$. Place in a clean graduated cylinder, and add sufficient boiled distilled water to make 25 mL of solution. Discard all but 2.5 mL of the solution, and add enough water to make 25 mL of final solution. Measure the pH as before.

Using the five indicators as described, measure the pH of the following solutions provided on the reagent table: 1.0 M $CaCl_2$, 0.50 M $ZnCl_2$, 0.0010 M $KAl(SO_4)_2$, 0.00010 M Na_2S.

(b)   Test 1.0 M $CaCl_2$ solution and 0.50 M $ZnCl_2$ solution for amphoterism as follows: Place 5 mL of the solution in a test tube, and add dropwise 6 M NaOH until an insoluble hydroxide precipitates. Split the precipitate into two portions in separate test tubes. To one, add dilute HCl; to the other, 6 M NaOH. Record your observations.

Test for amphoterism in the case of aluminum by first preparing an approximately 0.2 M Al^{3+} solution starting with solid $KAl(SO_4)_2 \cdot 12H_2O$ and then proceeding as above.

Test for amphoterism in the case of CO_2 as follows: Place about 50 mL of water in a plain cylinder. Add a few drops of alizarin yellow indicator, and then with stirring add 6 M NaOH dropwise until the color changes from yellow to red. Drop in a small piece of Dry Ice, and note what happens. In a second cylinder, place 50 mL of water, a few drops of methyl-orange indicator, and sufficient 6 M HCl to change the color from yellow to red. Add Dry Ice and observe.

Data

(a)   Record the indicator colors in each solution.

	Methyl Orange	Methyl Red	Bromthymol Blue	Phenolphthalein	Alizarin Yellow
H_2O	_____	_____	_____	_____	_____
NaCl	_____	_____	_____	_____	_____

	Methyl Orange	Methyl Red	Bromthymol Blue	Phenol- phthalein	Alizarin Yellow
$NaC_2H_3O_2$	____	____	____	____	____
$Cu(NO_3)_2$	____	____	____	____	____
$CaCl_2$	____	____	____	____	____
$ZnCl_2$	____	____	____	____	____
$KAl(SO_4)_2$	____	____	____	____	____
Na_2S	____	____	____	____	____

(b) Record pertinent observations.

Results

(a) Complete the following table (for the cations, assume that only one OH^- is picked up; for the anions, only one H^+):

	Conc.	pH	Hydroly- sis Const. K_h
NaCl soln	____	____	____
$C_2H_3O_2^-$	____	____	____
Cu^{2+}	____	____	____
Ca^{2+}	____	____	____
Zn^{2+}	____	____	____
Al^{3+}	____	____	____
S^{2-}	____	____	____

(b) Record your conclusions about amphoterism of $Ca(OH)_2$, $Zn(OH)_2$, $Al(OH)_3$, and CO_2.

Questions

E.32.1 Starting with the expressions for K_{diss} of a weak acid and K_w, show that $K_h = K_w/K_{diss}$.

E.32.2 Clearly explain how you can experimentally obtain hydrolysis constants by measuring the pH of the solutions in part (*a*).

E.32.3 Show that a 1.0 *M* NaCl solution results from dissolving 1.5 g NaCl in enough water to make 25 mL of solution.

E.32.4 Aqueous Al^{3+} ion is better described as $Al(H_2O)_6^{3+}$, an octahedral structure having aluminum at the center and six water molecules surrounding it at the corners of an octahedron. $Al(OH)_3$ is better described as $Al(H_2O)_3(OH)_3$. Draw structures for these octahedral molecules.

E.32.5 Describe the reaction of 1 mol of $Al(H_2O)_6^{3+}$ with 3 mol of OH^- using Brønsted acid-base terminology.

E.32.6 An explanation for the amphoterism of $Cr(H_2O)_3(OH)_3$ is that the Cr^{3+} is sufficiently positively charged to weaken (by withdrawing electrons) O—H bonds of the H_2O molecules bound to it to allow the $Cr(H_2O)_3(OH)_3$ molecule to donate protons; but the Cr^{3+} is not so strongly positive (electron-withdrawing) as to prevent protons from being added to the hydroxides (forming O—H bonds) to the Cr^{3+}. Use a parallel argument to explain why:

(*a*) $Cr(H_2O)_4(OH)_2$ is nonamphoteric, behaving only as a basic hydroxide.

(*b*) $Cr(OH)_6$ is nonamphoteric, behaving only as an acid (usually written as $H_2CrO_4 + 2H_2O$).

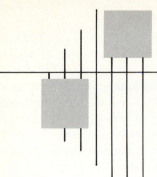

Experiment E.33

Separation of Iron from Aluminum Ore

Special Items

Fine filter paper
Thread
4 g Al(OH)$_3$/Fe$_2$O$_3$ mixture
15 mL distilled water
25 mL 2 *M* KOH
Ca. 30 mL 3 *M* H$_2$SO$_4$
Few drops 0.1 *M* KNCS

In the commercial production of aluminum by electrolysis, one of the problems is that of freeing the aluminum ore of iron. This is done by taking advantage of the fact that the hydroxide of aluminum is amphoteric whereas that of iron is not. Such a separation gives a product sufficiently pure for preparing aluminum metal. However, for the dyeing industry, it is necessary to exclude even trace amounts of iron from the aluminum salts used. Further separation of aluminum and iron can be achieved by crystallizing alum from solution. Such crystals are very free of iron impurity.

In this experiment, you will be given a mixture of Al(OH)$_3$ and Fe$_2$O$_3$. Your job will be to separate out most of the iron by solution in a strong base and then to remove the rest by crystallization of alum, KAl(SO$_4$)$_2$ · 12H$_2$O.

Procedure

■ WEAR YOUR SAFETY GOGGLES

Weigh out on a platform balance 4.0 g of the Al(OH)$_3$–Fe$_2$O$_3$ mixture [75% Al(OH)$_3$ by weight]. Place in a 200-mL beaker, and add 15 mL of water followed by 25 mL of 2 *M* KOH. Heat nearly to boiling while stirring constantly, keeping the mixture hot for 5 min so that the Al(OH)$_3$ dissolves.

■ CAUTION

Hot basic solutions are extremely corrosive to the skin. Do not boil, because violent bumping will occur, which will cause spattering and may even break the beaker.

Let the solution cool, and filter through a fine filter paper. To speed filtration, you may want to use suction, as shown in Figure A.14 in Part A of the manual.

Calculate the volume of dilute sulfuric acid (3 M) that now must be added to the solution to neutralize all the base [consider the original $Al(OH)_3$ as well as the original KOH]. Add this calculated volume of sulfuric acid plus 3 mL in excess (to suppress hydrolysis of Al^{3+}) to the filtrate in a small beaker. If a precipitate forms, heat with stirring to dissolve.

Tie a short piece of thread to a splint, and suspend the thread in the solution. Set aside in the desk (loosely covered with a filter paper) until the next laboratory period.

At that time, dissolve one of the alum crystals in water, and test the solution for the presence of Fe^{3+} by adding a few drops of KNCS solution. A pink color indicates the presence of $(FeNCS)^{2+}$.

Data and Results

Calculate the maximum yield of the product $KAl(SO_4)_2 \cdot 12H_2O$ expected from 4.0 g of starting mixture.

Questions

E.33.1 $Al(OH)_3$ is soluble in either acidic or basic solutions. Why is a basic, rather than an acidic, solution used to dissolve the $Al(OH)_3$ in this experiment?

E.33.2   Calculate the number of moles of excess hydroxide in the solution after all the $Al(OH)_3$ has been converted to $Al(OH)_4^-$.

E.33.3   Describe the reaction between $Al(OH)_3$ and OH^- using Brønsted terminology.

E.33.4   Explain why adding an excess of sulfuric acid to the solution from which crystallization of alum is to take place helps to suppress the hydrolysis of Al^{3+}.

E.33.5   What is the purpose of suspending the thread in the solution from which crystallization of alum is to take place?

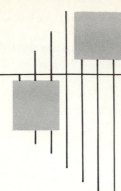

Experiment E.34

Separation and Analysis of a Mixture

Much of the work involved in conducting chemical research entails separating, purifying, and analyzing materials. To achieve efficient separations requires knowledge both of the properties of the substances and of the various techniques of separation. In this series of experiments you will separate and analyze the components of a mixture of an alkali metal sulfate and an organic acid. The entire sequence of experiments will require several laboratory periods for completion and will consist of the following:

(a) Separation of an alkali metal sulfate from an organic acid
(b) Drying and gravimetric analysis of the alkali metal sulfate
(c) Recrystallization and recovery of the organic acid
(d) Analysis of the organic acid by titration
(e) Analysis of the organic acid by freezing-point depression

You may also have the opportunity to analyze your alkali metal sulfate spectroscopically by the methods described in Experiment E.12.

Experiment (a)

Separation of an Alkali Metal Sulfate from an Organic Acid   You will be issued an unknown mixture weighing about 13 g. About 5 g (i.e., five-thirteenths of the mass) will consist of an alkali metal sulfate, one of Na_2SO_4, K_2SO_4, or Li_2SO_4; the remainder will consist of an organic acid. Your problem in this experiment will be to separate the sulfate from the organic acid as completely as possible using solubility differences and the techniques of decantation and filtration. In subsequent experiments you will purify and then analyze each component of the mixture.

The basis for the separation procedure lies in the fact that the alkali metal sulfate is quite soluble in both hot and cold water, whereas the organic acid is rather soluble in hot water but nearly insoluble in cold water. To achieve maximum separation you should add enough water to dissolve the sulfate but not so much that an appreciable amount of the organic acid dissolves along with it. To determine the appropriate quantity of water to add, you will first carry out solubility tests on 1-g samples of Na_2SO_4, Li_2SO_4, and K_2SO_4. Then you will know how much water to add to the unknown mixture containing 5 g of sulfate per 8 g of organic acid. A perfectly reasonable question you may ask at this point is, ''Why not simply look up the solubilities in tables and save a lot of time and trouble?'' The answer is that the values you obtain experimentally will describe how much of each salt will dissolve in a certain amount of water in a few minutes. The solubility values listed in tables, by contrast, refer to solubilities at equilibrium, which may not be reached for a long time—perhaps for days. You may occasionally become frustrated when something unexpected occurs. For example, a ''soluble'' sulfate may turn into a hard lump upon the addition of water. The fact that something is unexpected does not make it wrong or impossible to deal with. What you must do is find by experimentation a pragmatic solution to the problem and keep good notes for future reference. As you carry out the experimental work, be sure to observe what happens very carefully and enter your observations into your laboratory notebook.

Procedure (a)

■ WEAR YOUR
SAFETY
GOGGLES

Your laboratory instructor will show you how much Na_2SO_4, Li_2SO_4, or K_2SO_4 to place in a 15-cm test tube to give 1 g of material. Label three 15-cm test tubes, and place 1-g samples of the three sulfates in the test tubes. Fill your plastic wash bottle with distilled water, squirt 2 mL of water into your graduated cylinder, and transfer the water to a 15-cm test tube to get an idea of how much water is contained in 2 mL. To each of the three test tubes containing sulfates, add approximately 2 mL of water and tap the sides of the tubes briskly to mix. If solution is complete, note how long it took. Also, note whether each test tube cooled off or warmed up; compare the relative amounts of cooling or heating if possible. If solution is not complete, add another 2 mL of water, and repeat until the solid has entirely dissolved. When you have identified the least-soluble sulfate, obtain a second 1-g sample and recheck its solubility by adding slightly less than the total volume of water required the first time, attempting dissolution, and then adding more water if necessary. Calculate a value for the solubility of the least-soluble sulfate in units of grams per 100 mL of water at room temperature. Each member of your laboratory section should list his or her results on the chalkboard, and an average value for the solubility of each sulfate should be calculated. Both your value and the class average value should be entered into your laboratory notebook.

You are now in a position to carry out the actual separation. Weigh your unknown mixture plus its container on the triple-beam balance. Transfer the mixture to a large beaker, and reweigh the container to determine the mass of unknown. Calculate the volume of water that should be added to your approximately 13-g mixture, assuming that it contains five-thirteenths by mass of the *least*-soluble sulfate. Add the required volume of distilled water to the mixture, and stir briskly to dissolve the sulfate.

Read carefully the instructions for decantation and filtration given in Part A, Section 2*b* of this manual. Using qualitative filter paper, carry out the decantation and filtration. Collect the sulfate solution in your largest beaker. Rinse the solid in the filter paper with several small portions of chilled distilled water, and add the washings to the sulfate solution. Remove the filter paper containing the organic acid, place it in a clean labeled beaker, and keep it in your desk.

Support the beaker containing the sulfate solution on a ring stand equipped with a wire gauze, add two boiling chips, and heat the solution until it boils very gently. Continue heating until the water is almost gone. Do *not* heat the mixture to dryness, because it contains some of your organic acid (why?), which will smoke and produce acrid fumes in the laboratory. Set the hot beaker on a transite mat to cool somewhat. Then label it and leave it uncovered in your laboratory desk where it can dry until the next laboratory session.

Experiment (b)

Drying and Gravimetric Analysis of the Alkali Metal Sulfate   At this stage, your sulfate sample is contaminated to some extent with organic acid, and it also contains both ''chemically bound'' water (water of hydration) and water in the form of moisture. The desired end product in this experiment is pure anhydrous (no water of hydration) sulfate. To accomplish this, you will first heat the sample enough to decompose most of the organic acid impurity, converting it to volatile gases and particulate carbon. You will then add water, dissolve the sulfate, and filter out the carbon, leaving the purified sulfate in solution. This solution will be heated to dryness in a beaker in order to remove most of the water. The water of hydration will be driven off by vigorous heating of the sulfate sample in a crucible. After you have thereby obtained a pure anhydrous sample of your unknown alkali metal sulfate, you will determine if it is Li_2SO_4, Na_2SO_4, or K_2SO_4 by carrying out a gravimetric analysis.

Procedure (b)

■ WEAR YOUR SAFETY GOGGLES

Clean a crucible thoroughly. Remove the boiling chips from the damp sulfate in your beaker, and transfer the sulfate entirely to the crucible. Carry a ring stand to the hood. Set the crucible containing the sulfate on a wire triangle, and heat the crucible gently to drive off the water and other volatile materials. If, as you are heating the solid, decrepitation (popping) occurs, remove the heat for a bit and then continue. You should heat the solid until it is thoroughly dry and the organic material has been converted to gases and black carbon particles. When the smoking subsides, allow the crucible to cool. Transfer the sulfate to a small beaker, add 30 mL of distilled water, and heat gently to dissolve. It may be necessary to add a bit more than 30 mL of water to get all the sulfate to dissolve. Using qualitative filter paper, filter out the black impurities, collecting the filtered sulfate solution in a beaker. Add two boiling chips, and heat until almost all the water is gone.

Place the beaker containing the damp sulfate on a wire gauze on a ring stand. Carefully remove the boiling chips. Then heat the beaker very gently until the solid appears to be dry. Stir it around with a stirring rod. If decrepitation occurs, slow down the heating process. Using beaker tongs, transfer the beaker to a transite mat, and allow it to cool. Thoroughly clean your crucible. Supporting the crucible on a wire triangle on a ring, heat the crucible for about 5 min and allow to cool; then, using crucible tongs, obtain the mass of the crucible on the analytical balance. Transfer the sulfate to the crucible, place the crucible on a wire triangle, and heat it, first very gently and then as strongly as possible. Cool the crucible, and get the mass of the crucible plus solid on the analytical balance. (Be sure to use crucible tongs.) Return the crucible to the wire triangle, heat it strongly for several minutes, cool it, and reweigh. Continue this heating, cooling, and reweighing procedure until the sulfate appears to be anhydrous (free of water of hydration). Calculate the mass of anhydrous sulfate by difference. This final mass of sulfate should be at least 1 g, because you will need two 0.5-g samples for use in the gravimetric analysis. If you find you have less than 1 g, you should obtain some additional sample from your instructor and dry it. Store the anhydrous sulfate in a labeled glass vial. Calculate a percent recovery of sulfate, assuming that your original unknown was five-thirteenths sulfate by weight. Submit the labeled glass vial to your laboratory instructor for visual examination.

The alkali metal sulfate is now ready for gravimetric analysis. The procedure is given in Experiment E.35. You should carry out the analysis on two samples simultaneously. In addition to calculating the percentage of your unknown that is sulfate, you should determine if your unknown is Li_2SO_4, Na_2SO_4, or K_2SO_4.

Experiment (c)

Recrystallization and Recovery of the Organic Acid
In this experiment you will purify the organic acid component of your unknown. Recall that the organic acids in the unknowns are nearly insoluble in cold water but rather soluble in hot water. The object is to dissolve the organic acid in slightly more than the minimum amount of hot water, and then to cool the solution very slowly in order to form large, pure crystals, leaving impurities in the water.

Procedure (c)

■ **WEAR YOUR SAFETY GOGGLES**

Obtain a 250-mL Erlenmeyer flask, transfer the solid organic acid from the filter paper into the flask, and add 150 mL of distilled water. Then place the flask on a wire gauze on a ring stand, and heat it very gently. Check the temperature periodically with your thermometer. If the solid has not all dissolved at a temperature of 95°C, add more water and bring the temperature back to approximately 95°C. Do not go above 95° for two reasons: First, the solubility of the acids will increase so much that too little water will be used, resulting in impure crystals; second, the solution could easily start to boil and spillage would result. Continue this process until the solid dissolves entirely at approximately 95°C. Then add about 10 mL of water in excess. Remove the flask from the heat, and set it on a transite mat in your laboratory desk where it can cool until the next laboratory period. Cover the flask with a small beaker.

To recover the organic acid crystals, use the technique of suction filtration, as described in Part A at the beginning of this manual. When you have collected the crystals on the filter paper in the Büchner funnel, wash them with a few 5-mL portions of ice-cold water. Then allow the aspirator to run until the crytals are quite dry. Transfer the crystals, together with the filter paper, to a large, open beaker, where they may finish drying.

When the crystals are dry, find their mass on the analytical balance. Calculate a percent recovery of the organic acid, assuming that eight-thirteenths of your original unknown by weight was organic acid and five-thirteenths was sulfate. Place a small sample of your crystals in a small, labeled glass vial for evaluation by your laboratory instructor. Keep the remainder of your crystals for analysis by titration and by freezing-point depression.

Experiment (d)

Analysis of the Organic Acid by Titration   The objective in this experiment is to determine the gram molecular weight of your unknown organic acid. To make the determination you will titrate the acid with a solution of NaOH, which you must first standardize by titration with a standard acid solution. The procedure is described in Experiment E.36.

Procedure (d)

Your first problem will be to prepare a standardized solution of NaOH. The procedure for doing this is described in part (a) of Experiment E.36. Starting with a standardized solution of H_2SO_4, you will prepare a NaOH solution and standardize it. *Note:* You will *not* do part (b) of Experiment E.36. Instead, you will use the standardized NaOH solution to titrate a sample of your unknown organic acid, as described in Experiment E.36, part (c).

At the end of each titration, you will have used a measured volume of NaOH solution of known concentration, from which you can calculate the number of moles of NaOH added to neutralize the known mass of acid. Since each of the possible unknown acids is assumed to have only one neutralizable hydrogen atom per molecule of acid, you can calculate a value for the gram molecular weight of your organic acid.

Experiment (e)

Analysis of the Organic Acid by Freezing-Point Depression   This procedure is described in Experiment E.37. Take a 2-g sample of your unknown organic acid and determine its gram molecular weight.

Questions

E.34.1   Prepare a flowchart for each of experiments (a) through (e).

E.34.2   Suggest a reason for noting whether your test tube cooled off or heated up when you dissolved each of the known sulfates in water.

E.34.3   In determining the solubility of the alkali metal sulfates, what is the rationale for determining the solubility of the least-soluble sulfate more carefully than that of the more-soluble sulfates?

E.34.4   Obtain from your colleagues as many values for the solubilities of the alkali metal sulfates as possible, and calculate an average value and an average deviation and/or standard deviation for each average solubility value.

E.34.5   Boiling chips should never be added for the first time to a solution at or near the boiling point. Why should boiling chips always be added *before* heating a solution vigorously rather than *after* heating it?

E.34.6   How does a hydrated salt differ from an anhydrous salt?

E.34.7   Compare the technique of suction filtration with gravity filtration. Under what conditions is each method preferred? What are the advantages and disadvantages of each method?

E.34.8   In the procedure you have followed, you have lost a fair portion of both the sulfate and the organic acid. Suggest some ways one could achieve a separation without as much loss. Construct a flowchart showing your ideas in schematic form.

E.34.9   Tell how the expressions "equivalence point" and "end point" differ in meaning.

E.34.10   If your unknown acid is represented by the symbol HUn, where H stands for the neutralizable hydrogen atom, the titration reaction equation could be written as follows:

$$HUn + NaOH \rightarrow H_2O + NaUn$$

If 0.35 g of unknown acid, HUn, reacted exactly with 25.00 mL of 0.25 M NaOH solution, calculate the gram molecular weight of the acid. If there were actually two neutralizable hydrogen atoms per acid molecule (that is, H_2Un), what would be the gram molecular weight (use same data as before)?

E.34.11   Why would you expect the salt of your unknown organic acid to be more soluble in water than the organic acid itself?

E.34.12   Explain any differences in your results for the gram molecular weight of your organic acid as determined by titration from that found by freezing-point depression. Which method should give more reliable results?

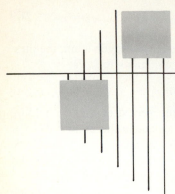

Gravimetric Analysis of a Sulfate

Special Items

200 mL distilled water
1 mL 3 M HCl
25 mL 0.25 M BaCl$_2$
0.5 g unknown sulfate
Ashless fine filter paper

One method for determining the quantity of a given ion in a soluble salt is to dissolve a weighed amount of the salt and then precipitate the ion as an insoluble salt. The insoluble salt can be filtered off, dried, and weighed. From the mass and known composition of the salt, the quantity of the ion in question in the original sample can be calculated. In this experiment, you will analyze a sample for sulfate content by precipitating and weighing the sulfate as barium sulfate (BaSO$_4$).

The principal sources of error in an analysis of this type are (1) incomplete precipitation, (2) precipitation of other substances along with the one desired, and (3) decomposition of the final precipitate on drying. To reduce errors of type 1, an excess of Ba^{2+} is added and the solution is kept near the boiling point long enough to allow the precipitation reaction to reach equilibrium. Keeping the solution hot also favors the formation of larger particles so that filtering and washing are easier. Formation of large particles also reduces the probability of adsorbing foreign ions on the surface of the precipitate. To reduce errors of type 2, the BaSO$_4$ is precipitated from a large volume of solution (so that the more soluble salts stay in solution), which is kept acidic (so that hydroxides and carbonates do not precipitate), and the final precipitate is washed well with water. Errors of type 3 can be minimized by carefully igniting the precipitate and filter paper so that the BaSO$_4$ is not chemically reduced by the carbon from the paper.

Procedure

The following procedure is for the analysis of a single sample of unknown sulfate. If possible, you should carry out two or

■ **WEAR YOUR
SAFETY
GOGGLES**

three simultaneous analyses. Weigh a clean, dry 200-mL beaker on an analytical balance to the nearest 0.0001 g. Add about 0.5 g of the solid unknown sulfate, and reweigh. Fill the beaker nearly half full of distilled water, add 1 mL of dilute HCl, and heat to boiling. While waiting for the solution to boil, thoroughly clean a crucible. Dry it first with cheese-cloth and then by holding it with tongs in the side of the burner flame. Then place the crucible on a wire triangle and heat it as strongly as possible for several minutes. Allow it to cool on your transite mat, and weigh it on the analytical balance. Repeat the heating, cooling, and weighing until two successive measurements agree to within ± 0.0005 g.

To the boiling solution, add very slowly from a graduated cylinder (pour down along a stirring rod) 25 mL of 0.25 M $BaCl_2$. Stir the solution well. Turn down the flame so that it just stays lighted (close off the air). The $BaSO_4$ precipitates in a very finely divided form. In fact, if the mixture were filtered immediately, much of the precipitate would pass right through the filter paper. To increase the particle size, digest (boil very gently) the mixture for a minimum of 30 min. Remove the flame, and allow the beaker to cool sufficiently so that you can hold it near the top. Meanwhile, prepare a filter as in Figures A.13 and A.14, using fine filter paper. (Use of quantitative, ashless filter paper such as Whatman 42 or Fisher quantitative is recommended. Otherwise you will have to correct later for the weight of the ash from the filter paper.) Pour a few milliliters of distilled water through the filter. Using a stirring rod to guide flow, decant most of the hot solution through the filter. Do not allow the level of the liquid in the funnel to reach the top of the paper. When only a few milliliters of liquid remain in the beaker, swirl it gently and pour the mixture into the funnel. Rinse down the sides of the beaker with a stream of distilled water from your wash bottle, swirl, and pour into the funnel. Repeat until the beaker and stirring rod are clean. Allow the filter to drain.

Fold over the top edges of the filter paper, carefully remove from the funnel, and place in the weighed crucible. Support the uncovered crucible on a wire triangle, and heat gently. It is desirable that the heating be done so as to char the paper without its bursting into flame. After the paper is charred, heat the crucible as hot as possible until the $BaSO_4$ is white. Heat 5 min more, cool, and weigh. Heat strongly for 5 min again, cool, and reweigh. Repeat until two successive weights differ by no more than ± 0.0005 g. (If you did not use ashless filter paper, subtract 0.001 g for the ash left from the filter paper.)

Data

Record the results of each of the weighings. Label each.

Results

From the final mass of $BaSO_4$, calculate the mass of sulfate in the original sample. Report to your instructor the percentage of the original sample that is sulfate.

Questions

E.35.1   Calculate the number of moles of barium ion you added to your sulfate sample in excess of the amount needed for complete precipitation of $BaSO_4$.

E.35.2   The rate of dissolving of a solid is proportional to the surface area per gram of the solid exposed to the liquid. How is this fact utilized in the digestion of the $BaSO_4$ precipitate in this experiment?

E.35.3   One can help avoid errors of type 2 by checking the filtrates after each individual washing by adding a few drops of $AgNO_3$ solution to the filtrates. Cloudiness of the filtered wash solution upon addition of the $AgNO_3$ would indicate a need for further washing of the precipitate. Explain.

E.35.4   A student using ashless filter paper (no ash left after ignition) mistakenly subtracted 0.001 g from the final weight of $BaSO_4$ because she didn't realize that her filter paper was ashless. Assuming her mass of $BaSO_4$ to be the same as yours, calculate the percent error this would cause in her reported mass of $BaSO_4$.

E.35.5   Find the percentage of sulfate in a mixture containing 5.00 g of Na_2SO_4 and 5.00 g of Li_2SO_4.

E.35.6   Suppose that you are given the information that your unknown is a pure sample of one of these: Na_2SO_4, K_2SO_4, or Li_2SO_4. Explain how you could use your result for the percentage of your unknown that is sulfate to determine which of the three salts you have. Be as quantitative as possible in your explanation.

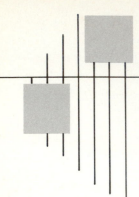

Experiment E.36

Acid-Base Titration

Special Items

300 mL distilled water
75 mL standardized H_2SO_4
   (ca. 0.05 M)
25 mL ca. 1 M NaOH
1 mL phenolphthalein
   solution
75 mL unknown H_2SO_4

One equivalent of any acid reacts with one equivalent of any base. Given an unknown solution of acid, you can determine its normality by titration—i.e., by measuring the volume of it that is required to react with a measured volume of a basic solution whose concentration is known. To find out at what point equivalent quantities have been mixed, you can use an indicator such as phenolphthalein, which changes from colorless in acid to pink in basic solution. If, as in this experiment, a base solution is progressively added to an acid solution in the presence of phenolphthalein, the point at which the first persistent pink coloration appears is taken as the equivalence point. At this point, sometimes called the *end point*, neglecting the small amount of reagent required in excess to bring about the color change, you can assume that equal numbers of equivalents of acid and base have been mixed. By measuring the volumes of the two solutions mixed and knowing the normality of one, you can calculate the normality of the other.

In this experiment, you will first prepare a solution of the base sodium hydroxide. You will then titrate it against a known acid and thereby determine precisely the normality of the base. In part (*b*) the standardized base will be used to determine the normality of a sulfuric acid solution. In part (*c*) the standardized base solution wil be used to titrate an unknown solid organic acid and to determine its molecular weight.

There are several things that may concern you in performing the titration of the unknown organic acid. One of these is that you don't know how many moles of hydronium (H^+ or H_3O^+) ion are released by one mole of the organic acid in the reaction with NaOH. Unless you are told differently by your instructor, you may assume that your unknown organic acid has only one neutralizable hydrogen atom per molecule. Something else of concern is that the unknown solid acids are quite insoluble in cold water. In theory, this should not be a serious problem, because the base solution will convert the insoluble acid to a soluble salt. In practice, however, it is found that the undissolved acid reacts very slowly with the base solution, causing problems. To dissolve the acid you will simply heat the flask immediately before starting the titration. This will have some effect on the equilibria, but not enough to cause appreciable error. A third potential problem is that at the equivalence point the solution will be slightly basic, due to the nature of the acids used as unknowns. This turns out not to be a real problem, however, because the indicator phenolphthalein changes color from colorless to pink under slightly basic conditions.

Procedure

(a)   Read Section 2*b* in Part A of this manual to review how to use a pipette and a burette. Clean and dry a beaker, and use it to obtain 100 mL of the standard sulfuric acid solution from the reagent table. Be sure to record the exact concentration written on the bottle.

■ **WEAR YOUR SAFETY GOGGLES**

Prepare a dilute solution of NaOH by placing about 30 mL of the concentrated ($\approx 1\ M$) NaOH solution from the side shelf in a clean 500-mL Florence flask and adding to it 210 mL of distilled water.* Swirl the solution vigorously to mix, and then cover the flask by placing a small beaker in an inverted position over the top.

Clean three Erlenmeyer flasks, and rinse them with distilled water. Clean and rinse a 25-mL volumetric pipette, first with distilled water and then with at least one 5-mL portion of the standard sulfuric acid solution. Using the pipette, place a 25-mL portion of the standard acid in each of the three Erlenmeyer flasks. Also add 2 drops of phenolphthalein indicator to each flask.

Clean and rinse your burette, first with distilled water and then twice with 5-mL portions of the dilute NaOH solution. Using a clean, dry funnel, fill the burette, checking for air bubbles in the tip, and record the initial burette reading.

Each of the portions of standard acid is to be titrated with the NaOH, but the first portion will be used only to locate approximately the end point for the other two. Add the NaOH rather rapidly to flask 1 with vigorous stirring (stir with the right hand while controlling the burette with the left) until a persistent pink color appears. This color is more easily seen

Note: The use of freshly boiled and cooled distilled water for all dilutions is strongly recommended to minimize errors caused by dissolved CO_2.

if a piece of white paper is placed under the flask. Note the burette reading. Now carefully titrate portions 2 and 3 as follows: Allow about nine-tenths of the estimated amount of needed NaOH to flow from the burette into the flask. Stir the acid solution well while adding the NaOH. Rinse down the inner walls of the flask with a jet of distilled water from your wash bottle. Add the last bit of NaOH needed dropwise, stirring between drops, until the pink persists. Record the burette readings.

(b)   Secure from your instructor a sample of sulfuric acid of unknown concentration. Clean and rinse two Erlenmeyer flasks with distilled water. Clean and rinse a 25-mL volumetric pipette with the unknown acid solution, and then place a 25-mL sample of the acid and 2 drops of phenolphthalein in each flask. Fill the burette with your standardized NaOH solution, and titrate each acid sample according to the procedure described in part (a). Report the normality of the unknown sulfuric acid to your instructor; you will be graded on its accuracy.

(c)   Clean and dry two 250-mL Erlenmeyer flasks. Weigh each flask on the analytical balance. Using the triple-beam balance, weigh out on a piece of notebook paper two samples of organic acid crystals. Each sample should have a mass of between 0.3 and 0.4 g. Transfer the samples to the titration flasks, and determine the exact masses on the analytical balance. Obtain the mass of each sample by subtraction.

To the first flask add around 40 mL distilled water and 2 drops of phenolphthalein. Clean a 50-mL burette, and rinse it with about 5 mL of your standardized NaOH solution. Support the burette on a ring stand, and fill it carefully with the standardized NaOH solution. After recording the burette reading, heat the flask, swirling the contents, until the crystals of your unknown organic acid dissolve. Place a sheet of white paper over the base of the ring stand, and titrate the organic acid slowly and carefully, swishing the titration flask continuously. When the pink color signaling the equivalence point starts to persist for roughly a second before disappearing, add the NaOH dropwise until there is a permanent pink color. It is a good practice to take a burette reading a number of times just before the end point. Record the burette reading, and then perform a second titration.

Data

(a) Normality of the standard acid _____

Volume of acid pipetted _____

	1	2	3
Initial burette reading for base	_____	_____	_____
Final burette reading for base	_____	_____	_____

(b) Volume of acid pipetted _____

	1	2
Initial burette reading for base	_____	_____
Final burette reading for base	_____	_____

	1	2
(c) Mass of flask plus acid	_____	_____
Mass of flask	_____	_____
Initial burette reading for base	_____	_____
Final burette reading for base	_____	_____

Results

	2	3
(a) Volume of acid used	_____	_____
Equivalents of acid used	_____	_____
Volume of base used	_____	_____
Normality of base	_____	_____
Average		_____

	1	2
(b) Volume of base used	_____	_____
Equivalents of base used	_____	_____
Volume of unknown acid	_____	_____
Normality of unknown acid	_____	_____
Average		_____

	1	2
(c) Volume of base used	_____	_____
Equivalents of base used	_____	_____
Mass of acid	_____	_____
Equivalent (molecular) weight of acid	_____	_____

Questions

E.36.1 It is advisable in performing acid-base titrations to use boiled distilled water in order to get rid of any dissolved CO_2 in the water. If this is not done, how will the presence of CO_2 in your solutions tend to affect your result? Explain.

E.36.2 Calculate an average deviation of your results for the molarity of the NaOH solution you determined in part (*a*). How many significant figures can you justify in reporting the value? Explain.

E.36.3 Distinguish between the terms "end point" and "equivalence point."

E.36.4 The molarity of H_3O^+ in a 1.00 *M* H_2SO_4 solution is considerably less than 2.00 *M*. Why is this the case; and why is it found, nonetheless, that a full 2 mol of NaOH are required to neutralize 1 mol of H_2SO_4?

E.36.5 At the beginning of the part (*c*) procedure you are directed to clean and dry two titration flasks. Why (or why not) is it important that they be dry?

E.36.6 In the titration of the unknown organic acid in part (*c*), the solution will be slightly basic at the equivalence point. Why?

E.36.7 If in part (*c*) your instructor told you your unknown organic acid was monoprotic, but it actually contained two neutralizable hydrogens per molecule, what error would result in your calculated value of the gram molecular weight?

E.36.8 In part (*c*), why would you expect the salt of your unknown organic acid to be more soluble than the organic acid itself?

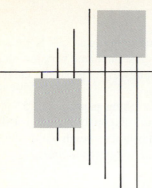

Experiment **E.37**

Molecular Weight from Freezing-Point Lowering

Special Items

Two 5-cm squares of
  weighing paper
20 g naphthalene
2 g powdered unknown
Wire stirrer

The addition of a solute to a solvent, in general, lowers the freezing point of the solvent. For a given solvent, the freezing-point lowering is directly proportional to the concentration of particles dissolved in it. For naphthalene, the solvent used in this experiment, the freezing point is lowered by 6.9°C for each mole of solute particles in 1 kg of naphthalene.

In this experiment, you will determine the molecular weight of an unknown solute dissolved in naphthalene by observing the freezing point of a solution that contains known masses of unknown and naphthalene. From the observed freezing-point lowering, you will be able to calculate the number of moles of dissolved particles per kilogram of naphthalene. Since you know the mass of unknown used per kilogram of naphthalene, you can calculate the apparent molecular weight of the solute.

Procedure

■ **WEAR YOUR
SAFETY
GOGGLES**

Assemble the apparatus shown in Figure E.37.1. Be careful when inserting the thermometer into the two-hole stopper. (Lubricate well with glycerine, use cheesecloth to protect your hands, and keep hands close together.) Make sure that the temperature scale is visible from 70° up. If necessary, slit the stopper with a razor blade.

On the platform balance, weigh out on a piece of creased weighing paper (previously weighed) about 20 g of naphthalene to the nearest 0.1 g. Carefully pour into the large test tube. Heat the water until the naphthalene melts.

Then remove the burner, and allow the naphthalene to cool. Stir continuously, and record the thermometer reading

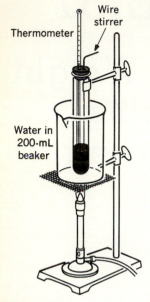

Thermometer

Wire stirrer

Water in 200-mL beaker

Figure E.37.1

every minute, starting at 85° and stopping at 75°. Replace the burner under the beaker, and adjust the flame very low so that the water stays hot while you are weighing out the unknown solute.

Take a piece of clean weighing paper about 5 cm on each edge, and crease it along the diagonals to form a shallow depression. Using the platform balance, weigh out on this paper about 2 g of unknown solute (neglect the mass of the paper).

Take another similar clean piece of creased weighing paper and the packet of unknown to the analytical balance. First weigh the clean paper to the nearest 0.001 g. Take the weighed paper out of the balance, carefully add the unknown and then reweigh.

Reheat the water in the beaker, if necessary, until the naphthalene has melted. Gently lift out the stopper assembly, and carefully add *all* the weighed unknown to the molten naphthalene. Replace the stopper assembly, and stir vigorously until all the unknown has dissolved. (If the unknown does not all dissolve, remove the thermometer-stirrer assembly and heat the mix carefully over a gas-burner flame until the cloudiness just disappears. Do not overheat; there is danger of fire from the naphthalene vapor, which is flammable. Replace the test tube containing the naphthalene solution, wait a minute or so, and replace the thermometer-stirrer assembly.)

■ CAUTION

Watch the thermometer scale and pull it out immediately if there is a danger that the mercury thread will go too high. If you break a thermometer, do not heat or touch the mercury. Notify your instructor.

Remove the burner, and allow the naphthalene solution to cool. Stir continuously, and record the thermometer reading every minute, starting at 85° and stopping at 70°.

(In cleaning out the test tube at the end of the experiment, heat the naphthalene until it just melts. This can be done in the yellow flame of a burner *if you take care not to heat the thermometer beyond its temperature range*. Remove the stopper, and pour the molten naphthalene onto a crumpled wad of paper. When the naphthalene has solidified, throw both the paper and solid naphthalene into a crock. Do not pour liquid naphthalene into the sink! The small amount of naphthalene left in the test tube can be dissolved with a few millimeters of xylene. Be careful with the xylene, because it is flammable.)

Data

Mass of paper .... _____

Mass of naphthalene plus paper .... _____

Temperature of the naphthalene at 1-min intervals
as it cools

Mass of paper .... _____

Mass of paper plus unknown solute .... _____

Temperature of the solution at 1-min intervals as
it cools

Results

Plot the temperature of the cooling naphthalene (on
vertical axis) vs. time (on horizontal axis).

Freezing point of pure naphthalene .... _____

Plot the temperature of the cooling unknown solute-
naphthalene solution vs. time.

Freezing point of solution (at the first sign of
crystallization) .... _____

Freezing-point lowering   _____

Moles of solute per kilogram of naphthalene

(= freezing-point lowering/6.9)   _____

Mass of naphthalene in solution   _____

Mass of unknown in solution   _____

Mass of unknown per kilogram of naphthalene   _____

Mass of unknown in 1 mol of unknown   _____

Questions

E.37.1   Estimate the absolute error and relative error (percent) in your measurement of the mass of naphthalene (≈ 20 g) and of the mass of unknown solute (≈ 2 g).

E.37.2   Sometimes as a solute dissolves in a solvent, the solute molecules dissociate to some extent; sometimes solute molecules associate to form larger molecular units, such as dimers or trimers. Discuss the effect of dissociation and of association on the apparent molecular weight of a solute.

E.37.3   Usually as a solution cools, the solvent will crystallize out of the solution as the pure solvent. What effect will this have on the concentration of the solution as solvent is solidifying?

E.37.4   Why is it important not to expose the open end of your test tube containing naphthalene to the flame from your gas burner?

E.37.5   The accepted value for the freezing point of naphthalene is 80.29°C. Find the percent error in your value. In this experiment, what advantage might there be in using your value for the freezing point of naphthalene, rather than the accepted value, in determining the freezing-point depression?

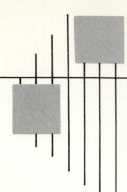

Experiment E.38

pH Titration of Phosphoric Acid

Special Items

pH meter
150 mL 0.100 M NaOH
 (standardized)
20 mL 0.100 M H$_3$PO$_4$
 (standardized)
20 mL unknown mixture of
 HCl and H$_3$PO$_4$
Methyl-orange indicator
Phenolphthalein indicator
10-mL volumetric pipette
50-mL burette
Combination pH electrode
Graph paper

This experiment has two parts: part (*a*), to prepare and describe a titration curve for a phosphoric acid–sodium hydroxide titration; part (*b*), to determine the molarity of an unknown HCl solution added to an equal volume of phosphoric acid solution of known concentration.

The structure of phosphoric acid may be represented as a central phosphorus atom double-bonded to an oxygen and single-bonded to three OH groups. Draw the structure, and explain why it is relatively easy to remove the first, considerably harder to remove the second, and extremely difficult to remove the third proton in chemical reactions. (*1*) (Where numbers in parentheses follow questions or directions, answers are to be written out on separate paper.)

Write the reactions and equilibrium expressions associated with K_I (dissociation of first proton), K_{II}, and K_{III}. (*2*)

Recall that, in general, for the reaction

$$H_2O + HX \rightleftharpoons H_3O^+ + X^-$$

$$pH = pK - \log \frac{[HX]}{[X^-]}$$

Using this relation, write pH expressions corresponding to pK_I, pK_{II}, and pK_{III} for H$_3$PO$_4$. (*3*)

Explain the concept of a buffer, relating it to the $[HX]/[X^-]$ ratio, pH, and pK. (*4*) In the case of H$_3$PO$_4$, pK_I = 2.16, pK_{II} = 7.16, and pK_{III} = 12.3. Recall that when the criteria for a buffer are met, additions of base will have less effect than under nonbuffer conditions. In titrating H$_3$PO$_4$ with NaOH, a number of "buffer situations" are created; in addition, a number of situations are created when very small additions of NaOH will cause great changes in pH. Draw a rough qualitative graph showing pH (*y* axis) vs. milliliters of NaOH added to an H$_3$PO$_4$ solution (*x* axis). Indicate those regions on the graph where buffering is most efficient and where it is least efficient. (*5*)

The "maximum buffer" point occurs when the pH equals the pK. Explain. (*6*)

The point in the titration when buffering is least efficient occurs at a *stoichiometric point*, or *equivalence point*, in the reaction. This point is often called a *point of inflection*, because there is a shift in curvature from convex to concave or from concave to convex at that point. Are there other inflection points in the titration curve? Explain. (7) Calculation of stoichiometric points of inflection is very useful in choosing indicators that will change color (end point) at the equivalence point. How does an end point differ from an equivalence point? (8) An approximate method of calculating the pH at the first stoichiometric point of the H_3PO_4 titration is to take an average of the pK_I and pK_{II}, values, that is, pH $\approx \frac{1}{2}(pK_I + pK_{II})$. In a similar manner, pH at the second stoichiometric point is an average of pK_{II} and pK_{III}.

Calculate the theoretical values for the pH at the first and second equivalence points in the H_3PO_4 titration. (9) Suppose that we chose methyl orange (changes from red to yellow at pH ≈ 4) and phenolphthalein (changes from colorless to red at pH ≈ 9) as indicators for the first and second stoichiometric points. Would they be reasonable choices? Explain. (10)

In part (a) of the experiment you will work in a group to prepare a titration curve for 0.100 *M* H_3PO_4 being titrated with 0.100 *M* NaOH solution. There will be as many groups as there are pH meters.

In part (b) of the experiment you will work on an individual basis to find the concentration of HCl added to an equal volume of 0.100 *M* H_3PO_4 solution. Before coming to the laboratory, you should spend some time thinking about the effect on the titration curve due to the addition of HCl. The general procedure will be nearly identical in the two parts of the experiment, except, of course, that the amount of NaOH required will be different in the second part.

Procedure

(a)   Your teaching assistant will divide your laboratory section into several small groups, each of which will carry out the phosphoric acid titration twice.

■ WEAR YOUR
SAFETY
GOGGLES

Use of the Fisher Model 230 pH Meter (Figure E.38.1)
To set up the pH meter for operation, the following steps must be taken:

1   Set the Operate switch to STBY.
2   Connect the power line plug into a power receptacle. The pilot

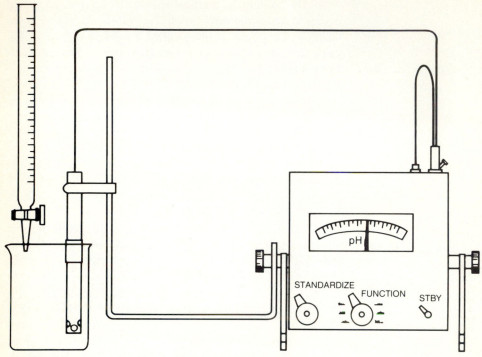

Figure E.38.1

lamp should light to indicate the "ready" status of the instrument.

3 Allow the instrument to warm up for approximately 10 min before using.

To perform pH measurements, the following steps must be carried out:

1 Immerse the electrode and a thermometer in the buffer solution, and stir the solution briefly.

2 Set the Function Switch on MAN TC.

3 Read the temperature, and set the Temperature control to the value observed.

4 Set the Operate Switch to USE, and adjust the Standardize control until the meter reads the pH of the buffer solution.

5 Record the Temperature control setting. Then set the Function switch to REF and record the meter reading. *Note:* If the Standardize control should be accidentally moved, you can re-standardize by simply setting the Function switch on REF and adjusting the meter needle to read the value just recorded.

6 With the Operate switch on STBY, remove the electrode from the buffer solution and rinse it with distilled water. Pour out your buffer solution, and rinse the beaker with distilled water.

Note: Do not adjust the Standardize control again unless it is moved accidentally. Remember, *the glass electrode is very fragile*, so handle it carefully. (*Always set the Operate switch on STBY before removing the electrode from the solution.*)

The pH meter is now ready to determine the pH of any solution.

Titration   To set up the titration, adjust the burette (using a ring stand and burette clamp) so that small amounts of NaOH may be released into the acid solution while the electrode is immersed in it. Fill the burette with ~45 mL of the 0.100 *M* NaOH solution, being certain to rinse the burette with about 5 mL of the 0.100 *M* NaOH first. Then, using a 10-mL pipette, put 10 mL of H_3PO_4 and 5 drops of methyl orange into a clean, dry beaker.

Place the burette above the beaker. Next, immerse the electrode. With the electrode immersed, set the Function switch to MAN TC and set the Operate switch to USE. Read and record the pH and the burette reading. Add the 0.100 *M* NaOH in 1-mL increments, stirring the solution after each addition (*don't touch the tip of the electrode*) and then taking the pH and burette reading after each addition. Also, note the color of the indicator.

Continue adding the NaOH in 1-mL increments and recording the information until the methyl orange turns from red to yellow. At this point, add 5 drops of phenolphthalein indicator. Then continue adding the 1-mL increments of NaOH until a total of 30 to 35 mL has been added, each time recording the information on the data sheet. [In part (*b*), you will add a different amount of NaOH, of course.]

Each group of students should carry out the standardization and titration procedure twice. Make up data sheets for known trial 1, known trial 2, unknown trial 1, unknown trial 2, using the following format:

Standard NaOH			
Added, mL	Color	pH	ΔpH
_____	_____	_____	_____
_____	_____	_____	_____

At the completion of each titration, turn the function knob to STBY. Then carefully remove the electrode from the solution and rinse with distilled water.

(b)   Obtain an unknown HCl + 0.100 *M* H_3PO_4 solution sample from your teaching assistant. You will be told the exact molarity of the H_3PO_4 before it was mixed with an equal volume of HCl solution to prepare your unknown. Using the same procedure your group used in part (a), each student should titrate the unknown two times. The problem will be to determine the molarity of the HCl added to the equal volume of 0.100 M H_3PO_4 solution. To make a precise determination, it will be necessary, in part (b), to add the standard NaOH solution dropwise (rather than a milliliter at a time) around one of the stoichiometric points, and to take additional data corresponding to that dropwise addition. Which stoichiometric point will this be? Explain. (11)

At the end of each laboratory period, the pH meters should be turned off. To turn off the pH meter, first rinse the electrode in distilled water, and then immerse it in fresh distilled water. Disconnect the line cord from the electric outlet. Place the dust cover over the instrument.

Treatment of the Data

(a) Prepare two graphs, one above the other, on the same sheet of graph paper. On the first, plot pH on the y axis vs. milliliters of 0.100 M NaOH added on the x axis for both trials. Use ○'s to identify one set of data points and x's to identify the other set. On your graph, label each inflection point.

On the second graph, plot change in pH per incremental volume ($\Delta pH/\Delta mL$) of 0.100 M NaOH added on the y axis for each trial. Are there any advantages in such a plot? Explain. Also label each inflection point.

How do your results compare with the "theoretical" values for the pK's and equivalence points? [Refer to the exercise followed by the number (9) earlier in this experiment.]

(b) Prepare graphs similar to those prepared in part (a). Using the information derived from the graphs, figure out the molarity of the unknown HCl added to the 0.100 M H_3PO_4 solution if the HCl solution was added to an equivalent volume of 0.100 M H_3PO_4 (that is, 50 mL of unknown HCl solution added to 50 mL of 0.100 M H_3PO_4 solution).

Questions

E.38.1   Sometimes the structure of phosphoric acid is written with single bonds to both the OH groups and to the oxygen atom;

other times the bond between the P and O is shown as a double bond. The "truth" probably lies somewhere between. Explain.

E.38.2   Show that the pH at the first stoichiometric point of the H_3PO_4 titration is mathematically equal to $\frac{1}{2}(pK_I + pK_{II})$.

E.38.3   In part (a), several small groups determined the concentration of a H_3PO_4 solution by titrating it with standardized NaOH solution. Obtain as many values as you can for the molarity of the H_3PO_4. Calculate an average value and an average deviation and/or standard deviation.

E.38.4   In part (b), your instructor gave you the value for the concentration of the H_3PO_4 before it was mixed with an equal volume of HCl solution. It should be theoretically possible for you to obtain the concentration of the H_3PO_4 from your titration graph. Explain. Using your titration graph, calculate a value for the H_3PO_4 concentration and compare it to the value given by your instructor.

E.38.5   A 50.0-mL sample of a solution that is 0.040 M in HCl and 0.080 M in H_2X is titrated with 0.20 M NaOH. H_2X is a weak diprotic acid having dissociation constants $K_I = 1.0 \times 10^{-4}$ and $K_{II} = 1.0 \times 10^{-9}$. How many milliliters of the 0.20 M NaOH are required to reach the point at which
(a) Essentially all the H_2X has been converted to HX^-?
(b) The second equivalence point of H_2X has been reached?
(c) A buffer solution containing equal concentrations of HX^- and X^{2-} has been created?

E.38.6   Why is the buffering action most efficient in the region where $[HX]$ and $[X^-]$ are about equal?

E.38.7   Indicate what ions are present in principal amounts at each buffer point and at each equivalence point.

E.38.8   What is the advantage of carrying out each titration twice?

E.38.9   How would your part (a) curves be changed if you used 1.0 M solutions instead of 0.1 M?

E.38.10   A student carried out an analogous titration using $H_2C_2O_4$ ($pK_I = 1.23$, $pK_{II} = 4.19$) instead of H_3PO_4. The shape of the curve for this diprotic acid is the same as that for the triprotic. Explain.

E.38.11   The amino acid glycine contains an NH_2 group on one end that acts like a base and a COOH group on the other end that acts as an acid. What sort of pH titration curve would you predict
(a) On stepwise addition of HCl to the approximately neutral material?
(b) On stepwise addition of NaOH to the approximately neutral material?

(c) On stepwise addition of HCl to the solution obtained from (b)?

E.38.12   Explain how your pH titration curve with HCl + H_3PO_4 is different from that for H_3PO_4 alone. What are the differences due to?

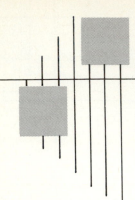

Experiment E.39

Enthalpy of Formation

Special Items

Styrofoam coffee cups
Scissors
Thermometer
450 mL 1.50 M NH$_3$
150 mL 1.50 M HCl
150 mL 1.50 M HNO$_3$
75 mL 1.50 M H$_2$SO$_4$
12 g NH$_4$Cl
18 g NH$_4$NO$_3$
15 g (NH$_4$)$_2$SO$_4$

The enthalpy of formation per mole of a compound is the ΔH accompanying formation of one mole of the compound from the elements in their standard states. Direct measurement of enthalpies of formation is often difficult experimentally, so indirect methods involving enthalpies of reaction are frequently used instead. In this experiment you will determine the heat of formation of various ammonium salts NH$_4$X(s) (where X is Cl, NO$_3$, or SO$_4$) by combining measurements of the heat for the neutralization reaction

$$NH_3(aq) + HX(aq) \rightarrow NH_4X(aq) \qquad \Delta \overline{H}_{neut}$$

and the heat of the dissolution reaction

$$NH_4X(s) \xrightarrow{H_2O} NH_4X(aq) \qquad \Delta \overline{H}_{diss}$$

with known heats of formation of NH$_3$(aq) and HX(aq). (The bar over the H means per mole.)

Hess's law states that the change in a thermodynamic property (such as enthalpy) depends only on the initial and final states and is independent of which combination of steps is taken. In other words, adding several consecutive reactions gives a net reaction for which the $\Delta \overline{H}$ is simply the sum of the $\Delta \overline{H}$'s for the component reactions. Thus, for example, the $\Delta \overline{H}$ of formation of *aqueous ammonia* is equal to the sum of the $\Delta \overline{H}$ of formation of *gaseous ammonia* plus the $\Delta \overline{H}$ that accompanies dissolving the gas in water:

$$\frac{1}{2}N_2(g) + \frac{3}{2}H_2(g) \longrightarrow NH_3(g) \qquad \Delta H \text{ of formation of } NH_3(g) = \Delta \overline{H}_1$$

$$NH_3(g) \xrightarrow{H_2O} NH_3(aq) \qquad \Delta H \text{ of dissolving} = \Delta \overline{H}_2$$

$$\frac{1}{2}N_2(g) + \frac{3}{2}H_2(g) \xrightarrow{H_2O} NH_3(aq) \qquad \Delta H \text{ of formation of } NH_3(aq) = \Delta \overline{H}_1 + \Delta \overline{H}_2$$

Given that $\Delta \overline{H}_1 = -45.8$ kilojoules per mol and $\Delta \overline{H}_2 = -35.4$ kilojoules per mol, we can calculate that the $\Delta \overline{H}$ of formation of NH$_3$(aq) is -81.2 kilojoules per mol. A similar calculation can be set up to get $\Delta \overline{H}$ of formation of each HX(aq). The main point to be careful about is to set up the steps correctly per mole so that the component steps do indeed add up to the final net equation.

The way in which the $\Delta\overline{H}$ of your reactions will be determined is to measure temperature changes of the system. The experimental setup consists of an insulated cup, a thermometer, and a water solution of reactants or products. To simplify matters, the system will be limited to only the water solution of reactants or products. The experiment is designed to make it possible to correct (by extrapolation) for any heat exchanges between the surroundings (cup, thermometer, air) and the system (water plus reactants of products). This will allow us to define the systems as adiabatic ($q = 0$; no heat exchange between system and surroundings). Since the system is adiabatic, and since each reaction is carried out at constant pressure, $q_p = \Delta\overline{H}_{\text{system}} = 0$. (The subscript p means at constant pressure.)

For n moles of reaction, n times the molar heat of reaction $\Delta\overline{H}_{\text{reaction}}$ must be considered. If heat is generated ($\Delta\overline{H}_{\text{reaction}} < 0$) by the reaction (exothermic), this heat must be absorbed entirely within the system itself, because the system (adiabatic) cannot exchange heat with the surroundings. The resulting increase in the temperature (ΔT) of the system will depend inversely on the heat capacity (heat required to increase the temperature of the system by $1.0°C$) of the entire system at constant pressure C_p. The larger the C_p, the smaller will be the ΔT necessary to absorb the heat from the reaction:

$$q_p = 0 = \Delta H_{\text{adiabatic system}} = n\Delta\overline{H}_{\text{reaction}} + C_p\Delta T$$

Therefore, $n\Delta\overline{H}_{\text{reaction}} = -C_p\Delta T$.

Likewise, if the reaction is endothermic ($\Delta\overline{H}_{\text{reaction}} > 0$), heat taken for the reaction must come from the energy of the system itself, resulting in a decrease in the temperature of the system. Since the system is mainly water, it is a very good approximation to assume that C_p per gram for the system is equal to C_p per gram for water, which is 4.184 joules per g-°C. For convenience, one more approximation will be made, namely, that the density of the solutions is equal to 1.00 g per mL. To find a given value of $C_p\Delta T$, then, you will simply multiply the temperature change ΔT by the number of milliliters of solution times 4.184 joules per g-°C:

$$C_p\Delta T = (\Delta T)(\text{mL solution})\left(\frac{1.00 \text{ g solution}}{1.00 \text{ mL solution}}\right)\left(\frac{4.184 \text{ J}}{\text{g-°C}}\right)$$

In order to measure ΔT, you must make a series of temperature-vs.-time measurements and plot them. From the rate

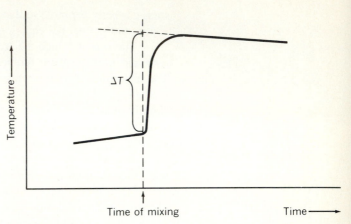

Figure E.39.1

of temperature drift before and after reaction occurs, you will be able to correct for any heat exchange between the system and the surroundings. Extrapolation to the time of mixing, as shown in Figure E.39.1, will give ΔT.

Procedure

■ WEAR YOUR SAFETY GOGGLES

Set up the apparatus shown in Figure E.39.2. The thermocup calorimeter is made of a Styrofoam cup or other well-insulating material.

Insert your thermometer in a stopper, and clamp it at a height to bring the thermometer bulb to the center of the liquid in the cup.

The experimental work will be done by students working in pairs or groups of three. Each pair or group of three students will be assigned by the teaching assistant to study one of the following:

1 $\Delta \overline{H}_{neut}$ for HCl + NH$_3$ and $\Delta \overline{H}_{diss}$ for NH$_4$Cl

2 $\Delta \overline{H}_{neut}$ for HNO$_3$ + NH$_3$ and $\Delta \overline{H}_{diss}$ for NH$_4$NO$_3$

3 $\Delta \overline{H}_{neut}$ for H$_2$SO$_4$ + 2NH$_3$ and $\Delta \overline{H}_{diss}$ for (NH$_4$)$_2$SO$_4$

Thus, one-third of the laboratory section will use HCl; one-third, HNO$_3$; and one-third, H$_2$SO$_4$. Data for the two portions that your section will not do will be obtained from other groups of students in your section.

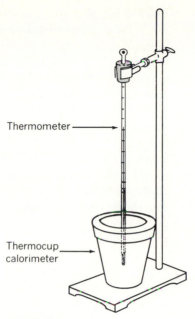

Thermometer ———

Thermocup calorimeter ———

Figure E.39.2

Procedure

(a) **Neutralization** Obtain two Styrofoam cups. In the first cup place 50 mL of 1.50 M NH_3. Next, calculate the volume of your acid (1.50 M HCl, 1.50 M HNO_3, or 1.50 M H_2SO_4) required to neutralize the NH_3 in the first cup, and place that volume of acid in the second cup. Place a thermometer in the cup containing the NH_3, and collect temperature (nearest 0.1°C) vs. time readings at 30-sec intervals for 5 min prior to mixing. Immediately following the last temperature-time reading, add the acid solution all at once to the NH_3 and swirl to mix. Continue taking temperature-time data for 10 min, swirling the solution occasionally. Repeat the procedure to obtain a second set of data.

(b) **Dissolving** In part (a), the neutralization reaction will produce NH_4Cl, or NH_4NO_3, or $(NH_4)_2SO_4$. The final volume of solution after adding the acid solution to the NH_3 solution will depend on which acid was used. Calculate the mass of salt [NH_4Cl, NH_4NO_3, or $(NH_4)_2SO_4$] produced in your neutralization reaction in part (a), and weigh out that mass of salt into a clean, dry beaker. Place a volume of distilled water equal to the final volume of solution from part (a) in a Styrofoam cup, and collect temperature-time data for 5 min prior to mixing. Immediately add the weighed amount of salt, swirl to dissolve (use stirring rod if necessary), and continue taking temperature-time data for 10 min. Repeat the procedure to obtain a second set of data.

Your laboratory instructor will allocate one chalkboard in your laboratory for $HCl + NH_3$ and NH_4Cl data, one chalkboard for $HNO_3 + NH_3$ and NH_4NO_3 data, and a third chalkboard for $H_2SO_4 + 2NH_3$ and $(NH_4)_2SO_4$ data. Students should enter their time-temperature data on the appropriate chalkboards while they are doing the experiment. Each student should copy at least one set of data on each neutralization and one set of data on each dissolving process into a laboratory notebook.

Treatment of the Data   Plot your data, and determine $\Delta \overline{H}_{neut}$ for (a) and $\Delta \overline{H}_{diss}$ for (b). Using -81.2 kilojoules per mol for the $\Delta \overline{H}$ of formation of 1.5 M NH_3 and -165.1 kilojoules per mol for the $\Delta \overline{H}$ of formation of 1.5 M HCl, calculate the $\Delta \overline{H}$ of formation of $NH_4Cl(s)$. Do the same thing for $NH_4NO_3(s)$, using -206.0 kilojoules per mol for the $\Delta \overline{H}$ of formation of 1.5 M HNO_3. Finally, calculate the $\Delta \overline{H}$ of formation of $(NH_4)_2SO_4(s)$, using -884.2 kilojoules per mol for the $\Delta \overline{H}$ of formation of 1.5 M H_2SO_4. [Note that $\Delta \overline{H}$'s of formation are per mol (not per 1.5 mol).]

Questions

Prelaboratory

E.39.1   Prepare three different flowcharts for the procedural portions of this experiment, the first corresponding to HCl, NH_3, and NH_4Cl; the second for HNO_3, NH_3, and NH_4NO_3; and the third for H_2SO_4, NH_3, and $(NH_4)_2SO_4$.

E.39.2   Calculate the volumes of each of the different acid solutions (1.50 M HCl, 1.50 M HNO_3, 1.50 M H_2SO_4) required in part (a) of the procedure. Show your reasoning.

E.39.3   Calculate the masses of NH_4Cl, NH_4NO_3, and $(NH_4)_2SO_4$ required in part (b) of the procedure. Show your reasoning.

E.39.4   Explain in stepwise fashion (number the steps) how you will determine the $\Delta \overline{H}$ of formation of $(NH_4)_2SO_4$ from your data and information given in the experiment. Use actual numbers whenever possible.

E.39.5   Explain how the design of the experiment permits the system to be described as adiabatic.

Postlaboratory

E.39.6   Explain why in Figure E.39.1 the straight line on the left has a rising slope whereas the straight line on the right has a falling slope. What can you deduce about the initial temperature of the starting materials?

E.39.7   How would the form of the curve in Figure E.39.1 appear for an endothermic reaction?

E.39.8   How is the heat of dissolution of the salt $NH_4Cl(s)$ related to the lattice energy of $NH_4Cl(s)$ and the hydration energy of $NH_4^+(g)$ plus $X^-(g)$?

E.39.9   Consider the following standard molar heats of formation, $\Delta \overline{H}_f^\circ$: $NH_3(g)$, -46.1 kilojoules per mol; $NO(g) + 90.2$ kilojoules per mol; $H_2O(l)$, -286 kilojoules per mol; $O_2(g)$, 0 kilojoules per mol. Find ΔH° per mole of reaction for $4NH_3(g) + 5O_2(g) \rightarrow 4NO(g) + 6H_2O(l)$.

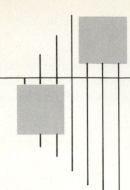

Experiment E.40

Determination of the Faraday

Special Items

Anode of 4 × 1 cm copper sheet with end folded over and pinched to bare copper wire

Cathode of 25-cm piece of insulated copper wire with 1 cm bared at each end

Current source (dry cells or battery charger, ammeter, switch)

Laboratory barometer

Watch

100 mL distilled water

50 mL 3 M H_2SO_4

The faraday is defined as the amount of electricity needed to produce one equivalent of product at each electrode during an electrolysis. In other words, a faraday is equal to the total charge carried by the Avogadro number of electrons. In this experiment, you will determine the value of the faraday by measuring the amount of charge needed to reduce the Avogadro number of H^+ (or H_3O^+) ions. Electric charge is conveniently measured in coulombs. One coulomb (1 C) is the amount of electricity that is transferred when a current of one ampere (1 A) flows for 1 sec; therefore, the current in amperes multiplied by time in seconds is equal to the number of coulombs.

You will electrolyze a solution of sulfuric acid and measure the volume of hydrogen produced by a measured current in a measured period of time. At the same time that hydrogen gas is being evolved at the cathode, the copper anode is being oxidized to copper ions in solution. By measuring the mass loss of the copper anode, you will be able to find the mass of copper oxidized per faraday of electricity, i.e., the mass of one equivalent of copper in this reaction.

Procedure

WEAR YOUR SAFETY GOGGLES

Fill your 200-mL beaker half full of distilled water, and add 50 mL of dilute (6 N) sulfuric acid. Stir to mix. Set your gas-measuring tube in the beaker, as shown in Figure E.40.1. Empty your wash bottle, and take off the nozzle. Slip a pinch clamp over a piece of rubber tubing, and connect the tube between the tip of the gas-measuring tube and the wash bottle. Suck on the mouthpiece of the wash bottle to draw solution to the very top of the gas tube. Close this pinch clamp. (The wash bottle merely serves as a safety reservoir to keep acid out of the mouth.)

Weigh the copper plate anode on an analytical balance to the nearest 0.001 g. Insert it and the insulated cathode into

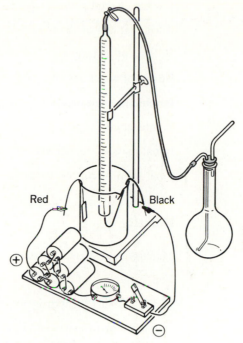

Figure E.40.1

the beaker as shown. Connect the copper plate to the positive terminal and the insulated cathode to the negative terminal of a dc source (less than 10 volts) with a switch and an ammeter in series as shown in Figure E.40.1. Close the switch, noting the time and ammeter reading. Be very careful not to move the electrodes during the electrolysis, since this would give a different current for part of the measured time. Allow the electrolysis to proceed until about 20 mL of hydrogen has collected (about 2 min). Note the ammeter reading. Disconnect, and record the time. Measure the height of the water column (with a meter stick), gas volume, temperature of solution, and barometric pressure.

Rinse the copper anode gently in a beaker of distilled water, blot it dry with filter paper, and weigh it on the analytical balance. Wash and dry the other electrode, and return it to the reagent table.

Data

Time electrolysis begins   _____

Time electrolysis ends   _____

Current at start of electrolysis   _____

Current at end of electrolysis   _____

Volume of hydrogen   _____

Height of water column   _____

Barometric pressure   _____

Temperature   _____

Initial mass of copper anode   _____

Final mass of copper anode   _____

Results

Mercury equivalent of water column   _____

Vapor pressure of water   _____

Partial pressure of hydrogen   _____

Volume of hydrogen at STP   _____

Moles of H_2 formed   _____

Moles of H^+ reduced   _____

Average current during electrolysis   _____

Elapsed time (in seconds) in electrolysis   _____

Coulombs transferred in electrolysis   _____

Value of faraday (coulombs per mole of H^+)   _____

Mass loss of copper anode   _____

Equivalent mass of copper (mass loss
per faraday)   _____

Questions

E.40.1   Why should the pinch clamp be applied as near to the top of the gas-measuring tube as possible?

E.40.2   Use your data to obtain an atomic weight for copper. Calculate the percent error in your value, given that the atomic weight is known to be 63.546 g.

E.40.3   Trace the motion of electric charges throughout the circuit. Label the anode and cathode, and write a balanced half-reaction for the process that occurs at each.

E.40.4   Suppose that some of the Cu^{2+} ions generated at the anode migrated over to the insulated cathode and were reduced there to copper metal. What effect would this have on your calculated value of the faraday?

E.40.5   How would your results be affected if, unknown to you, someone moved the electrodes closer together for the middle part of the electrolysis?

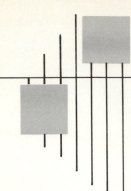

Experiment E.41

Electrode Potentials

Special Items

Zinc electrode, tin electrode,
    inert (nichrome wire)
    electrode
Voltmeter
10 mL distilled water
150 mL 1 M $ZnSO_4$
150 mL 1 M $SnCl_2$
50 mL 1 M NH_4NO_3
150 mL 0.2 M Fe^{2+}/Fe^{3+}
10 mL cyclohexane
1 mL chlorine water
1 mL iodine water
1 mL bromine water
Few crystals of NaCl, NaBr,
    NaI

NOTE: Students work in
    pairs.

Any oxidation-reduction reaction can be split into two half-reactions, one showing the oxidation and the other the reduction. In order to compare half-reactions, it is convenient to write them all as reduction reactions. It is then possible to assign to each a voltage (the reduction potential) that indicates the relative tendency of that half-reaction to occur. The actual values assigned are relative values because all that can be observed is the complete cell voltage, that is, the difference between the two reduction potentials. Once we decide how much of the voltage observed for the complete cell is to be assigned to each component half-reaction, we can use that value in other cells having the same component.

In part (*a*) of this experiment, you will measure the voltage associated with various cells. Since these cells will have half-reactions in common, you will be able to show that the same voltage can be assigned to a given half-reaction no matter what other half-reaction it is coupled with. Your measurements will allow you to rank the various half-reactions according to their tendency to occur.

In part (*b*), you will investigate the relative tendency of various half-reactions to occur by noting whether the oxidizing agent of one half-reaction reacts with the reducing agent of another. If reaction occurs spontaneously, it implies that the oxidizing agent of the first half-reaction is stronger than the oxidizing agent of the second. In other words, the reduction potential of the first is greater than the second. The relative rankings you thereby get in part (*b*) can be tied in with those from part (*a*) by trying appropriate additional chemical reactions.

Procedure

(a) Prepare the setup shown in Figure E.41.1 as follows: Fill one 200-mL beaker three-fourths full of 1 M $ZnSO_4$ and the other to exactly the same height with 1 M $SnCl_2$. Fill a U tube *com-*

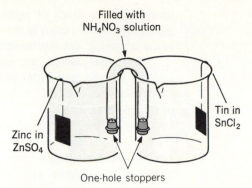

Filled with
NH$_4$NO$_3$ solution

Tin in
SnCl$_2$

Zinc in
ZnSO$_4$

One-hole stoppers

Figure E.41.1

pletely with 1 *M* NH$_4$NO$_3$, push one-hole stoppers into both ends so that the holes are filled *completely*, stuff the holes with cotton, and invert the bridge into the beakers as shown. Make sure that there are no air bubbles trapped in the stopper holes. Insert a zinc electrode into the ZnSO$_4$ solution and a tin electrode into the SnCl$_2$ solution. Attach the electrode leads to a voltage-measuring meter provided by your instructor, and measure the voltage. (Hook it up so that electrons from the zinc enter the meter at the negative terminal.)

Carefully remove the U tube. Store the SnCl$_2$ solution and the tin electrode in a spare beaker. Rinse the 200-mL beaker, and refill it with the solution that contains Fe^{2+} and Fe^{3+} at equal concentrations. Insert an inert electrode from the reagent table. Replace the U tube. The zinc electrode is to be hooked to the same terminal as before. Measure the voltage.

Repeat the operation, this time measuring the voltage of the Fe^{3+}/Fe^{2+} electrode vs. the Sn^{2+}/Sn electrode. Decide from the two preceding cells how the meter should be connected.

(b)   Determine the color of cyclohexane solutions of I$_2$, Br$_2$, and Cl$_2$ by shaking a few drops of each aqueous solution with 1 mL of cyclohexane in a test tube. These colors (in cyclohexane) are characteristic and will serve to identify the elemental halogens.

Cyclohexane is flammable. Keep away from open flames.

Place in a test tube a few small crystals of sodium iodide, 5 mL of distilled water, and 1 mL of cyclohexane. Tap the side of the test tube briskly with the finger to dissolve;

then add a few drops of aqueous Cl_2 solution. Tap briskly again, and note the color in the cyclohexane layer. Repeat the entire test using aqueous bromine in place of aqueous Cl_2.

In similar fashion, test sodium bromide separately with aqueous Cl_2 and then with aqueous I_2.

Likewise, test NaCl for reaction with aqueous solutions of Br_2 and then with I_2.

On the basis of these chemical tests, decide which of the following is the best reducing agent: I^-, Cl^-, or Br^-. To rank these reducing agents with those of part (a), test the best of the reducing agents (I^-, Cl^-, or Br^-) with each of the oxidizing agents from part (a) (Zn^{2+}, Fe^{3+}, and Sn^{2+}). Where a positive test is obtained, try with the same oxidizing agent the second-best reducing agent of the list I^-, Cl^-, or Br^-. Repeat with the third-best reducing agent if necessary.

At the end of the experiment, wash and dry all three of the electrodes, and return them to the reagent table in their proper places.

Data

(a)   Voltage

Zn^{2+}/Zn versus Sn^{2+}/Sn _____

Zn^{2+}/Zn versus Fe^{3+}/Fe^{2+} _____

Fe^{2+}/Fe^{2+} versus Sn^{2+}/Sn _____

In each case, circle the cathode.

(b)   Colors

I$_2$ in cyclohexane _____

Br$_2$ in cyclohexane _____

Cl$_2$ in cyclohexane _____

In the space below, note each of the reactions tried and the results observed.

Results

(a)   Write the three half-reactions tested in order of decreasing reduction potential. Assume that the Sn^{2+}/Sn contributes zero voltage, and assign a set of consistent reduction potentials to the other two. Repeat the assignment, this time using -0.14 volt for $Sn^{2+} + 2e^- \rightarrow Sn$.

(b)   Write all six half-reactions tested in order of decreasing reduction potential.

Questions

E.41.1   The U tube filled with 1 M NH_4NO_3, called a salt bridge, allows electrical neutrality to be preserved as the cell reaction occurs. Propose an explanation for *how* the salt bridge functions; and tell what would happen, and why it would happen, if the salt bridge were not present.

E.41.2   A cell in which the anode reaction is $Zn \rightarrow Zn^{2+} + 2e^-$ and the cathode reaction is $Cu^{2+} + 2e^- \rightarrow Cu$ may be represented symbolically as

$$Zn \left| Zn^{2+} (1\ M) \right\| Cu^{2+} (1\ M) \left| Cu \right.$$

Note that phases (e.g., solid vs. solution) are separated by a single vertical line. The anode portion is separated from the cathode portion by a double vertical line. Concentrations and pressures are given in parentheses. Write such symbols for each of the cells in part (*a*) of the procedure.

E.41.3   Why is it not possible to measure the reduction potential of a half-reaction independently; that is, why must one half-reaction always be coupled with another half-reaction to generate a measurable voltage?

E.41.4   If, in part (*a*) of this experiment, the concentration of $ZnSO_4$ were 0.5 *M* and that of $SnCl_2$ were 1 *M*, would you expect the voltage to be less than or greater than the voltage observed when both concentrations were 1 *M*? Explain.

E.41.5   With reference to the reduction potential table you prepared in part (*b*) of the procedure, choose the set of two half-reactions that if taken together would give the largest overall cell reaction potential. Identify the cathode reaction and the anode reaction, and give the balanced equation for the overall cell reaction.

E.41.6   Suppose in the second and third portions of part (*a*) that you had used pure Fe^{2+} solutions (not really achievable in practice) instead of solutions in which you had both Fe^{2+} and Fe^{3+} present in equal concentrations. Predict how this would affect your measured voltages.

E.41.7   With reference to Figure E.41.1, suppose that you used a platinum electrode in place of the tin electrode but kept everything else unchanged. What effect would this have on the voltage measured for the cell? Explain.

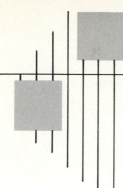

Experiment **E.42**

Redox Titration of Ferrous Ion with Ceric Ion

Special Items

Potentiometer
Saturated calomel electrode
Platinum electrode
50-mL burette
10-mL volumetric pipette
50-mL volumetric flask
5 g solid ferrous salt
   (unknown)
50 mL 0.10 M
   $Ce(NH_4)_2(NO_3)_6$
   (standardized) in 1 M
   H_2SO_4
5 mL 10 M H_2SO_4
Graph Paper

The purpose of this experiment is to determine the percentage of Fe in a soluble sample containing Fe^{2+} by dissolving the sample and titrating it potentiometrically against a standard solution of ceric ammonium nitrate, $Ce(NH_4)_2(NO_3)_6$.

The titration is carried out in an acidic (1 M H_2SO_4) medium to prevent formation of hydroxides and hydrated oxides of iron and cerium, which have limited solubility in neutral solution.

When ferrous ion (Fe^{2+}) is titrated with ceric ion (Ce^{4+}) in the sulfuric acid solution, the reactions that occur can be written in simplified form as follows:

$$Fe^{2+} \rightarrow Fe^{3+} + e^- \qquad E° = -0.68 \text{ volt } (1\ M\ H_2SO_4)$$

$$\underline{Ce^{4+} + e^- \rightarrow Ce^{3+}} \qquad E° = +1.44 \text{ volts } (1\ M\ H_2SO_4)$$

Overall reaction $Ce^{4+} + Fe^{2+} \rightarrow Fe^{3+} + Ce^{3+}$ $\qquad E°_{reaction} = 1.44 - 0.68 = 0.76$ volt

The titration is carried out by adding Ce^{4+} solution directly to Fe^{2+} solution; the two half-reactions ($Ce^{4+} \rightarrow Ce^{3+}$ and $Fe^{2+} \rightarrow Fe^{3+}$) are not physically separated as would be the case in a cell. The reaction comes to equilibrium as fast as the Ce^{4+} is ad to the Fe^{2+} solution, and one may regard the reaction to be *at* equilibrium at all times during the titration. At equilibrium, the voltage $E_{reaction}$ is equal to zero.

The setup you will use to conduct the titration is shown in Figure E.42.1.

At the start of titration, the beaker contains ferrous (Fe^{2+}) ion solution. Dipping into the solution are the platinum electrode and the saturated (with KCl) calomel reference electrode.

The saturated calomel reference electrode that is used to

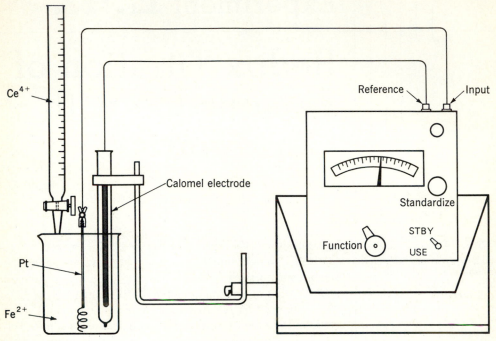

Figure E.42.1

monitor the titration behaves as an anode. Its constant potential corresponds to the reaction

$$Hg(s) + Cl^- \rightleftharpoons \tfrac{1}{2}Hg_2Cl_2(s) + e^- \qquad E_{REF} = -0.246 \text{ volt}$$

The voltage read on the potentiometer, E_{POT}, during the titration equals the sum of the saturated calomel reference half-cell potential E_{REF} and the potential developed at the platinum electrode (cathode) E_{Pt}.

What is the potential associated with the platinum electrode, E_{Pt}? At the beginning of the titration, when the solution contains Fe^{2+} plus a very small amount of Fe^{3+} but no Ce^{4+} or Ce^{3+}, E_{Pt} is given by

$$E_{Pt} = E_{Fe^{3+} \rightarrow Fe^{2+}} = E^\circ_{Fe^{3+} \rightarrow Fe^{2+}} - 0.0591 \log \frac{[Fe^{2+}]}{[Fe^{3+}]}$$

Note that since the platinum electrode is the cathode, the half-reaction $Fe^{3+} \rightarrow Fe^{2+}$ is written as a reduction, opposite that for the titration half-reaction $Fe^{2+} \rightarrow Fe^{3+}$. This does not interfere with the titration, since no appreciable amount of the $Fe^{3+} \rightarrow Fe^{2+}$ reaction occurs, because the potentiometer draws essentially no current.

If the $[Fe^{2+}]/[Fe^{3+}]$ ratio determines E_{Pt} before any Ce^{4+} is added, what determines E_{Pt} after some Ce^{4+} has been added to the Fe^{2+} solution? Looking at the reaction

$$Ce^{4+} + Fe^{2+} \rightarrow Ce^{3+} + Fe^{3+}$$

one can see that the addition of a small amount of Ce^{4+} converts an equivalent amount of Fe^{2+} to Fe^{3+}. At equilibrium (reached immediately), $[Fe^{3+}] = [Ce^{3+}]$; $[Ce^{4+}]$ is infinitesimally small; and $[Fe^{2+}]$ is still fairly large. E_{Pt} may be considered to be determined by the $[Fe^{2+}]/[Fe^{3+}]$ ratio, given by the Nernst equation as before, or it can be considered to be determined by the $[Ce^{3+}]/[Ce^{4+}]$ ratio in the solution. The Nernst equation for this half-reaction is

$$E_{Pt} = E_{Ce^{4+} \rightarrow Ce^{3+}} = E^{\circ}_{Ce^{4+} \rightarrow Ce^{3+}} - 0.0591 \log \frac{[Ce^{3+}]}{[Ce^{4+}]}$$

It makes no difference which of the two half-reactions we choose to calculate E_{Pt}. This can be seen as follows: Recall that the titration reaction consists of an oxidation process and a reduction process,

$$E_{reaction} = E_{oxidation(Fe^{2+} \rightarrow Fe^{3+})} + E_{reduction(Ce^{4+} \rightarrow Ce^{3+})}$$

At equilibrium, $E_{reaction} = 0$; so

$$E_{Ce^{4+} \rightarrow Ce^{3+}} = -E_{Fe^{2+} \rightarrow Fe^{3+}}$$

$$E_{Ce^{4+} \rightarrow Ce^{3+}} = E_{Fe^{3+} \rightarrow Fe^{2+}}$$

Since the potentials of the two half-reactions are equal, it makes no difference which we choose to calculate E_{Pt}. Thus, both these equations must be true:

$$E_{Pt} = E_{Fe^{3+} \rightarrow Fe^{2+}} \tag{I}$$

$$E_{Pt} = E_{Ce^{4+} \rightarrow Ce^{3+}} \tag{II}$$

Summing the two equations, we may state that

$$2E_{Pt} = E_{Fe^{3+} \rightarrow Fe^{2+}} + E_{Ce^{4+} \rightarrow Ce^{3+}}$$

or

$$E_{Pt} = \tfrac{1}{2} \left[(E_{Fe^{3+} \rightarrow Fe^{2+}}) + (E_{Ce^{4+} \rightarrow Ce^{3+}}) \right] \tag{III}$$

The potential of the platinum electrode may be calculated using any of the three equations, I, II, or III, whichever is most convenient to use with the available information. Because $[Ce^{4+}]$ before the equivalence point in the titration is extremely small, it is best to relate E_{Pt} to $[Fe^{2+}]$ and $[Fe^{3+}]$, both of which are relatively large and easily calculated. E_{POT} before the equivalence point is then determined by the relation

$$E_{POT} = E_{REF} + E_{Fe^{3+} \to Fe^{2+}}$$

$$E_{POT} = -0.246 + \left(0.68 - 0.0591 \log \frac{[Fe^{2+}]}{[Fe^{3+}]} \right)$$

An important point in the titration occurs when enough Ce^{4+} has been added to convert half the original Fe^{2+} to Fe^{3+}. Recalling that $\log 1 = 0$ and using the equation above, calculate E_{POT} for that point in the titration, that is, when $[Fe^{2+}] = [Fe^{3+}]$. (1) (Where numbers in parentheses follow questions or directions, answers are to be written out on separate paper.)

The most important point in the titration is the equivalence point. The potential of the platinum electrode at that point is most easily calculated from equation III:

$$E_{Pt} = \tfrac{1}{2} \left[(E_{Fe^{3+} \to Fe^{2+}}) + (E_{Ce^{4+} \to Ce^{3+}}) \right]$$

Substituting in the appropriate Nernst equations,

$$E_{Pt} = \frac{1}{2} \left[\left(E^{\circ}_{Fe^{3+} \to Fe^{2+}} - 0.0591 \log \frac{[Fe^{2+}]}{[Fe^{3+}]} \right) + \left(E^{\circ}_{Ce^{4+} \to Ce^{3+}} - 0.0591 \log \frac{[Ce^{3+}]}{[Ce^{4+}]} \right) \right]$$

$$E_{Pt} = \frac{1}{2} \left(E^{\circ}_{Fe^{3+} \to Fe^{2+}} + E^{\circ}_{Ce^{4+} \to Ce^{3+}} - 0.0591 \log \frac{[Fe^{2+}][Ce^{3+}]}{[Fe^{3+}][Ce^{4+}]} \right)$$

At the equivalence point, $[Fe^{3+}] = [Ce^{3+}]$ and $[Fe^{2+}] = [Ce^{4+}]$, so that $\log [Fe^{2+}][Ce^{3+}]/[Fe^{3+}][Ce^{4+}] = \log 1 = 0$. Therefore, it must be true at the equivalence point that

$$E_{Pt} = \tfrac{1}{2} \left(E^{\circ}_{Fe^{3+} \to Fe^{2+}} + E^{\circ}_{Ce^{4+} \to Ce^{3+}} \right)$$

Substitute in the E° values and calculate the E_{Pt} at the equivalence point. (2) Also, using the relation $E_{POT} = E_{REF} + E_{Pt}$, calculate the voltage you should read on the potentiometer at the equivalence point. (3)

After the equivalence point, $[Fe^{2+}]$ will be extremely small, and it becomes best to relate E_{Pt} to $[Ce^{3+}]$ and $[Ce^{4+}]$, both of which are large and easily defined.

The titration curve will have the general appearance shown in Figure E.42.2.

Note that, as with a pH titration curve, there is a large change in potential near the equivalence point. Explain. (4) Also note that the shape of the curve is symmetrical. This is because the oxidation-state change for both species, Ce^{4+} and Fe^{2+}, involves a one-electron transfer per ion. If the oxidation-state change had involved different numbers of electrons transferred per ion for the oxidizing and the reducing agents,

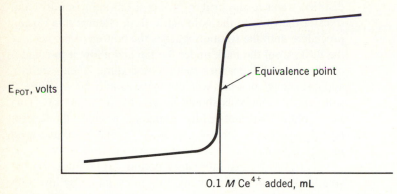

E_{POT}, volts

Equivalence point

0.1 M Ce^{4+} added, mL

Figure E.42.2

the curve would have been unsymmetrical (e.g., titration of Fe^{2+} with MnO$_4^-$).

Your problem in this experiment is to carry out a potentiometric titration of an unknown Fe^{2+} sample, to prepare a graph (or graphs) of the titration, and to use the graph(s) to calculate the percentage by weight of Fe in your unknown.

Note: The potentiometer is not standardized, because it is not necessary to know the voltages on an absolute scale in order to identify the equivalence point. The equivalence point in the titration, where voltage changes rapidly with added Ce^{4+}, is identical to that found with a standardized instrument. The observed E_{POT} at the equivalence point, however, may be slightly different from the calculated value.

Procedure

(a) **Preparation of Fe^{2+} Solution (Unknown)**   Because Fe^{2+} in solution is oxidized readily to Fe^{3+} by atmospheric oxygen, it is necessary to have a freshly prepared solution. Obtain approximately 5 g of solid unknown containing Fe^{2+} from your instructor in a labeled, clean, dry 10-cm test tube. The test tube should be about two-thirds full to contain 5 g.

■ **WEAR YOUR SAFETY GOGGLES**

Find the mass of the container (test tube) plus unknown; pour the solid carefully into a clean (not necessarily dry) 50-mL beaker; then find the mass of the empty container to determine the exact mass of your unknown. Add approximately 20 mL of hot distilled water to the beaker, and swirl it to dissolve as much of the solid as possible. Add approximately 5 mL of 10 M H$_2$SO$_4$, and swirl until the salt has dissolved completely. Carefully transfer the solution from the beaker to a clean 50-mL volumetric flask. Add approximately 5 mL of distilled water to the beaker, swish it around carefully, and transfer the rinse to the volumetric flask. Repeat this rinsing procedure until the solution reaches the bottom of the neck of the flask. Cool the flask under the tap and allow it to stand 5 or 10 min until it reaches room temperature. Then carefully add just enough distilled water to make *exactly* 50 mL of Fe^{2+} solution (the meniscus should be exactly at the mark). (This Fe^{2+} solution will not oxidize appreciably within the laboratory period, but it must not be made up one week and used the next; therefore, if you do not have the time to do the titration within a given laboratory period, do *not* make up the solution.) Stopper the flask and mix the solution by inverting the flask, righting it, inverting, etc., at least 50 times. Pour the Fe^{2+} solution into a clean, *dry* container so that someone else may use the 50-mL volumetric flask. Again swish the solution some more to ensure complete mixing.

(b)   **Setting Up Apparatus**   Obtain a 10-mL pipette, pipette bulb, and a clean 200-mL beaker; attach a clean 50-mL burette to a ring stand with a burette clamp. Attach a platinum electrode to the INPUT receptacle and a calomel electrode to the REF receptacle on a pH meter (which will be used as a potentiometer). *Make sure that the meter is on STBY.* Rinse the 10-mL pipette with a small amount of the Fe^{2+} solution, and using the pipette, place exactly 10.00 mL of the Fe^{2+} solution in the beaker. Add about 15 mL of distilled water.

Immerse the electrodes in the solution, and set the Operate toggle switch to USE. Set Function switch on ZERO, and turn the Standardize dial until the meter reads zero (midscale). Turn the Function switch to the ± 1400 mV (millivolt) range, and take a reading. Turn toggle switch to STBY.

Rinse the burette with a bit of the ceric solution; then fill it and record the burette reading. (Be careful not to spill the Ce^{4+} solution on your hands. It is 1 M in H_2SO_4.) Place the burette above the beaker of ferrous solution, and add 1 mL of Ce^{4+} solution. Turn the toggle switch to USE and take a reading. Before taking readings, stir the solution thoroughly with a stirring rod.

Repeat the procedure of adding 1 mL and taking a reading until the change in voltage begins to increase (near the equivalence point). At this point decrease the volume added before each reading to approximately 0.1 mL. When the voltage change again levels off, increase the increment until you have gone about 10 mL past the equivalence point. Repeat the procedure, using a second aliquot of Fe^{2+} solution. It will save time if you do the first trial quite rapidly; then do the second trial very carefully.

Record your results in two parallel columns—"mL Ce^{4+} added" and "E_{POT}."

Treatment of Data

Plot the data on a graph, placing "E_{POT} (volts)" on the y axis and "mL 0.1 M Ce^{4+} added" on the x axis. Identify the equivalence point, and then calculate the percentage of Fe by weight in your unknown sample.

Data

mL Ce^{4+} Added	E_{POT}, volts	mL Ce^{4+} Added	E_{POT}, volts	mL Ce^{4+} Added	E_{POT}, volts

Results

Percentage of Fe by weight in unknown _____ .

Questions

E.42.1 Consider a potentiometric titration of a sample of chromous (Cr^{2+}) salt with a standardized 0.1000 M Ce^{4+} solution in 1 M H_2SO_4 using the same setup as used in this experiment. $E°$ for $Cr^{3+} + e^- \rightarrow Cr^{2+} = -0.41$ volt. Suppose that the beaker contains a 10.00-mL aliquot of 0.300 M Cr^{3+} solution and that you are adding 0.1000 M Ce^{4+} solution progressively from a burette.

(a) Predict the potential observed on the potentiometer (using saturated calomel reference electrode)

    (i) After adding 10.0 mL of 0.1000 M Ce^{4+}

    (ii) After adding 15.0 mL of 0.1000 M Ce^{4+}

    (iii) At the equivalence point of the titration

(b) Calculate values for $E°$, $\Delta G°$, and K for the reaction of Ce^{4+} reacting with Cr^{2+} to give Ce^{3+} and Cr^{3+}.

E.42.2 The potentiometric titration curve for the oxidation of Fe^{2+} by Ce^{4+} is symmetrical, but the curve for the titration of Fe^{2+} using $Cr_2O_7^{2-}$ in acidic solution is unsymmetrical. Explain.

E.42.3 Suppose that you used a standard hydrogen electrode (SHE) in place of the saturated calomel electrode. How would your data have been different?

E.42.4 If E of the titration reaction is always zero during the titration, and E of the saturated calomel electrode is always -0.246 volt, why is the voltage read on the potentiometer not always -0.246 volt?

E.42.5 The $E°$'s given here for the $Fe^{2+} \rightarrow Fe^{3+} + e^-$ and the $Ce^{4+} + e^- \rightarrow Ce^{3+}$ half-reactions are not the same in magnitude as those normally given in a table of standard reduction potentials. Explain.

E.42.6 Given that $E_{reaction} = E°_{reaction} - (0.0591/n) \log Q$, where Q, the mass-action expression, is given by $[Fe^{3+}][Ce^{3+}]/[Fe^{2+}][Ce^{4+}]$, calculate K_{eq} for the overall reaction $Ce^{4+} + Fe^{2+} \rightarrow Fe^{3+} + Ce^{3+}$, using data given earlier in this experiment.

E.42.7 What function is served by the calomel electrode in this experiment? How does the value of E_{REF} affect the determination of volume Ce^{4+} added at the equivalence point?

E.42.8 Estimate the precision with which you have determined the percentage of Fe by weight in your unknown in this experiment?

E.42.9   What is the fundamental reason why either the $[Fe^{2+}]/[Fe^{3+}]$ or the $[Ce^{3+}]/[Ce^{4+}]$ ratio can be used to fix the voltage of the platinum electrode during the titration?

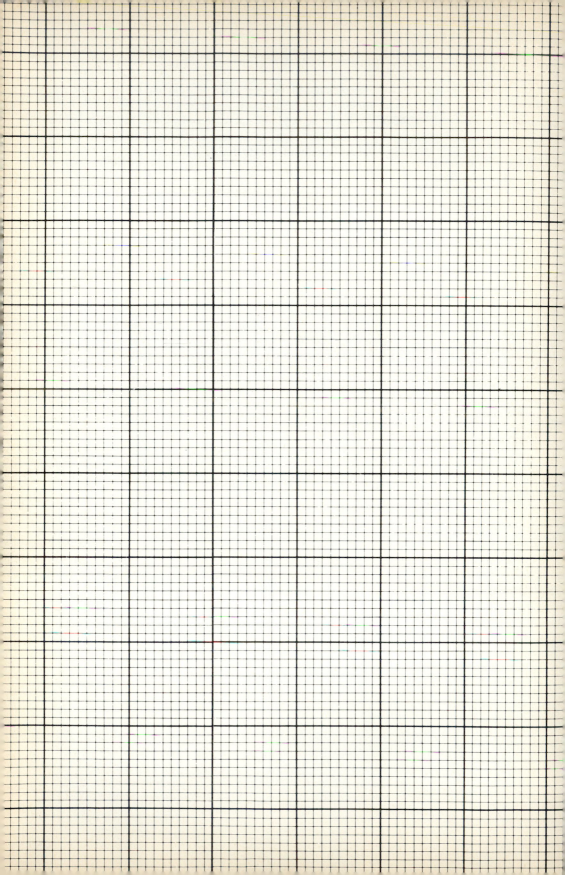

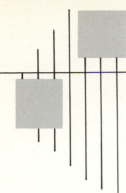

Hydrogen and Preparation of Pyrophoric Iron

Special Items

Boiling chips
60-cm length of glass tubing
10 g mossy zinc
Few drops $CuSO_4$ solution
15 g calcium chloride lump
2 g steel wool
30 mL 3 M H_2SO_4

NOTE: Students work in pairs.

Since iron and water are cheap, hydrogen is sometimes made industrially by reduction of steam with hot iron. In this experiment, you will carry out this reaction. By increasing the concentration of hydrogen above the product and removing steam, you will find that the reaction can be reversed. In fact, this is one way of making pure iron. The iron thus prepared is in a state of high subdivision, and its great surface area makes it extremely reactive. Because it may burst into flame when heated in air, it is called "pyrophoric."

Procedure

(a)

■ **WEAR YOUR SAFETY GOGGLES**

Set up the apparatus shown in Figure E.43.1. Have the 500-mL flask about half full of water. Add a boiling chip to it to prevent bumping. Secure from the stockroom 60 cm of glass tubing, which is to extend almost to the bottom of the flask and serve as a safety tube. The test tube should be tilted as shown so that any water that may condense will drain away from the heated iron. Weigh the test tube on an analytical balance to the nearest 0.001 g. Place a small loose wad of steel wool in the test tube, and weigh it again. Heat the water to boiling. When boiling starts, turn down the flame so as to maintain steam evolution at a low steady rate.

Heat the test tube gently at first; then step up the intensity to full heat. Continue to heat for about 10 min. Then by water displacement collect a test tube full of the gas coming out of the delivery tube. Test it with a lighted splint.

While still continuing the steam evolution, disconnect the rubber tubing and turn off the burner. Note what has hap-

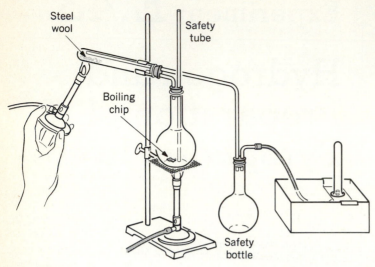

Steel
wool

Safety
tube

Boiling
chip

Safety
bottle

Figure E.43.1

pened to the iron. Note also what is present in the safety bottle.

(b)   When the test tube is cool, disconnect it and, using cheesecloth, wipe out as much of the condensed water as possible without disturbing the product. Weigh the test tube and contents, and then use to build the apparatus shown in Figure E.43.2. Have about 10 g of mossy zinc plus a few drops of copper sulfate solution in the bottle. The thistle tube should extend nearly to the bottom of the bottle. Wrap the bottle in

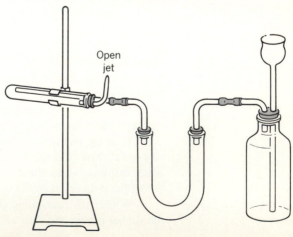

Open
jet

Figure E.43.2

a towel. Pour about 30 mL of water and then about 30 mL of dilute sulfuric acid through the thistle tube.

Let the generator run for at least 1 min to sweep out the air. Collect a test tube full of the gas issuing from the jet, carry it to a burner several feet away, and try to light it. Repeat until no pop is heard. Quickly return the quietly burning hydrogen to the jet to ignite it.

■ CAUTION

Do not light the jet in any other way. Students who blow up a hydrogen generator will be expelled from the laboratory immediately.

Once the jet is burning quietly, heat the iron oxide in the test tube gently at first and then with full flame. Continue heating for about 10 min. Remove the burner, and let the test tube cool while the hydrogen flow continues. Pour water into the generator to kill the evolution of hydrogen. Disconnect the test tube, stopper it at once, and weigh it.

Data and Results

Note here the observations you have made in parts (*a*) and (*b*) and your interpretations of them.

On the basis of your data, what is the simplest formula of the product formed by your oxidation of iron by steam? Assume that the product consists of only iron and oxygen.

Questions

E.43.1   With reference to Figure E.43.2, explain why it is important to have the bottom of the thistle tube extend below the surface of the liquid in the hydrogen generator.

E.43.2   Why in part (a) is it possible to collect the hydrogen gas in an inverted test tube?

E.43.3   The iron produced in part (b) has roughly the same composition as many other iron samples. Why, then, is it pyrophoric?

E.43.4   Calculate the volume of hydrogen required at STP to reduce the iron oxide in part (b) to elemental iron.

E.43.5   Write balanced equations for each of the reactions that occur in this experiment.

E.43.6   What is the function of the copper sulfate solution in part (b)?

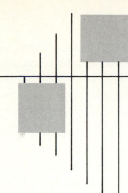

Experiment **E.44**

Structure of Ice

Special Items

Framework molecular model
kit

The purpose of this special project is to examine structural features and symmetry of ordinary ice and to calculate the density of ice from unit-cell dimensions. You will first construct a representation of the lattice structure of ice using framework molecular-model tubes and metal connectors. Differently colored tubes will be utilized in order to point out structural characteristics, not to distinguish between bonds. You will then use this model to help you visualize structural characteristics and symmetry elements. Finally, you will make measurements of the model to determine the density of ice.

Construction of the Ice I Structure

Step 1   Attach four black-and-white striped tubes, one to each prong of a tetrahedral connector. Make two other identical tetrahedral structures. Place all three on a flat surface and align them so that one leg of each points inward toward a central point. Using another tetrahedral connector, connect the three tetrahedral units together at the central point. If you look directly down on it, you should be able to see that the structure you have made can be inscribed within a regular hexagon. Draw it. (*1*) (Where numbers in parentheses follow questions or directions, answers are to be written out on separate paper.)

Step 2   Attach a tetrahedral connector to every vertex (nine in all) of the hexagonal structure. Examine the central vertex (joining the three tetrahedral units), and attach to it a black-and-blue striped tube. Compare the orientation of the central tetrahedral unit relative to the three original ones. (*2*) To each of the six vertices that outline the hexagon, attach a white tube, and align the white tubes so that they are parallel to the central black-and-blue striped tube. Place the structure on your desk so that it stands on the white tubes.

Step 3   At this point the structure should have three black-and-white striped tubes pointing up (vertical), each with a metal tetrahedral connector at the top. To each of these tetrahedral connectors attach one yellow and two red tubes, and align them so that the three yellow tubes point in to a central point and the three pairs of red tubes point outward. Using one metal tetrahedral connector, attach the three yellow tubes together. The structure should now be complete.

The lattice you have put together is a representation of the structure of ordinary ice. There are many different forms of ice (at least nine), but the most common form (ice I) has a wurtzite or tridymite structure as shown in Figure E.44.1. Compare the lattice you have constructed with Figure E.44.1.

In the ice I structure (also called hexagonal ice), the oxygen atom in each water molecule resides at the center of a tetrahedron formed by four adjacent oxygen atoms, each of which is approximately 0.276 nm (2.76 Å) away.

Except for a negligible fraction of ionized molecules, there is evidence that the water molecules are intact in the ice structure and that they retain essentially the same shape as water molecules in the vapor state. Recall that in the vapor state the geometry of the water molecule and its unshared electron pairs corresponds to a slightly distorted tetrahedron, with the oxygen atom at the center, hydrogen atoms along two of the axes separated by an angle of 104°27′, and electron pairs pointing along the other two axes. Draw a picture of a water molecule, together with the "clouds" representing the unshared electrons. (3) In the ice structure, adjacent water molecules are oriented in such a way that only one hydrogen atom lies (approximately) along the axis between any two adjacent oxygen atoms. Using a metal tetrahedral connector and four differently colored plastic tubes (blue, red, black, green), construct a tetrahedral shape. Next, figure out all the different

Figure E.44.1

possible ways of orienting a single water molecule on it. Describe each of the different orientations, using the colors (e.g., one blue and one red) in your descriptions. (*4*) In an ice crystal, the water molecules are oriented in every one of the possible ways in a random fashion.

Each water molecule in an ice crystal is hydrogen-bonded to its four nearest neighbors. Each of the two hydrogen atoms on a particular oxygen atom is directed toward a lone pair of electrons on a neighboring oxygen atom; each of the lone pairs on the oxygen atom is directed toward a hydrogen atom bonded to a neighboring oxygen atom. Thus, there are four hydrogen atoms, two covalently bonded and two hydrogen-bonded, surrounding each oxygen atom. One $O-H$ distance is 0.101 nm (1.01 Å), and the other is 0.175 nm (1.75 Å). Which corresponds to the covalent bond, and which to the H bond? Explain. (*5*)

Now refer to the model of the ice structure you have constructed. One thing you should notice is the existence of hexagonal "puckered"rings. If you take an organic chemistry course you will get accustomed to seeing these six-membered rings made up of tetrahedrally bonded atoms (especially carbon). The two important low-energy configurations of hexagonal rings are called the "boat" and "chair" configurations. In the boat form, atoms on opposite sides of the hexagon either both point up or both point down. In the chair form, opposite atoms are configured differently; if one points up, the other points down. Draw diagrams showing the two configurations of hexagonal rings. (*6*)

Hold your model in such a way that the white tubes point down. In what configuration are the vertical rings? (*7*) How about the horizontal rings? (*8*)

Symmetry of the Ice I Structure

The symmetry of the ice I structure is described in the Hermann-Mauguin (H-M, or international) symbolism as $P6_3/mmc$. The precise meaning of the symbols will be made clear as we progress. The structure is very symmetrical. In fact, the symmetry of hexagonal close-packing is also $P6_3/mmc$. Look at the model you have constructed and see if you can see any similarities between it and hexagonal close-packing arrangement. The symbol $P6_3/mmc$ indicates that the unit cell is primitive (P), that the principal symmetry axis is a sixfold screw axis (6_3), that there is a mirror plane ($/m$) perpendicular to the 6_3 axis, and that one of the sets of planes parallel to the 6_3 axis is a set of glide planes, with translation in the same direction (c) as the 6_3 axis.

6_3 Screw Axis   Like any other symmetry operation, the operation done with respect to the screw axis results in a configuration identical to the original. A 6_3 screw axis means that the rotational unit is $\frac{360}{6}$, or $60°$, and in twisting $60°$ a point will move three-sixths of the way from the bottom to the top of the unit cell. (Similarly, a 4_3 screw axis means that in twisting $90°$ a point will move three-fourths of the way up the axis.) In the model you have constructed you can, by looking down on the top (provided the yellow tubes are on top), see three partially completed hexagonal channels running vertically. The top of each of the channels is defined by two yellow and two red tubes. Take two black tubes and some tetrahedral connectors and complete one of the three top partial hexagons. Take two more black tubes and complete the lower hexagon (same channel) also. Then connect the two horizontal hexagons together with a fifth black tube placed vertically. Now, you should be able to see the complete channel. Passing down through the center of the hexagonal channel is the 6_3 axis. Using the differently colored tubes in your description, explain why this is a 6_3 screw axis. (*9*)

Mirror Plane (*m*)   Recall that a mirror plane is a plane through which all elements on one side are mirrored (reflected) on the other side. See if you can find a mirror plane perpendicular to the 6_3 axis. Describe its location. (*10*)

Glide Plane   A glide plane is the result of two consecutive symmetry operations, reflection and translation. Consider a glide plane parallel to the 6_3 axis in your model. A glide plane is similar to a mirror plane, except that instead of reflecting directly across the plane to the opposite side, the reflection is moved up or down, usually by half the height of the unit cell. To find a glide plane, look for mirror images that are not directly across from each other but are across *and* up or down from each other. Try to find a glide plane in your model, and describe its location. (*11*)

Density Calculation

The fact that the structure of ice is very open, with hexagonal channels running through it, helps to explain why solid ice is less dense than liquid water (at 0°C). It is believed that in the liquid state, as in the solid state, each oxygen atom is surrounded by four other oxygen atoms in tetrahedral directions but that there is some interpenetration of the structural cavities by water molecules in the liquid to give the increased density.

The model you have put together, if it represents the true structure of ice, should permit you to calculate the density of ice, provided you are given the distance separating oxygen atoms in neighboring water molecules. This distance is 0.276 nm (2.76 Å).

Your first problem will be to determine the unit cell (repeating unit) for ice and then to figure out how many oxygen atoms and hydrogen atoms are contained in that cell. The model you put together contains three unit cells, which can be seen if you first remove the five black tubes you added to see the 6_3 axis and then turn the model upside down so the white tubes point up. The unit cell is a rhombohedral (diamond-shaped sides) six-sided cell whose boundaries are defined by three white tubes and the blue-and-black striped tube. Recall that in step 1 you inscribed the base of your structure within a hexagon. Do the same thing again, except this time show the boundary of one unit cell. (*12*)

To figure out the number of atoms in the unit cell, first stand the structure on the white tubes. Put metal tetrahedral connectors at the tips of the three red tubes in the unit cell to represent oxygen atoms. If you have any metal connectors at the bottoms of the white tubes, take them off. Take your neighbor's model and stack it on top of yours so that the unit repeats vertically. Look at the "oxygens" at the top tips of the red tubes and the one at the apex of the yellow tubes. Are these atoms the same or different from the atoms on the bottoms of the white tubes? (*13*) Should they be counted once or twice per unit cell? (*14*) How many other unit cells share each of the four top oxygen atoms (three red + one yellow)? (*15*) You should be able to see it from the number of unfilled "valences" on each "red" tetrahedral connector (oxygen atom). Of these four "top" oxygen atoms, then, how many actually "belong" to one unit cell? (*16*)

In the other horizontal "layer" of four oxygen atoms (top of blue-and-black tube and top of three white tubes), the oxygen atoms are also shared partially by other cells. Of these four, how many "belong" to one unit cell? (*17*)

There are two additional oxygen atoms. Describe their location in the unit cell, and determine how many of them should be considered part of one unit cell. (*18*) What is the total number of oxygen atoms in each unit cell? (*19*)

How many hydrogen atoms must then occupy each unit cell? (*20*)

Measure the vertical height of the unit cell and its other dimensions. Also measure the length of one of the plastic tubes. List your measurements in centimeters. (*21*) You were given the information that the oxygen-oxygen distance (length of tube) is 0.276 nm. Use this information to convert (by ratio) the dimensions to nanometers. (*22*) Note that the base of the cell is a rhombus (two equilateral triangles). Calculate the volume of the unit cell in cubic centimeters, using the relation that volume equals the area of the base times the height, and the conversion factor $1 \text{ nm} = 10 \text{ Å} = 10^7 \text{ cm}$. (*23*)

Calculate the density of ice with the data you have gathered and the use of Avogadro's number, which says that there are 6.02×10^{23} molecules in 1 mol of water (18.015 g). (*24*)

Questions

E.44.1   In the ice structure, a hydrogen can move from being closer to one oxygen to being closer to an adjacent oxygen. Relate this to the graph in Figure E.44.2, which describes the potential energy of a system of two oxygens and a hydrogen in the ice structure.

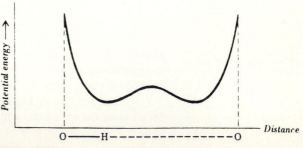

Figure E.44.2

E.44.2   In general, substances soluble in liquid water are not soluble in ice. Ammonium fluoride, however, has considerable solubility in ice. Suggest a reason why this may be the case.

E.44.3   In aqueous solutions of acids, hydronium ions exist as H_3O^+, $H_5O_2^+$, $H_7O_3^+$, etc. Propose a structure for $H_7O_3^+$.

E.44.4   Assuming all the distances stay the same, what would happen to the density of ice if all the H atoms were replaced by D (deuterium) atoms?

E.44.5   In terms of your structure, indicate how one could explain that liquid water is more dense than ice at $0°C$, but that its density increases and then decreases as the temperature is raised.

E.44.6   Tabulate the symmetry elements of the tetrahedral structures. Show how these symmetry elements are related to the 6_3 screw axis of Figure E.44.1.

E.44.7   Suppose that you have two isolated six-membered rings made up of tetrahedrally bonded atoms—one in the boat form, the other in the chair form. Which probably has the higher potential energy? Explain.

E.44.8   Show that a hexagonal close-packed array of identical spheres has symmetry describable as $P6_3/mmc$. Locate each of the symmetry elements specifically.

E.44.9   Draw a sketch to show how the location of points related by a normal sixfold rotation axis differs from the location of points related by a 6_3 screw axis. Show that if there is a 6_3 screw axis, there must also be a normal threefold axis.

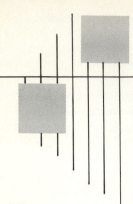

Experiment **E.45**

Fixation of Nitrogen

Special Items

0.4 g Mg ribbon
10 drops distilled water

If magnesium metal is heated in nitrogen gas, reaction occurs to form magnesium nitride (Mg_3N_2). Since air is mostly nitrogen, it seems possible to prepare this nitrogen compound by heating magnesium in air. However, air also contains oxygen, which reacts with magnesium to form magnesium oxide (MgO). Although air contains about four times as much nitrogen as oxygen, it is not necessarily true that the product of burning Mg will contain N and O in the same ratio as in air. The composition of the product will be governed by the relative rates with which nitrogen and oxygen react with Mg and the relative stabilities of Mg_3N_2 and MgO with respect to the elements. In this experiment, you will determine the relative amounts of Mg converted to Mg_3N_2 and MgO on burning a known mass of Mg. Then you will heat the product in the presence of water to see if all the Mg_3N_2 can be converted to MgO.

Procedure

■ WEAR YOUR SAFETY GOGGLES

(a) Dry with a cheesecloth a clean crucible and cover, and then hold them with tongs in the side of a burner flame. Allow to cool, and weigh the crucible and its cover on an analytical balance to the nearest 0.0001 g. Place in it about 0.4 g of Mg ribbon (about 0.5 m), cut into small pieces. Reweigh. Support the crucible on your triangle with the cover cocked so that a small crack is left for air to enter. Heat with the *hottest flame* of your burner for at least 15 min. Remove the cover, and heat 10 min longer. Cool, and reweigh crucible with cover.

(b) Add about 10 drops of distilled water to the contents of the crucible. By waving your hand over the top of the crucible and wafting the odor cautiously toward your nose, note any odor produced. Reheat gently with cover completely on. Increase the heat to full flame for about 5 min. Cool, and then reweigh crucible and cover.

Data

Results

(a)   To calculate the percentage of Mg converted to Mg_3N_2, proceed as follows: Let x be the moles of Mg_3N_2 in the product and y be the moles of MgO. From the formula weights of Mg_3N_2 and MgO, write an equation expressing the final mass of product in terms of x and y. Since $3x$ moles of Mg atoms is needed to produce x moles of Mg_3N_2 and y moles of Mg atoms is needed to produce y moles of MgO, you can set up a second equation, using the atomic weight of Mg, expressing the initial mass of Mg in terms of x and y. Solve the two simultaneous equations for x and y, and calculate the percentage of the initial Mg that is converted to Mg_3N_2.

(b)   From the initial mass of Mg and the final mass of product after ignition with water, decide if the Mg_3N_2 has been completely converted to MgO.

Questions

E.45.1   Recalling that air is 80 percent nitrogen and 20 percent oxygen, what can you infer from your data and calculations about the relative tendency of magnesium to combine with oxygen vs. nitrogen?

E.45.2   Write electron dot structures for the molecules O_2 and N_2, the ions O^{2-} and N^{3-}, and the ionic compound Mg_3N_2.

E.45.3   Calcium metal reacts with N_2 and O_2 to form a nitride and an oxide like those of magnesium. How many grams of product would be formed if 0.40 g of calcium were reacted with (*a*) pure nitrogen and (*b*) pure oxygen?

E.45.4   What are likely the initial products formed when water (*Hint:* HOH) is added to Mg_3N_2? What happens to these products when they are heated in a crucible?

E.45.5   What procedural errors would lead to a reported result for the percentage of Mg converted to Mg_3N_2 that is (*a*) too low and (*b*) too high?

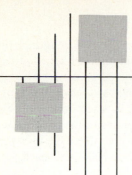

Experiment E.46

Some Elements of Group V

Special Items

5 mL distilled water
12 mL 3 M H_2SO_4
3 g $NaHCO_3$
20 mL 0.1 M $MnSO_4$
1 g NaI
Few grains sodium
  bismuthate
50 mL 3 M HCl
50 mL 3 M NaOH
3 mL each "PCl$_3$ plus water,"
  "AsCl$_3$ plus water,"
  "SbCl$_3$ plus water," and
  "BiCl$_3$ plus water"
8 mL each 0.1 M solutions of
  sodium phosphate, sodium
  arsenate, and sodium
  antimonate

In this experiment, you will investigate the chemical behavior of the group V elements P, As, Sb, and Bi, with particular reference to acid-base properties and oxidation-reduction. For the acid-base properties, you will start with aqueous solutions made from the trichlorides, check for hydrolysis, and then add base dropwise to see what happens. The ultimate purpose will be to decide if the corresponding oxides are acidic, basic, or amphoteric. For the oxidation-reduction, you will be mainly interested in the change from 5+ to 3+ states; and you will investigate what happens to compounds of 5+ states when mixed with some simple reducing agents. In this way, you will be able to rank the corresponding half-reactions in order of decreasing reduction potential.

Procedure

■ CAUTION

Compounds of As, Sb, and Bi are very poisonous. Take care to keep these compounds and their solutions out of cuts. Wash your hands thoroughly before leaving the laboratory.

■ WEAR YOUR SAFETY GOGGLES

(a) In a clean test tube, place a few milliliters of the solution labeled "PCl$_3$ plus water," and test with litmus paper. Then add dilute NaOH dropwise until the solution is definitely basic. Finally, add dilute HCl dropwise until the solution becomes acidic.

Examine closely the solution labeled "AsCl$_3$ plus water." Test a few milliliters of it with litmus, NaOH, and HCl, as above.

Repeat for "SbCl$_3$ plus water" and "BiCl$_3$ plus water."

(b) In a clean test tube, mix 5 mL of 0.1 M sodium phosphate solution with 1 mL of dilute (3 M) sulfuric acid. Add 0.2 g of NaI. Warm and observe. Save the solution.

Repeat separately with 5 mL of 0.1 M sodium arsenate and with 5 mL of 0.1 M sodium antimonate. Save the solutions.

To test with sodium bismuthate, place 1 mL of dilute sulfuric acid and 0.2 g of NaI in a test tube with 5 mL of distilled water. Drop in a few grains of the solid sodium bismuthate. Warm slightly and observe. Save the mixture.

Weigh out four portions of $NaHCO_3$, each weighing 0.7 g. Add one portion of the $NaHCO_3$ very slowly to each test tube containing the saved solutions. (*Note:* Addition of HCO_3^- neutralizes most of the acid and therefore may change the oxidizing strength of an oxyanion.)

In each of four test tubes, place 5 mL of 0.1 M manganous sulfate and 2 mL of 3 M sulfuric acid. To the first three tubes, add separately 3-mL portions of the 0.1 M solutions of phosphate, arsenate, and antimonate. To the fourth, add a few grains of sodium bismuthate. Shake the mixtures, and heat gently. Allow any solids to settle, and observe the colors of the solutions. If a centrifuge is available, it is helpful to centrifuge each of the solutions after shaking and heating the mixtures. Dispose of all reaction products in the waste containers provided; do not throw down the sink drain.

Data

Results

(a)   On the basis of your observations, what do you conclude about the acid-base behavior in group V?

(b)   Write half-reactions for each of the $3+$ to $5+$ conversions, arranging them in order of decreasing reduction potential. Assume species to be as follows: H_3PO_3, H_3PO_4; As $(OH)_2^+$, H_3AsO_4; $Sb(OH)_2^+$, Sb_2O_5; $Bi(OH)_2^+$, $NaBiO_3$.

Questions

E.46.1   In part (*a*), why is the hydrolysis product of $PCl_3 + H_2O$ better described as H_3PO_3 than as $P(OH)_3$?

E.46.2   Write the hydrolysis reaction of $AsCl_3$ using
(*a*) Brønsted terminology
(*b*) Lewis acid-base terminology

E.46.3   Given that the oxidation product of I^- is I_2 and that of Mn^{2+} is MnO_4^-, write balanced equations for each of the $3+$ to $5+$ conversions in acidic solution of group V elements in part (*b*).

E.46.4   With reference to the reduction-potential table you prepared in part (*b*), identify the strongest
(*a*) Reducing agent
(*b*) Oxidizing agent

E.46.5   Propose an explanation for what happened when $NaHCO_3$ was added to the acidic reaction products of sodium arsenate and sodium iodide. Write a balanced equation for the reaction.

Experiment **E.47**

Thiosulfate

Special Items

Burette
65 mL distilled water
6 g $Na_2S_2O_3 \cdot 5H_2O$
1 mL 3 M HCl
70 mL standardized $Na_2S_2O_3$
  (ca. 0.1 M)
0.4 g iodine
25 mL 0.2 M KI
5 mL 1 M acetic acid
1 g each of NaCl, NaBr, NaI
6 mL 1% $AgNO_3$
15 mL 3 M NH_3

When powdered sulfur is boiled with a solution of sodium sulfite (Na_2SO_3), the sulfur reacts with SO_3^{2-} to form thiosulfate ion ($S_2O_3^{2-}$). The structure of $S_2O_3^{2-}$ is similar to that of SO_4^{2-} except that one of the oxygen atoms has been replaced by a sulfur atom. In this experiment, you will study some of the properties of thiosulfate ion. You will first find out what happens to it in acid solution. Then, by titrating against iodine, you will determine the change in oxidation state that $S_2O_3^{2-}$ undergoes when it reacts with I_2 to form I^- in acid solution. Finally, you will compare the complexing ability of $S_2O_3^{2-}$ with that of NH_3 for Ag^+ ion.

Procedure

(a) Dissolve about 1 g of sodium thiosulfate in 25 mL of water. Split the solution into equal portions in three test tubes. In test tube 1, add 1 drop of dilute HCl; in 2, 2 drops; in 3, 4 drops. Stir, and note the time required for reaction to occur.

■ WEAR YOUR
SAFETY
GOGGLES

(b) Set up your burette, and after rinsing it twice with 5-mL portions, fill with the standard solution of sodium thiosulfate provided. Note the concentration on the bottle.

Weigh your clean, dry 125-mL flask on an analytical balance. Carefully add about 0.3 to 0.4 g of iodine, and reweigh.

■ CAUTION

Iodine is poisonous and corrosive to the skin.

Add 25 mL of 0.2 M KI solution, and stir to dissolve. Add 5 mL of 1 M acetic acid. Titrate until the solution is colorless. Use a white sheet of paper under the flask.

(c) Place small amounts of solid NaCl, NaBr, and NaI in three clean test tubes. Add a few milliliters of distilled water to dissolve each solid. Add 2 mL of 1% silver nitrate solution to precipitate the silver halides. Attempt to dissolve each precipitate with dilute aqueous ammonia. Make up a solution of sodium thiosulfate by dissolving 5 g of solid in 15 mL of distilled water. Add a small portion of this solution to each of the silver halides that did not dissolve in the ammonia.

Data

Results

(a)   What do you conclude about the rate of the decomposition of $S_2O_3^{2-}$ by acid to S and H_2SO_3 and about its dependence on H_3O^+ concentration?

(b)   What is the oxidation state of the sulfur in the product formed by oxidation of $S_2O_3^{2-}$ by I_2? If the product is a dinegative ion containing only four sulfur atoms and some oxygen atoms, what is its formula?

(c)   How does the complexing ability of $S_2O_3^{2-}$ for Ag^+ compare with that of ammonia?

Questions

E.47.1   Sulfate ion, SO_4^{2-}, is tetrahedral, with the sulfur atom at the center. Thiosulfate ion, $S_2O_3^{2-}$, is similar, except for the fact that one of the oxygens is replaced by a sulfur atom. In peroxydisulfate ion, $O_3S—O—O—SO_3^{2-}$, the geometry of the bonds around each sulfur is tetrahedral, and there is a peroxide structure ($—O—O—$) between the two sulfurs. Tetrathionate ion ($S_4O_6^{2-}$) is similar to peroxydisulfate, except that the peroxide group is replaced by a disulfide ($—S—S—$) group. Draw electron dot structures for each of these ions, and determine the oxidation state of sulfur in each.

E.47.2   The solubility of iodine is much higher in aqueous potassium iodide than it is in pure water.

(*a*) Explain this enhanced solubility, given the reaction
$$I_2(aq) + I^-(aq) \rightleftharpoons I_3^-(aq).$$

(b) Write a balanced equation for the oxidation of $S_2O_3^{2-}$ by I_2 in acidic solution.

(c) Write a balanced equation for the oxidation of $S_2O_3^{2-}$ by I_3^- in acidic solution.

E.47.3 Given the following reduction potentials, show that $S_2O_3^{2-}$ should disproportionate to $S(s)$ and SO_2 in acidic solution:

$$6H^+ + S_2O_3^{2-} + 4e^- \rightarrow 2S(s) + 3H_2O \quad E° = +0.50 \text{ volt}$$

$$2H^+ + 2SO_2 + 4e^- \rightarrow S_2O_3^{2-} + H_2O \quad E° = +0.40 \text{ volt}$$

E.47.4 Given the relation $\Delta G° = -n\mathfrak{F}E°$, where the faraday ($\mathfrak{F}$) = 96,500 joules per volt-equivalent and n is the number of electrons transferred per mole of reaction, find $\Delta G°$ for the disproportionation reaction described in question E.47.3. Also, find K for the reaction, given that $\Delta G° = -RT \ln K$, where $R = 8.31$ joules per deg-mol. Assume $T = 298$ K.

E.47.5 Based on your results in part (c), order the following constants in decreasing order of magnitude:

K_{sp} for $\qquad$ $AgCl(s) \rightleftharpoons Ag^+ + Cl^-$

K_{sp} for $\qquad$ $AgBr(s) \rightleftharpoons Ag^+ + Br^-$

K_{sp} for $\qquad$ $AgI(s) \rightleftharpoons Ag^+ + I^-$

K for $\qquad$ $Ag^+ + 2NH_3 \rightleftharpoons Ag(NH_3)_2^+$

K for $\qquad$ $Ag^+ + 2S_2O_3^{2-} \rightleftharpoons Ag(S_2O_3)_2^{3-}$

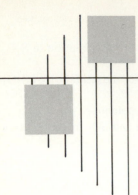

Chlorine Chemistry

Special Items

15 g manganese dioxide
350 mL distilled water
15 mL 12 M KOH
100 mL 3 M HCl
10 g NaOH

In the commercial preparation of $KClO_3$, a hot aqueous solution of KCl is electrolyzed with vigorous stirring so that the Cl_2 generated at the anode mixes with the OH^- produced at the cathode. In basic solution, Cl_2 disproportionates into Cl^- and ClO^- (hypochlorite); but in hot basic solution, ClO^- disproportionates further into Cl^- and ClO_3^- (chlorate). Since $KClO_3$ is only moderately soluble, it will precipitate if the final mixture is cooled. In this experiment, you will carry out all these reactions stepwise. Chlorine will be generated by oxidizing HCl with MnO_2. The chlorine formed will be dissolved in aqueous KOH to form hypochlorite and then heated so that ClO^- disproportionates. On cooling, crystals of $KClO_3$ will separate. You will then decompose the $KClO_3$ and test for oxygen.

Procedure

Chlorine is poisonous. Do not inhale it.

■ CAUTION

■ WEAR YOUR SAFETY GOGGLES

Set up the chlorine generator shown in Figure E.48.1. Place about 15 g of manganese dioxide in the flask. Have the thistle tube extend nearly to the bottom of the flask. The first cylinder is empty and serves as a safety trap. The second cylinder contains about 50 mL of water to trap any HCl gas carried over. The large test tube supported in a plain cylinder contains 15 mL of 12 M KOH. Note that the delivery tube extends 2 cm below the level of the KOH solution. Have the setup checked by your instructor.

Pour about 100 mL of 3 M hydrochloric acid through the thistle tube. Warm the flask gently to start the chlorine evolution. Continue the generation for about 15 min or until the solution is saturated with chlorine. Remove the burner and the receiving tube. Dissolve about 10 g of NaOH in 300 mL of H_2O, and pour this into the thistle tube to kill the generator and dissolve chlorine gas. Dismantle the apparatus, and pour

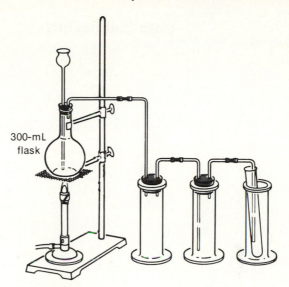

300-mL
flask

Figure E.48.1

the solution from the 300-mL flask and the cylinder into the
sink.

Carefully heat the solution in the test tube, but do not
boil.

■ CAUTION

Hot KOH is extremely corrosive. Avoid bumping.

Stir, and keep the solution hot for about 5 min. Cool under
the water tap with vigorous stirring to promote crystallization.
Filter off the crystals, and wash once with a very small portion
of cold water. Press the crystals dry of water between filter
papers.

Transfer some of the dry crystals to a small dry test tube.
Add a sprinkle of manganese dioxide (catalyst), and shake to
mix.

■ CAUTION

**Do not grind $KClO_3$ mixtures! Place any filter papers
that came into contact with $KClO_3$ in the waste container
provided.**

Heat very gently over a flame to drive off any water, and then
heat more intensely. Test the gas given off with a glowing
splint.

Weigh the rest of your product, place it in a labeled test
tube, and turn it in to your instructor.

Data and Results

Questions

E.48.1   Write balanced equations for
(a) The disproportionation of Cl_2 in basic solution
(b) The disproportionation of ClO^- in basic solution

E.48.2   List each of the safety precautions to be observed in the generation of chlorine gas, and give the reasons for each precaution.

E.48.3   How many moles of chlorine gas could be produced by oxidation of half of the HCl added to the generator, if MnO_2 is present in excess and the reduction product of MnO_2 is Mn^{2+}?

E.48.4   What is the maximum number of grams of NaOCl that could be produced by reacting 10.0 g of NaOH with excess chlorine gas?

E.48.5   Give the oxidation state of chlorine in (a) Cl_2, (b) ClO^-, and (c) ClO_3^-.

E.48.6   Write a balanced equation for the reaction that occurs when $KClO_3(s)$ is heated in the presence of $MnO_2(s)$ catalyst.

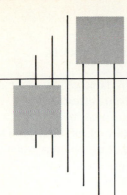

Experiment E.49

Chemistry of Iodine

Special Items

Pyrex glass wool
100 mL distilled water
25 mL 0.1 M Na_2SO_3
15 mL 0.1 M NaI
10 mL 0.1 M KIO_3
15 mL saturated aqueous iodine
Small crystal I_2
0.6 g NaI
3 mL cyclohexane
3 mL 3 M H_2SO_4
3 mL 5 M NaOH
6 mL fresh chlorine water
Ca. 4 g KIO_3
Ca. 2 g $NaHSO_3$

The principal oxidation states of iodine are -1, 0, and $+5$, as represented by the species I^-, I_2, and IO_3^-. In this experiment, you will investigate the relations between these species. In part (*a*), you will study the equilibrium of I^- and I_2 with I_3^- (triiodide ion). In part (*b*), you will investigate the behavior of the various oxidation states of iodine with the oxidizing agent chlorine and the reducing agent H_2SO_3 (sulfurous acid) in an attempt to rank the various half-reactions with respect to reduction potential. Finally, in part (*c*), you will apply some of the chemistry learned to the preparation of solid iodine from sodium iodate, a naturally occurring iodine compound.

Procedure

■ CAUTION

Iodine is poisonous and corrosive to the skin.

(a) In each of three test tubes, place 5 mL of saturated aqueous iodine. To number 1, add a small crystal of solid iodine. To 1 and 2, add approximately 0.3-g portions of solid NaI. Stopper, and shake. Compare the colors in 1 and 2 with that in 3.
Add 1 mL of cyclohexane to each of the three test tubes.

■ WEAR YOUR SAFETY GOGGLES

■ CAUTION

Cyclohexane is flammable, so keep it away from open flames.

Tap the side of the test tube briskly with the finger, and note what happens.

(b) To 10 mL of 0.1 M NaI in a small beaker, add 5 mL of 0.1 M KIO_3. Stir. Add 1 mL of dilute H_2SO_4(3 M). Note the result. Now add 3 mL of 5 M NaOH.

■ CAUTION   **Do not use NH$_3$ (as a substitute for NaOH).**

Observe what happens. (A faint yellow color may be due to IO$^-$, which is slow to disproportionate.)

To 5 mL of 0.1 M NaI in a small beaker, add 1 mL of chlorine water. Stir, and then add 1 mL of dilute H$_2$SO$_4$. Note result, and then add 5 mL more of the chlorine water.

Prepare a solution of sulfurous acid (H$_2$SO$_3$) by adding 1 mL of dilute H$_2$SO$_4$ to 25 mL of 0.1 M Na$_2$SO$_3$. Add this solution gradually with stirring to 5 mL of 0.1 M KIO$_3$ in a small beaker. Note all changes.

(c)   Prepare 50 mL of 0.4 M KIO$_3$ and 20 mL of approximately 1 M NaHSO$_3$. [If the solid furnished is Na$_2$S$_2$O$_5$, note that Na$_2$S$_2$O$_5$(s) + H$_2$O → 2NaHSO$_3$(s).] Fill your burette with the 0.4 M KIO$_3$, and place the 20 mL of 1 M NaHSO$_3$ in a small flask together with a drop of dilute sulfuric acid. Titrate *slowly* until a permanent yellow color appears. (The yellow color appears only after all the HSO$_3^-$ has been oxidized by the reaction 3HSO$_3^-$ + IO$_3^-$ → I$^-$ + 3HSO$_4^-$ and an additional drop of iodate solution reacts with I$^-$ to form I$_2$. You can enhance the sensitivity of the titration by adding some soluble starch to help detect I$_2$ by formation of the deep-blue I$_2$-starch complex.) Calculate the amount of IO$_3^-$ required to convert all the I$^-$ to I$_2$, and add this amount. Stopper the flask, and cool under running water.

Prepare a filter by placing a small plug of glass wool in your funnel. Filter off the solid iodine, and wash with a few milliliters of water. Then transfer it and the glass wool to a small beaker. Set your evaporating dish full of cold water (but dry on the outside) on top of the beaker. Warm the iodine gently until a deposit collects on the underside of the dish. Collect the product, weigh it, and turn it in to your instructor.

Data and Results

Questions

E.49.1   Show how the Le Châtelier principle applied to the equilibrium $I^- + I_2 \rightleftharpoons I_3^-$ accounts for the observations in the three test tubes in part (a).

E.49.2   Write a balanced equation for the reaction of I^- with IO_3^- in acidic solution to form I_2. Show what happens to this reaction when H^+ is destroyed by addition of OH^-.

E.49.3   Write balanced net equations for all the reactions that occur in part (b).

E.49.4   Show that the oxidation state of sulfur in $Na_2S_2O_5$ is the same as that in $NaHSO_3$.

E.49.5   How many milliliters of 0.10 M KIO_3 would you have to add to 1.00 g of $NaHSO_3$ to produce a yellow color due to I_2?

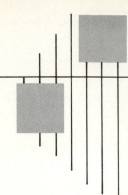

Experiment E.50

Reduction of Permanganate

Special Items

Burette
80 mL distilled water
80 mL standardized KMnO₄
  (ca. 0.01 M)
100 mL standardized NaHSO₃
  (ca. 0.01 M)
20 mL 3 M H₂SO₄
5 mL 12 M NaOH

In many cases the reduction of an oxidizing agent produces different products, depending on the pH of the solution in which the reaction is carried out. In this experiment, you will find out (by titration) what happens to permanganate ion when it is reduced in acidic, basic, and neutral solutions. When permanganate ion acts as an oxidizing agent, it picks up electrons. Because the product formed in acidic, basic, and neutral solutions may be different, the number of electrons picked up per MnO_4^- may differ. From the amount of reducing agent used and the number of moles of MnO_4^- reduced, you can decide the number of electrons each MnO_4^- ion picked up. Since manganese in MnO_4^- has an oxidation number of $7+$, you will be able to determine the oxidation number of Mn in the product in each case.

The reducing agent you will start with is sodium bisulfite ($NaHSO_3$). In each case, acidic, basic, and neutral, the sulfur changes its oxidation number from $4+$ to $6+$. However, the sulfur-containing species is different for each case. In acidic solution, H_2SO_3 (produced by the reaction of the acid with $NaHSO_3$) is oxidized to HSO_4^-. In basic solution, SO_3^{2-} is oxidized to SO_4^{2-}. In neutral solution, HSO_3^- is oxidized to SO_4^{2-}.

Procedure

(a) **Acidic** Set up your burette (see Part A, Figure A.8 or A.9). After rinsing once with distilled water and twice with 5-mL portions of the solution, fill the burette with the standard solution of KMnO₄ from the reagent shelf. (Record the KMnO₄ molarity shown on the bottle.) With your graduated cylinder, measure out into a clean beaker 25 mL of the standard NaHSO₃ solution (record its molarity). Add about 5 mL of dilute sulfuric acid. Titrate, with stirring, to the appearance of a pink color that persists for at least 30 sec. The pink color,

■ WEAR YOUR SAFETY GOGGLES

due to excess MnO_4^-, can be seen better if you have a white sheet of paper under the beaker.

(b)   **Neutral**   Refill the burette, and repeat the titration exactly as in (*a*) except with omission of the acid. During the titration, a brown precipitate will be formed that may first appear yellow in color. This is not the end point. Add $KMnO_4$ until a permanent pink color persists. To facilitate seeing the pink color in the presence of the brown precipitate, use a capillary rod to periodically withdraw small portions of the solution. Look at the solution in the capillary. You will find it relatively free of precipitate. Blow the solution out of the capillary back into the beaker before taking the next sample.

(c)   **Basic**   Allow 10.0 mL of the $KMnO_4$ solution remaining in the burette to run into a clean beaker. Empty the burette, and rinse it thoroughly with distilled water. Rinse it twice, and then fill it with the standard $NaHSO_3$ solution. Add to the $KMnO_4$ solution in the beaker 5 mL of 12 *M* NaOH and 15 mL of distilled water. Titrate until the solution has turned a clear green with no trace of violet. (Just to see what happens, when the titration is finished, pour some dilute sulfuric acid into the beaker.)

Data

Record the concentrations of solutions used, the volumes measured with the graduated cylinder, and the initial and final burette readings.

Results

Calculate from the volume and concentration of $NaHSO_3$ the moles of electrons furnished by the reducing agent in each case. Then, knowing that this number of electrons is picked up by MnO_4^-, calculate the moles of electrons picked up per mole of MnO_4^- in each case. Assign an oxidation number to the manganese in each of the products.

Questions

E.50.1 Why is it that you need progressively more and more of the standard $KMnO_4$ solution for titration as you go from acidic to neutral to basic conditions?

E.50.2 Give one reason why titration of $KMnO_4$ in acidic solution gives more precise results than does titration of $KMnO_4$ in basic solution.

E.50.3 Permanganate reacts rapidly with sulfite or bisulfite, but it does not react so fast with all reducing agents. How could a slow reaction lead to erroneous results in titrating with permanganate?

E.50.4 Suppose that as you are titrating the $KMnO_4$ in acid, some of the H_2SO_3 breaks down into $H_2O + SO_2$ and the SO_2 leaves the solution as a gas. How will this affect your determination of the number of moles of electrons picked up per mole of MnO_4^-?

E.50.5 There is a general rule about oxyanions that says that high oxidation states are stabilized by basic solution and low oxidation states are stabilized by acidic solution. Apply the rule to your observations in this experiment.

E.50.6 It is found in an experiment that in an acidic solution, 20.0 mL of 0.300 M H_2SO_3 solution reacts completely with 10.0 mL of 0.200 M $Cr_2O_7^{2-}$ solution. If the H_2SO_3 goes to HSO_4^-, what is the final oxidation state of the chromium atom?

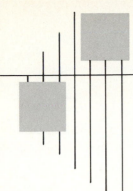

Synthesis of Coordination Isomers

Special Items

Filtering flask
Büchner funnel, No. 2
Aspirator or vacuum pump
1 kg ice
50 g NaCl
36 g KCl
25 g $Na_3Co(NO_2)_6$
11.5 g $Co(NO_3)_2 \cdot 6H_2O$
6 g $NaNO_2$
50 mL ethyl alcohol
55 mL 10% ethylenediamine solution
20 mL ethylenediamine–nitric acid solution (6.85 g 75% ethylenediamine + 10 mL H_2O + 3 mL conc. HNO_3)

As indicated in Experiment E.14, isomers having different properties result when it is possible to have two or more spatial arrangements of the same atoms. Of particular interest for the chemistry of the transition elements is the fact that many of the complex compounds formed by those elements can exist in isomeric forms. In this experiment you will attempt to synthesize and distinguish two geometrical isomers of cobalt, *cis*- and *trans*-dinitrobisethylenediamine cobalt(III). In its trivalent state, cobalt forms a host of such octahedral complexes. In this one, two of the octahedral corners are each occupied by a nitro (NO_2) group, and pairs of the other octahedral corners are occupied by the bidentate ligand ethylenediamine ($H_2NCH_2CH_2NH_2$). Ethylenediamine, usually abbreviated as *en* and frequently shown in structures as a simple arc, has two points of attachment to cobalt, since each N of *en* has an unshared electron pair that it can donate to one of the vacant cobalt orbitals. The two possible isomers are shown in Figure E.51.1. The formula for both is $Co(NO_2)_2(NH_2CH_2CH_2NH_2)_2{}^+$, which can also be written $Co(NO_2)_2en_2{}^+$.

One of the isomers can be made by the reaction

$$Co(NO_2)_6{}^{3-} + 2en \rightarrow Co(NO_2)_2en_2{}^+ + 4NO_2{}^- \qquad \textbf{(I)}$$

and the other by reaction

$$4Co^{2+} + 8NO_2{}^- + O_2 + 8en + 4H_3O^+ \rightarrow$$
$$4CO(NO_2)_2en_2{}^+ + 6H_2O \qquad \textbf{(II)}$$

In the first, ethylenediamine replaces nitro groups in hexanitritocobaltate; in the second, cobaltous ion is oxidized by O_2 in the presence of the ligands. Since in both cases we will

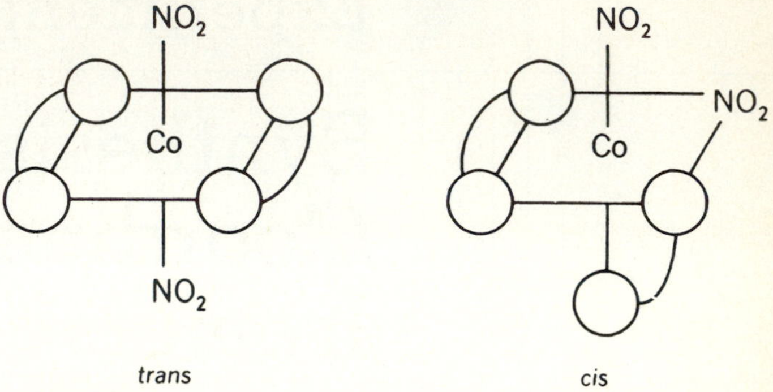

trans   cis

Figure E.51.1

want to end up with the nitrate salts, the full stoichiometry can be shown as follows:

Compound I $K_3Co(NO_2)_6 + 2en \rightarrow [Co(NO_2)_2en_2]NO_2 + 3KNO_2$

$2[Co(NO_2)_2en_2]NO_2 + 2HNO_3 \rightarrow$

$$2[Co(NO_2)_2en_2]NO_3 + NO_2 + NO + H_2O$$

Compound II $4[Co(NO_3)_2 \cdot 6H_2O] + 8NaNO_2 + 4en + 4(en \cdot HNO_3)$

$$+ O_2 \rightarrow 4[Co(NO_2)_2en_2]NO_3 + 8NaNO_3 + 26H_2O$$

The only additional minor complication is that $K_3Co(NO_2)_6$ is not available as a starting material, so you will have to make it from commercially available $Na_3Co(NO_2)_6$.

At the end of the synthesis you can discriminate your products as *cis* or *trans* by using the fact that the *cis* material precipitates with potassium chromate, ammonium oxalate, or sodium thiosulfate, whereas the *trans* does not.

Procedure

■ WEAR YOUR SAFETY GOGGLES

Get a filtering flask and a Büchner funnel from the stockroom. A filtering flask is a kind of heavy-walled Erlenmeyer flask with a sidearm at the neck that can be connected to a vacuum hose. A pinch clamp on the hose will enable you to regulate the vacuum in the flask. A Büchner funnel is a special type of porcelain funnel with a flat bottom full of tiny holes on which the filter paper can be laid flat.

Weigh out 25 g of $Na_3Co(NO_2)_6$, and transfer the material to a 600-mL beaker. Weigh out 36 g of KCl, and add it to the beaker. Assemble your filtration setup by putting the Büchner stem through the rubber adapter and inserting it in the mouth of the filtering flask. Hook the sidearm to the vacuum (pump or aspirator). Lay a filter paper of a diameter equal to the inside of the Büchner flat on the bottom so as to cover the holes. Squirt some distilled water on the paper and suck it through so the paper makes a complete seal. After the setup is assembled, add 100 mL of distilled water to the 600-mL beaker. Stir the mixture with a spatula so as to form a smooth slurry. After 2 or 3 min have elapsed, filter the slurry through the funnel. (*Note:* The solution must be filtered within 3 min of adding the water.) Let the suction suck the solid dry for a few minutes, and then pour 25 mL of ethyl alcohol into the funnel so as to cover the solid. Suck dry for an additional 2 min. Clamp off the vacuum hose, detach it from the flask, and remove the funnel. Scrape off the bright-yellow solid with your spatula onto a large piece of filter paper. You can store it in your desk until the next period, or proceed with the synthesis.

(a)   **Compound I**   Place the $K_3Co(NO_2)_6$ previously prepared in a 125-mL Erlenmeyer flask. Working in the hood, add 55 mL of ethylenediamine solution to the flask and stir.

■ CAUTION

Care must be taken in handling ethylenediamine because of its caustic nature and the irritating properties of its vapor. Work in the hood.

Clamp the Erlenmeyer for steaming, as shown in Figure E.51.2, over 300 mL of water in a 600-mL beaker. Heat the beaker of water with a moderate flame so the temperature of the solution in the Erlenmeyer rises slowly to 75°C. This should take about 15 min. Stir the slurry of $K_3Co(NO_2)_6$ in ethylenediamine constantly with a glass rod to prevent any solid from forming a layer on the bottom of the flask. Take the temperature of the reaction mixture every few minutes until the temperature reaches 70°C. A reaction soon occurs as is shown by dissolving the yellow $K_3Co(NO_2)_6$ to form a dark-brown solution. Make sure that the temperature remains at 75°C or below. (Be ready to remove the Erlenmeyer from the steam bath or turn off the burner to make sure that the temperature stays below 75°C. Above 75°C, your isomer will be destroyed.)

Filter the dark-brown solution while hot through a Büchner funnel into a clean filtering flask. Pour the filtrate into a 100-mL beaker, and set it in a salt-ice mix in a 600-mL beaker (one-third full of ice plus 50 mL of salt). When the solution in the beaker reaches about −10°C, a brown precip-

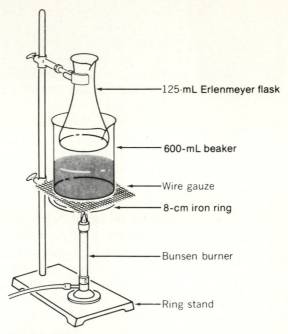

125-mL Erlenmeyer flask

600-mL beaker

Wire gauze

8-cm iron ring

Bunsen burner

Ring stand

Figure E.51.2

itate should appear. If it does not, vigorously scratch the inside wall of the beaker with a glass rod. After precipitation is complete, filter the solid and transfer to a 50-mL beaker. Dissolve in a *minimum amount* of water at 50°C, and then add from a dropper 1 mL of nitric acid. A vigorous reaction will occur, and a dark-yellow solid should precipitate. Filter and dry by suction filtration. This is your isomer I.

(b)   **Compound II**   Dissolve 11.5 g of $Co(NO_3)_2 \cdot 6H_2O$ and 6.0 g of $NaNO_2$ in 20 mL of water in a 125-mL Erlenmeyer flask. Swirl the flask vigorously to ensure that all the solid material has dissolved. Working in the hood, add to the mixture 20 mL of the ethylenediamine–nitric acid solution. Stir or shake rapidly until a yellow precipitate begins to appear. Cool in an ice bath (600-mL beaker, one-third full of ice, 50 mL of salt) for 20 min and then suction-filter through a Büchner. Wash the yellow crystalline material on the filter funnel with 25 mL of ethanol, and suction air dry for 3 min. Recrystallize the small yellow golden plates by dissolving in as little boiling water as possible and cooling. Dry the crystals further, if necessary, by spreading them out on a large piece of filter paper. This is your isomer II.

(c) **Isomer Test**   Add a small amount (just enough to cover the bottom of the tube) of each solid to a 150-mm test tube. Dissolve in a minimum amount of water (about 10 mL), and add a concentrated solution of K_2CrO_4, $(NH_4)_2C_2O_4$, or $Na_2S_2O_3$ to each test tube. Look for the precipitate that indicates the *cis* complex.

Questions

E.51.1   Prepare flowcharts for each preparation, showing in schematic form what reagents are added, what species are formed, and what reaction conditions (e.g., temperature) are involved at each stage.

E.51.2   Make a sketch of a vacuum filtration setup, and explain how to carry out such a separation properly.

E.51.3   In using the Büchner funnel for filtration, it is good practice to decant most of the supernatant liquid through the funnel first, before getting much solid on the filter paper. Suggest a reason for this procedure.

E.51.4   Temperature control is very critical in the preparation of isomer I. Suggest a reason for this.

E.51.5   Why do you use ethyl alcohol instead of water to wash the $K_3Co(NO_2)_6$ precipitate? Suggest two reasons.

E.51.6   Calculate the theoretical yield of $K_3Co(NO_2)_6$ you could prepare from the starting materials given.

E.51.7   How many grams of compound II could you make from 11.5 g of $Co(NO_2)_6 \cdot 6H_2O$?

E.51.8   Explain why the *cis* isomer of $[Co(NO_2)_2en_2]^+$ reacts with K_2CrO_4, $(NH_4)_2C_2O_4$, or $Na_2S_2O_3$ but the *trans* isomers does not.

E.51.9   What are the symmetry elements of each of the isomers shown in Figure E.51.1? Make models if necessary.

E.51.10   Both *cis* and *trans* isomers turn out to be diamagnetic (i.e., no unpaired electrons). Describe the electronic distribution in the central cobalt atom using the ideas of crystal-field theory.

E.51.11   How would you expect the dipole moment of the *cis* and *trans* isomers to differ?

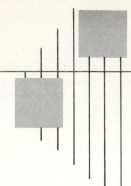

Cation Analysis Using Paper Chromatography

Special Items

Whatman No. 1 chromatography paper
Scissors
Ruler
Capillary tubes (1.6 to 1.8 × 100 mm)
800-mL beaker with watch glass to cover
20-gauge copper wire
Desiccator with shelf
Petri dishes
Solutions (1 g/5 mL H_2O) of $MnCl_2$, $FeCl_3$, $CoCl_2$, $NiCl_2$, $CuSO_4 \cdot 5H_2O$
Unknown mixtures of metal ions
Concentrated $NH_3(aq)$
0.25 M NaOH
1% dimethylglyoxime in isopropanol
Developing solution (19 parts acetone: 4 parts conc. HCl: 2 parts H_2O by volume)

Chromatography is a term applied to several separation techniques based on differential migration. In this experiment, one of these techniques, paper chromatography, will be used to separate a mixture of transition metal ions so that they may be identified. It will become apparent that the identification of individual ions is much easier after they have been separated.

In any differential migration technique, three things are required: First, there must be a migration medium, a place for the separation to occur. Second, there must be a driving force to move the species to be separated along the migration medium. Third, there must be a selective resistive force. It is this last force, the selective resistive force, that causes the separation of the chemical species under consideration.

If a mixture of chemical species is applied to a point in a migration medium, the driving force will tend to make the species move away from the starting point. The selective resistive force, however, holds the species back, so that they do not move along as quickly as they would if there were no resistive force. The resistive force is selective in that it retards the movement of each of the species by a different amount, so that each moves away from the starting point, or origin, at a different rate. This difference in rate of movement away from the origin is the basis of the term *differential migration*.

In this experiment, the differential migration method is paper chromatography. The migration medium is a strip of chromatography paper. The driving force is a flow of liquid solvent, also called the *mobile phase*, along the paper strip. The selective resistive force is not so easily identified. In fact it is likely that there is more than one type of selective resistive force in paper chromatography. A major one may be identified when it is realized that paper contains a great deal of water tightly bound to it. One may imagine little immobile "pools" of water. As the chemical species being separated move along the mobile phase, they encounter these regions of tightly bound water. They may even move into this immobile, or stationary, phase. Since the fraction of time each individual species spends in the stationary phase depends on its relative solubility in both the stationary phase and the mobile phase, and these solubilities depend on the identity of the chemical species, this describes a selective-resistive action. When a species is in the mobile phase, it moves with the velocity of the mobile phase. When the species is in the stationary phase, it does not move. The different chemical species in a given sample are therefore separated as they move at different rates away from the origin.

It is apparent that for a given pair of mobile and stationary phases, the fraction of the distance the chemical species moves with respect to the distance the mobile phase moves will be a constant. This fraction, called the R_f value of the species, may then be used to compare two chromatograms, even though the mobile phase may have moved a different distance in the two.

Although paper chromatography is a powerful tool for the separation of chemical species, the species are often not visible on the chromatogram. For this reason it is often necessary to treat the chromatogram with reagents that will form colored compounds on reaction with the species.

In this experiment Mn^{2+}, Fe^{3+}, Co^{2+}, Ni^{2+}, and Cu^{2+} will be separated and identified. First, chromatograms of known solutions will be made. These chromatograms will be treated with reagents (1) to make visible those ions that do not leave a visible spot on the chromatogram and (2) to identify positively each of the spots as one of the five possible ions. In addition, the R_f value for each ion will be measured. Then, a chromatogram of an unknown will be made and subjected to the same chemical tests. By comparison of the colors and R_f values of the spots on the chromatogram of the unknown solution with the colors and R_f values of the spots on the chromatograms of the known solutions, the composition of the unknown will be found.

Equipment you will need for this experiment includes 1-cm wide Whatman No. 1 chromatography paper, scissors, a small ruler, capillary tubes (1.6 to 1.8 × 100 mm), an 800-mL beaker with watch glass to cover, bendable wire (such as 20-gauge copper), a desiccator with shelf, and petri dishes. You will also need metal ion solutions containing 1 g per 5 mL of Mn^{2+}, Fe^{3+}, Co^{2+}, Ni^{2+} chlorides, $CuSO_4 \cdot 5H_2O$, and metal-ion solution mixtures (composed of equal volumes of solutions) of each of the following pairs of ions: (1) Mn^{2+}, Fe^{3+}; (2) Fe^{2+}, Co^{2+}; (3) Ni^{2+}, Mn^{2+}; (4) Ni^{2+}, Cu^{2+}; (5) Co^{2+}, Cu^{2+}. Finally, you will also need concentrated $NH_3(aq)$, 0.25 M NaOH, 1% dimethyglyoxime in isopropanol, and a *developing solution* composed of 19 parts acetone, 4 parts concentrated $HCl(aq)$, and 2 parts H_2O by volume.

Procedure

■ **WEAR YOUR SAFETY GOGGLES**

Your teaching assistant will divide the class into five groups. Using the procedure below, each group will prepare a chromatogram of a mixture of two of the possible ions and chromatograms of each of the two ions separately. Then each group will prepare chromatograms of each person's unknown. Using the information collected by all the groups on the chromatograms of known solutions, each person will identify his or her unknown.

Obtain an 800-mL beaker and a watch glass to cover it. Bend a piece of wire about 20-cm long, and hook it across the inside of the beaker so that strips of chromatography paper may be hung from it. Cut a piece of chromatography paper about 12 cm long, and bend it on the top end so that the bottom end just clears the bottom of the beaker when the paper is hung from the wire. Remove the paper strip from the beaker, and pour a few milliliters of the solvent (acetone/concentrated HCl/H_2O, 19:4:2 by volume) into the beaker. There should be enough solvent so that the paper strip just touches it when hung from the wire. Cover the beaker with the watch glass, and set it aside.

Using the measured strip prepared above, cut and fold additional strips so that there are enough strips for the three known solutions and all the unknown solutions in your group. With a ruler, measure up 1 cm from the bottom of each strip, and mark this distance with a pencil line. This line will mark the origin. With a capillary tube, apply enough solution to each strip to make a spot about 0.5 cm in diameter. The spot should be in the center of the strip on the pencil line. If the spot is misplaced or too large, the strip should be discarded. Allow the spots to dry thoroughly; then carefully hang the strips in the beaker. You should be able to get four strips in the beaker at once. The spots should be entirely above the surface of the solvent. If they are not, the strips should be discarded and new ones made.

Allow the chromatograms to develop for at least $\frac{1}{2}$ h, then remove them and re-cover the beaker. Quickly mark the solvent front with pencil before the solvent dries; then allow the chromatograms to dry thoroughly. Note the colors and determine the R_f values for any spots that are visible.

Pour a few milliliters of concentrated NH_3 into the bottom of a desiccator. The desiccator should have a ceramic shelf above the $NH_3(aq)$ so that the paper strips may be placed in the desiccator without getting wet. Expose each chromatogram to $NH_3(g)$ for several minutes. Note the appearance, disappearance, or change of color, and determine pertinent R_f values.

In a petri dish place a few milliliters of a solution of dimethylglyoxime and cover it with a second petri dish. Quickly dip each chromatogram in the solution, then allow it to dry. Do not soak the strips in the solution, because the spots may smear. Again make observations.

Last, place several milliliters of a solution of NaOH in a petri dish, and dip each chromatogram in it. When they are dry, note any changes.

By comparison of the R_f values and color changes, identify the ions in the unknown solution.

Questions

E.52.1   Briefly define:
(a) Differential migration
(b) R_f value
(c) Stationary phase
(d) Solvent front

E.52.2   Suppose that you discovered that two different ions gave the same R_f value? What could you try in order to separate them?

E.52.3   Obtain from your classmates about 10 R_f values for Ni^{2+}. Calculate an average value and an average deviation.

E.52.4   In hanging the strips in the developing chamber (beaker), why is it important not to allow the spots to become immersed in the solvent?

E.52.5   What is the purpose of exposing the developed chromatograms to NH_3, dimethylglyoxime, and NaOH?

E.52.6   Why is the length of time allowed for the chromatograms to develop not critical?

E.52.7   Suggest chemical formulas for the various colored species observed throughout this experiment.

E.52.8   How might you explain the fact that the R_f value obtained for a single ion by itself may differ slightly from the R_f value observed for that same ion when it is in a mixture of ions?

E.52.9   Why do you use a pencil line and not an ink line to mark the origin?

E.52.10   Why cover the beaker with a watch glass while the chromatograms develop?

E.52.11   How might you use this technique to make a *quantitative* separation of the ions in a mixture?

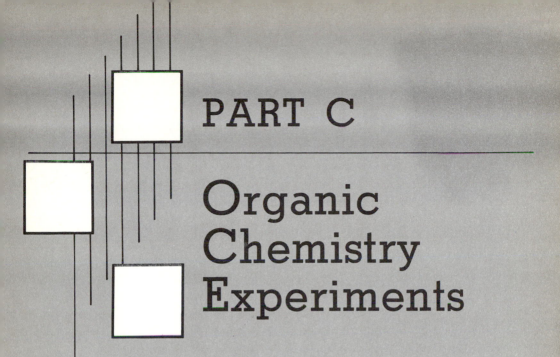

PART C

Organic Chemistry Experiments

The term "organic chemistry" once applied only to compounds composing or derived from living systems. Modern organic chemistry, however, is the study of virtually all of the compounds containing carbon. Carbon can bond to itself to form long chains of carbon-carbon bonds in molecules; it also can form not only single, but also double and triple, bonds to itself as well as other elements. Because of these properties, there is an infinite variety of carbon compounds. In these experiments, you will learn something about most of the more important classes of organic compounds through both synthesis and analysis. In doing so, you will also have a chance to develop some of the technical skills used in the organic chemistry laboratory.

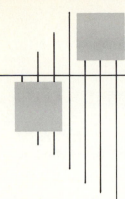

Experiment O.1

Molecular Models I: Structural Isomers and Geometric Isomers

Special Item

Sargent-Welch molecular model kit

Isomers are molecules having the same molecular formula but differing from one another by virtue of either the order in which the atoms are connected to each other or the arrangement of the atoms in space. Usually the term *structural isomers* is applied to molecules differing only in their connectivity, while the term *stereoisomers* is used to describe molecules in which the atoms have the same connectivity but different spatial arrangements. In this exercise you will investigate a number of types of both structural isomers and stereoisomers, and you will get some practice in naming some of them.

Perspective Formulas

Most organic compounds are three-dimensional, due to the tetrahedral nature of carbon. *Perspective formulas* communicate this by representing bonds in the plane with line segments; bonds coming up, out of the plane, with wedges; and bonds going down, below the plane, with dashed lines. Figure O.1.1 shows a perspective formula for methane, together with a sketch of a model of a methane molecule in the same orientation as that represented by the perspective formula.

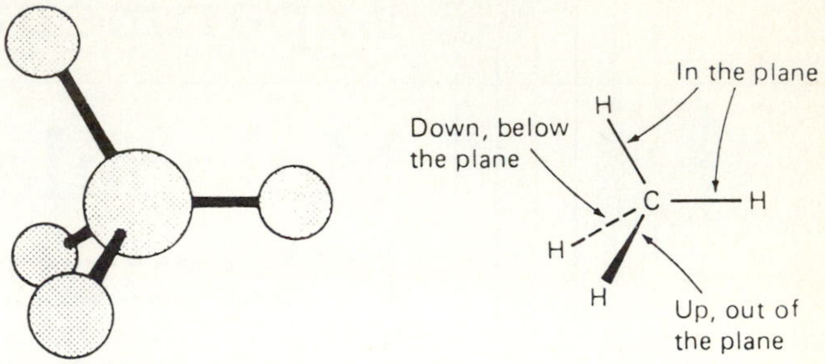

Figure O.1.1

Conformational Stereoisomers and Configurational Stereoisomers

Because there is virtually free rotation about most single carbon-carbon bonds, a single structural isomer of an organic molecule might assume an infinite variety of different conformations, each of which is, technically, a different stereoisomer. Such conformations are definitely *not* different structural isomers. If two conformations can be interconverted by a simple rotation requiring little or no expenditure of energy, the stereoisomers are known as *conformational isomers, conformations, or conformers.* More often than not, conformers are *not* considered full-fledged isomers; check with your instructor.

If the rotation about a carbon-carbon bond is hindered for some reason, such as by the presence of a double bond, there may be configurations that are *not* easy to interconvert; in this case, the different stereoisomers are called *configurational* isomers. Unfortunately, the dividing line between conformational isomers and configurational isomers is not always obvious. Usually, if it is possible physically to separate two such isomers, they are considered to be configurational isomers; if not, they are regarded as conformers.

Equivalent and Nonequivalent Hydrogens

Equivalent hydrogen atoms in a molecule have identical environments; conversely, nonequivalent hydrogens have different environments. For example, in the molecule propane there are two "kinds of hydrogen": hydrogens on terminal carbons (three on each carbon), and hydrogens on the central carbon (two). The six hydrogens on terminal carbons are equivalent to each other, but each is nonequivalent to either of the two hydrogens on the central carbon. Replacing any one of a group of equivalent hydrogens in a molecule to give a monosubstituted product gives only one product. For example, there is only one chloromethane, because all of the four hydrogens in methane are equivalent. There are, however, two possible chloropropanes, 1-chloropropane and 2-chloropropane, because there are two nonequivalent kinds of hydrogens.

Structural Isomerism

One of the simplest types of structural isomerism, often called *chain isomerism*, exists because carbon chains may be either branched or unbranched. For example, *n*-heptane, 3-methylhexane, 2-methylhexane, and 2,4-dimethylpentane, as well as many others, are structural (chain) isomers of C_7H_{16}. A second type of structural isomerism, usually called *positional isomerism*, results from having *substituent groups* (groups or atoms other than hydrogen) located at different positions. Examples of positional isomers include 2,3-dichloroheptane, 1,4-dichloroheptane, and 2,2-dichloro-3-methylhexane, all of which are structural isomers of $C_7H_{14}Cl_2$.

Structural Isomerism and Geometric (*cis-trans*) Stereoisomerism in Compounds with Double Bonds   The presence of a double bond in a molecule creates the possibility both of structural isomerism and of configurational stereoisomerism. If the only difference between a pair of isomers is the position of the double bond in a chain of carbon atoms, the two isomers are structural isomers (positional isomers). For example, 1-butene is a positional isomer of 2-butene.

$$CH_2{=}CH{-}CH_2{-}CH_3 \qquad CH_3{-}CH{=}CH{-}CH_3$$
1-Butene   2-Butene

The possibility of stereoisomerism about a double bond is a consequence of the highly restricted rotation characteristic of the double bond; rotation can occur only at the expense of breaking the pi bond. Groups attached to the double bond may be on the same side, *cis* to each other; or they may be on opposite sides, *trans* to each other. This type of configurational stereoisomerism is called *geometric isomerism*. To decide whether a structure is *cis* or *trans*, simply note the configuration of the longest continuous carbon chain as it passes through the double bond; if the two groups from *that* chain are *cis* to each other, the configuration of the molecule as a whole is taken as *cis*; if the two groups from the longest chain are *trans*, the molecular configuration is considered *trans*. The structural isomer 3-methyl-2-pentene, for example, has *cis* and *trans* geometric isomers.

$$
\begin{array}{cc}
\underset{H}{\overset{H_3C}{\diagdown}}C=C\underset{CH_3}{\overset{C_2H_5}{\diagup}} & \underset{H}{\overset{H_3C}{\diagdown}}C=C\underset{C_2H_5}{\overset{CH_3}{\diagup}} \\
cis\text{-3-Methyl-2-pentene} & trans\text{-3-Methyl-2-pentene}
\end{array}
$$

Procedure

(a) Perspective Formulas; Conformers and Structural Isomers

Ethane   Construct a model representing ethane, $H_3C—CH_3$. Use black spheres for carbon atoms and yellow spheres for hydrogen atoms. The conformation of ethane having the lowest potential energy is called the "staggered" conformation. To get your model into this conformation, twist it about the C—C bond until one hydrogen on one carbon points "up" and one hydrogen on the other carbon points "down" as much as possible. Set the model on the desk top, and draw a perspective formula representing staggered ethane. The conformation of ethane with the highest potential energy is called the "eclipsed" conformation. To get your model into this conformation, twist it about the C—C bond until one hydrogen attached to each carbon points "up." Now, when you set it on the desk top it should rest on four hydrogens, two from each carbon. Draw a perspective formula for eclipsed ethane. Offer an explanation for the fact that the staggered form is of lower potential energy than the eclipsed form of ethane.

Chloroethane   To make a model of chloroethane, replace one of the hydrogens of ethane with a green "chlorine atom." Use the model to help in drawing perspective formulas representing staggered and eclipsed chloroethane.

Dichloroethanes   There are two different structural isomers of $C_2H_4Cl_2$. Build models of each of them, and draw two perspective formulas for each, one representing the highest energy (chlorines eclipsed) conformation, and one the lowest potential energy conformation. Give the proper IUPAC name of each structural isomer. Identify the numbers of equivalent and non-equivalent hydrogens in each isomer.

Propane   Construct a model representing propane, $H_3C-CH_2-CH_3$. Twist the model until you achieve the lowest energy conformation. If you do this correctly, when you set the model on the desk top, two hydrogens from each terminal carbon should rest on the desk top. Pick any two adjacent carbon atoms, and note the spatial arrangement of hydrogens on the two carbons (eclipsed or staggered). Draw a perspective formula for propane in this conformation, and indicate the relative spatial arrangements of hydrogens on adjacent carbons on your drawing.

Hexanes   Construct models for all of the structural isomers having the formula C_6H_{14}. For each chain of carbon atoms exceeding three carbons, twist the model until the chain, when viewed from the top, sitting on the desk top, is straight (C—C—C—C—C, *not* C—C⟍ ⟋C, etc.). Draw a
⟍ ⟋
C—C
perspective formula (viewed from the top, looking down) for each isomer, and give each its proper IUPAC name.

(b)   Geometric Isomerism

Ethylene   Using two springs to represent a double C=C bond, connect two black spheres (carbon atoms) together, and then, using sticks and yellow spheres (hydrogen atoms), complete a model of ethylene (ethene), $H_2C=CH_2$. Recall that the double bond actually consists of a sigma bond and a pi bond. Notice that a plane going through the two springs is perpendicular to a plane including the four hydrogens. Sketch an ethylene molecule, showing the sigma and pi bonds, such that the pi bond is "up and down," that is, so the hydrogens appear to go down into and come up out of the plane of the paper. Also, sketch the molecular model from the same perspective.

Dichloropropenes   Build models of all possible structural (positional) isomers and *cis-trans* geometric isomers of $C_3H_4Cl_2$ containing one double bond. Give a perspective formula, and attempt to assign a proper IUPAC name to each isomer. (*Note:* There is a problem in naming the geometric isomers of 1,2-dichloropropene. Explain.)

Pentenes   Build models of all possible structural isomers and *cis-trans* geometric isomers of molecules containing one double bond and having the molecular formula C_5H_{10}. Give a perspective formula for each isomer, and name it according to IUPAC rules. Also, identify the number of different types of (non-equivalent) hydrogens there are in each isomer.

When you have finished using the molecular models, disassemble all models and put the parts away neatly in the box. Your instructor will check over your model kit before you leave.

Data and Results

(a)   Perspective Formulas; Conformers and Structural Isomers
Ethane

Chloroethane (staggered and eclipsed)

Dichlororethanes

Propane

Hexanes

(b)   **Geometric Isomerism**
Ethylene

Dichloropropenes

Pentenes

Questions

O.1.1   Draw three perspective formulas for each of the following molecules, each formula corresponding to a different conformation:
(a) n-Octane (C_8H_{18})
(b) 3-Ethylhexane

O.1.2   For ethane, there are only two important conformers, the staggered and the eclipsed. For 1,2-dichloroethane, there are several additional important conformers. Explain by giving perspective sketches.

O.1.3   Write out structural formulas for 2,3-dichloroheptane, 1,4-dichloroheptane, and 2,2-dichloro-3-methylhexane.

O.1.4   Give perspective sketches for cis- and trans-2-butene and for cis- and trans-4-ethyl-3-methyl-3-heptene.

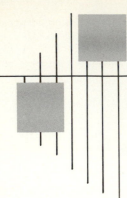

Experiment O.2

Molecular Models II: Optical Isomers

Special Items

Sargent-Welch molecular
 model kit
Prentice-Hall framework
 molecular model kit

Stereoisomers have the same connectivity between atoms but differ in their spatial arrangements. Simple rotations about carbon-carbon single bonds generate different *conformational stereoisomers* (conformers). Conformers are not ordinarily regarded as being different isomers, because they usually interconvert so freely that they cannot be physically separated. If different configurations of the same structural isomer do *not* interconvert easily, they *are* considered to be full-fledged isomers and are called *configurational stereoisomers*. Unless stated to the contrary, the term "stereoisomers" usually is taken to mean the same thing as "configurational stereoisomers." Stereoisomers may be split into two groups, optical isomers and geometrical (*cis-trans*) isomers. Optical isomers may be further divided into *enantiomers* and *diastereomers*. In this exercise you will investigate optical isomerism, as well as important conformations of cyclic molecules.

Optical Isomerism

Optical isomerism in organic molecules is a consequence of the tetrahedral geometry characteristic of carbon in its compounds. If a carbon atom in a molecule has four different groups (or atoms) bonded to it, it is said to be a *chiral carbon*. A *chiral molecule* is a molecule that cannot be superimposed on its mirror image. The mirror image of a chiral molecule is called its *enantiomer*. Chiral carbons in a molecule often are identified by placing an asterisk (*) next to each of them. Chiral molecules are often most conveniently represented using *Fischer projections*. In these representations, bonds going down, into the paper, are shown in the vertical direction, while bonds coming up, out of the plane of the paper, are horizontal. Enantiomers have identical chemical and phys-

Enantiomers

$$
\begin{array}{ccc}
\overset{\displaystyle Cl}{\underset{\displaystyle I}{H \blacktriangleright \overset{*}{C} \blacktriangleleft Br}} & = & \overset{\displaystyle Cl}{\underset{\displaystyle I}{H - \overset{*}{C} - Br}}
\end{array}
\qquad
\overset{\displaystyle Cl}{\underset{\displaystyle I}{Br - \overset{*}{C} - H}}
$$

Perspective formula

Mirror

Fischer projections

ical properties, except for the direction in which they rotate the plane of plane-polarized light and for the way in which they behave chemically with other chiral molecules. A *d*-en-antiomer, also called the (+) enantiomer, rotates the plane of plane-polarized light clockwise (as viewed by the observer); an *l*-enantiomer, or (−) enantiomer, rotates the light by the same amount, except in the counterclockwise direction. The *d, l* terminology must not be confused with the D, L terminology (relative configuration), which is also used to describe many enantiomers, especially sugars and amino acids. The symbols D and L do *not* specify the rotation of plane-polarized light; rather, they describe the *configuration* of an enantiomer "relative" to D-(+)-glyceraldehyde or L-(−)-glyceralde-

$$
\begin{array}{ccc}
\overset{\displaystyle CHO}{\underset{\displaystyle CH_2OH}{H - \overset{*}{C} - OH}} & & \overset{\displaystyle CHO}{\underset{\displaystyle CH_2OH}{HO - \overset{*}{C} - H}}
\end{array}
$$

D-(+)-Glyceraldehyde   Mirror   L-(−)-Glyceraldehyde

hyde. If, in the Fischer projection of a simple sugar molecule, the −OH group on the carbon adjacent to the terminal −CH$_2$OH group is on the right, as it is in D-(+)-glyceral-dehyde, the relative configuration of the simple sugar mole-cule is given the designation "D"; if it is on the left, it is "L." It is only by coincidence that D-glyceraldehyde happens to rotate plane-polarized light to the right, and L-glyceralde-hyde, to the left. Unlike relative configuration, optical rota-tion is an empirical property; that is, it can be determined only by conducting an experiment.

For molecules having more than one chiral carbon, there

is the possibility of having not only enantiomers but also *dia-stereomers*. Two optical isomers are diastereomers if they are not mirror images (enantiomers) of one another. Diastereomers have identical chirality at one or more chiral carbon atoms, but they also have the opposite chirality at one or more chiral carbon atoms. Diastereomers differ from one another not only in their optical properties but also in terms of ordinary physical properties, such as melting point and boiling point, and in their chemical reactivity. More often than not, diastereomers are given completely different names. The terms "enantiomer" and "diastereomer" are relative terms; a particular optical isomer is an enantiomer with respect to its mirror image but at the same time is a diastereomer with respect to an optical isomer that is not its mirror image. A special type of diastereomer, called a *meso form*, arises when there is an internal mirror plane; such a molecule is *not* chiral, because it *is* superimposable on its mirror image. Take, for example, 2,3-dibromobutane (Figure O.2.1). Structures A and B are enantiomers of each other and diastereomers of structure C. Structure C is a *meso* form, because it has an internal mirror plane. Note that even though structure C contains two chiral carbon atoms, the molecule itself is *achiral*; therefore it will not rotate the plane of plane-polarized light.

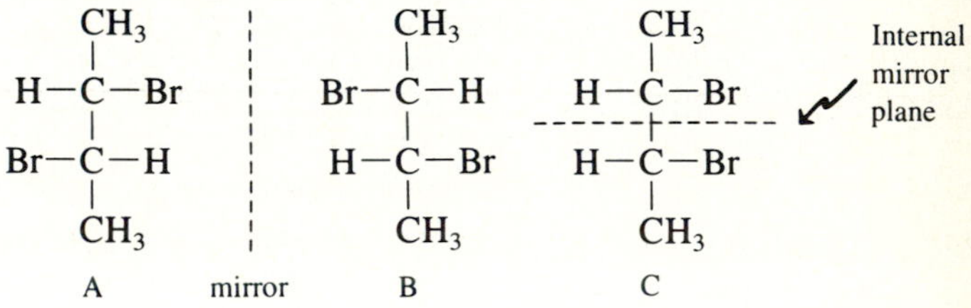

Optical isomers of 2,3-dibromobutane

Figure O.2.1

Conformers and Stereoisomers of Cyclic and Heterocyclic Molecules

There is a huge variety of *cyclic* (containing only carbon atoms in the ring) and *heterocyclic* (containing other atoms, such as oxygen or nitrogen, in addition to carbon in the ring) organic compounds. You will consider only a few examples of each in this exercise. The simplest cyclic compounds, the cycloalkanes, consist of carbon atoms (actually $-CH_2-$ units) joined together with single bonds in the form of a ring. The presence of the ring severely hinders the freedom of rotation

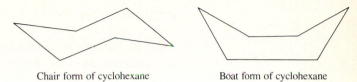

Chair form of cyclohexane   Boat form of cyclohexane

Figure O.2.2

about single carbon-carbon bonds, as compared with the free rotation characteristic of C—C bonds in acyclic (open-chain) hydrocarbons. Conformations of rings are often represented by using *sawhorse projections*. Figure O.2.2 shows sawhorse projections for the more stable "chair" conformation and the less stable "boat" conformation of cyclohexane. Note that carbon atoms and hydrogen atoms are not usually shown in sawhorse projections.

The *monosaccharides*, or simple sugars, exist in both open-chain and ring forms. The ring form is heterocyclic, the ring being made up of one oxygen atom and the remainder carbon atoms. A very convenient representation of a simple sugar in its ring form is a *Haworth projection* (Figure O.2.3). Side groups shown on the right in a Fischer projection point down in a Haworth projection; groups on the left in a Fischer projection point up in the Haworth projection. Shown in Figure O.2.3 are Fischer, Haworth, and sawhorse projection formulas for α-D-glucose, which is a simple hexose, or six-carbon monosaccharide. The carbons are numbered so as to give the carbon in the terminal —CH₂OH group the highest number, 6 in this case; so the other carbons run from 1 to 5. Notice that the oxygen on carbon 5 is bonded to carbon 1 in the heterocyclic ring structure.

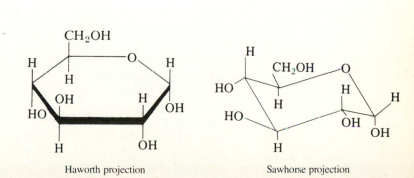

Fischer projection   Haworth projection   Sawhorse projection

Figure O.2.3

Procedure

Using the model kits, build models as directed, and use them to answer the questions that are asked.

Cyclobutanes   Using your framework model kit, assemble a model of cyclobutane, C_4H_8. Use tetrahedral (four-pronged) metal connectors for carbon atoms, black tubes for carbon-carbon single bonds, and black-and-white tubes for carbon-hydrogen single bonds. Imagine a hydrogen atom at the end of each C—H single bond. Draw a sawhorse projection formula for the molecule. Set the model on the desk top, and draw a sketch of it as seen from above, looking straight down on it. Although cyclobutane is a stable molecule, the C—C bond is easier to break than the C—C bond in *n*-butane. Suggest a reason. Are all the hydrogens in cyclobutane equivalent?

Replace two of the black-and-white C—H bonds on one of the carbon atoms with green tubes, representing C—Cl bonds, in order to make 1,1-dichlorocyclobutane. Set the model on the desk top, and use it to help you in drawing a perspective formula of the molecule, as viewed from above.

Rearrange the tubes representing 1,1-dichlorocyclobutane in order to construct a model of 1,2-dichlorocyclobutane. Notice that there are two possible configurational stereoisomers that you can make. One, the *cis* form, has both chlorines on the same side of the ring; the other, the *trans* form, has the chlorines on opposite sides of the ring. Draw sawhorse projection formulas for the two stereoisomers. Using another set of tubes and metal connectors, construct a mirror image of each (one at a time) of the two stereoisomers. Attempt to superimpose each stereoisomer on its the mirror image, and note your observations. Identify each chiral carbon atom with an asterisk (*). Following the same approach, build models and analyze the structures of *cis*- and *trans*-1,3-dichlorocyclobutane. Provide sawhorse projection formulas for the entire group of dichlorocyclobutanes, and describe the relationships between the different isomers. Which are structural isomers, which are enantiomers, and which are diastereomers? Also, identify any *meso* forms.

Cyclohexanes   Using six metal tetrahedral connectors for carbon atoms, black tubes for carbon-carbon bonds, and black-and-white tubes for carbon-hydrogen bonds, construct a framework model of cyclohexane, C_6H_{12}, in the chair conformation, and set it on the desk top. Note that each pair of adjacent carbon-carbon bonds forms a V; three V's are "right side up," and three V's are inverted. Each of the three "hydrogen atoms" on which the model is resting is at the base of one of the V's that are "right side up." Those three hydrogens are perpendicular to the average plane of the molecule and are called *axial* hydrogens. There are three other hydrogens that are perpendicular to the average molecular plane, those that point straight up; they are also axial. The remaining six hydrogens lie approximately in the average molecular plane and are called *equatorial* hydrogens. When a V is "right side up," does the equatorial hydrogen at the center of the V point slightly above, or does it point slightly below the average molecular plane?

Sketch a sawhorse projection formula for cyclohexane in the chair conformation, and label each axial and equatorial hydrogen atom. Sight down one of the six carbon-carbon bonds; are the hydrogens on adjacent carbons staggered, or are they eclipsed? Select another pair of adjacent carbons and repeat. Does it matter what pair you choose? Twist the model to get it into the slightly higher-energy boat conformation. Give a sawhorse projection formula of the boat conformer, and number the carbons 1 to 6, starting anywhere, on your drawing. For each pair of adjacent carbon atoms (e.g., C-1 and C-2), describe the orientation of the hydrogens (staggered or eclipsed). Explain why the chair conformer is of lower energy than the boat.

Return the model to the chair conformation, and replace one of the axial black-and-white C—H bonds with an orange tube, representing a C—Br bond. You have made a model of bromocyclohexane. Give a sawhorse projection formula for the molecule in this conformation. Twist the model such that you reverse the orientation of all the V's; that is, invert those that are "right side up," and vice versa. This should result in another chair conformation, different from the first. Observe and note what happens to the axial bromine atom and to all of the axial hydrogens in going from one chair to the other. Draw the sawhorse projection for this conformer. It turns out that there is very little energy required to go from one chair to the other. Explain why there is only one bromocyclohexane, not two different configurational stereoisomers. The chair conformer in which bromine is in the equatorial position is of lower potential energy than that having bromine in the axial position. Suggest a reason for this.

Glucose   Using the Sargent-Welch ball-and-stick model kit, construct a model of D-glucose in the open-chain form. Use black spheres for carbon, yellow for hydrogen, and red for

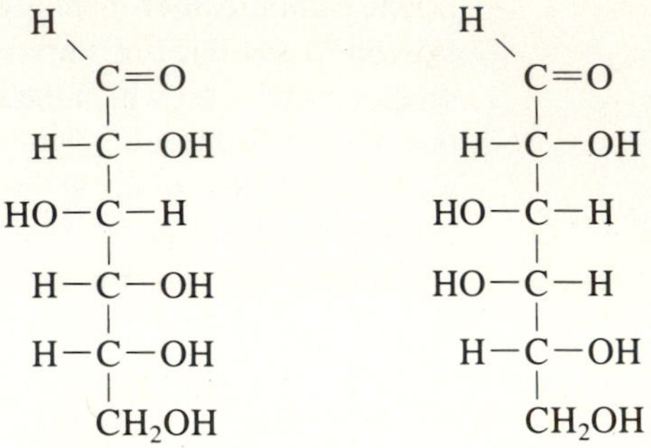

$$
\begin{array}{c}
H \\
\ \ \ \ \backslash \\
\ \ \ \ C=O \\
\ \ \ \ | \\
H-C-OH \\
| \\
HO-C-H \\
| \\
H-C-OH \\
| \\
H-C-OH \\
| \\
CH_2OH
\end{array}
\qquad
\begin{array}{c}
H \\
\ \ \ \ \backslash \\
\ \ \ \ C=O \\
\ \ \ \ | \\
H-C-OH \\
| \\
HO-C-H \\
| \\
HO-C-H \\
| \\
H-C-OH \\
| \\
CH_2OH
\end{array}
$$

D-Glucose   D-Galactose

Figure O.2.4

oxygen atoms. Use the longer wooden pegs for C—C bonds and the shorter pegs for the remaining single bonds. The double C=O bond in the aldehyde group should be made using springs. Shown in Figure O.2.4 are Fischer projections of D-glucose and D-galactose. Reproduce each Fischer projection on your answer sheet, and identify each chiral carbon with an asterisk. Twist the model of D-glucose about C—C bonds until it conforms to the Fischer projection. Do you get a ring? Sketch your result. Attempt to change the model of D-glucose into a model of D-galactose by rotating about carbon 4. Describe what happens. Are these molecules conformers, or are they distinct stereoisomers? If they are stereoisomers, are they enantiomers or diastereomers?

α-D-**Glucose and** β-D-**Glucose**   Change the model of open-chain D-glucose into a model of α-D-glucose (also called α-D-glucopyranose), one of the two possible ring forms (hemiacetal forms) of D-glucose. The oxygen that ends up in the six-membered ring comes from the —OH group on carbon 5; the oxygen on carbon 1 in the open chain form takes on a hydrogen, forming an —OH on carbon 1 in the ring. Note that there are two possible ways to orient the —OH group on carbon 1, one giving a Fischer projection having the —OH group on the right (the α form), the other having the —OH group on the left (the β form). α-D-Glucose and β-D-glucose are called *anomers*, and carbon 1 is called the *anomeric* carbon. Write out a Fischer projection, a Haworth projection, and a sawhorse projection for each anomer of D-glucose. Identify each chiral carbon atom with an asterisk.

When you have finished using the molecular models, please

disassemble all models and put the parts away neatly in the boxes. Your instructor will check over your model kits before you leave.

Data and Results

Cyclobutanes
Cyclobutane

1,1-Dichlorocyclobutanes

Cyclohexanes
Cyclohexane

Bromocyclohexane

Glucose
D-Glucose and D-galactose

Anomers of D-glucose

Questions

O.2.1   The following is a Fischer projection of a general formula for an L-amino acid.

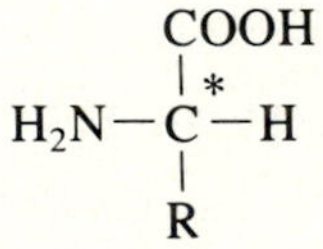

$$
\begin{array}{c}
\text{COOH} \\
| \; * \\
\text{H}_2\text{N}-\text{C}-\text{H} \\
| \\
\text{R}
\end{array}
$$

The R stands for one of a large number of possible groups, such as an alkyl group. Note that, for an L-amino acid, when the R group is ''down'' in the Fischer projection, the $-\text{NH}_2$ group is on the left. Provide Fischer projections for both the D and the L enantiomers of alanine (R group $= -\text{CH}_3$), and cysteine (R group $= -\text{CH}_2\text{SH}$).

O.2.2   The Fischer projection for one enantiomer of xylose is shown below. When this molecule is treated with HNO_3, both the

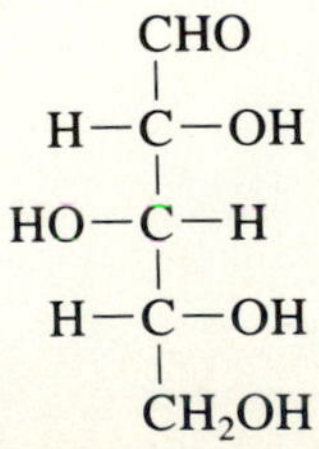

$$
\begin{array}{c}
\text{CHO} \\
| \\
\text{H}-\text{C}-\text{OH} \\
| \\
\text{HO}-\text{C}-\text{H} \\
| \\
\text{H}-\text{C}-\text{OH} \\
| \\
\text{CH}_2\text{OH}
\end{array}
$$

aldehyde group on carbon 1 and the alcohol group on carbon 5 are converted into acid groups ($-COOH$). Such a molecule is called a *glycaric acid*. Write out Fischer projection formulas for and identify (D or L) the two enantiomers of xylose. Label all chiral carbons with an asterisk (*). Also write out Fischer projection formulas for the glycaric acid produced upon oxidation of each enantiomer with HNO_3, identify all chiral carbons in each, and explain whether or not each of the glycaric acid molecules should be chiral.

O.2.3   Consider 1,3-dibromocyclohexane. Sketch Haworth projection formulas for all possible stereoisomers. Identify all chiral carbons (*), and point out the isomeric relationships (diastereomers, enantiomers, etc.) between the different structures.

O.2.4   The relationship between the open-chain and ring (hemiacetal) forms of D-xylose is analogous to that of D-glucose. In D-xylose, the oxygen that ends up in the five-membered ring comes from carbon 4; that same oxygen bonds to carbon 1 in forming the ring. Draw Haworth projections for the α and β anomers of both D-xylose and L-xylose.

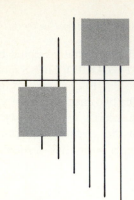

Experiment O.3

Hydrocarbons

Special Item
Sunlamp

The hydrocarbons are compounds consisting of only carbon and hydrogen. Among them are the alkanes and cycloalkanes, which have only single carbon-carbon bonds; the alkenes and cycloalkenes, which have one or more double bonds; the alkynes, having triple bonds; and the aromatic hydrocarbons, which contain special types of ring systems. In this experiment you will investigate some characteristic reactions of each of these groups of hydrocarbons.

Reactions of Hydrocarbons with Bromine

Different classes of hydrocarbons react differently with bromine; thus the bromine reaction is very useful in classifying and distinguishing between different hydrocarbons. Saturated hydrocarbons, including alkanes and most cycloalkanes, react very slowly with bromine unless heated or exposed to light. The reaction with alkanes is a chain reaction, the first step of which involves splitting the bromine molecule into two atoms, each of which is a *radical* because it has an unpaired electron. Although this step requires some input of heat (Δ) or light ($h\nu$) energy, the overall reaction is quite exothermic. The net reaction, which involves a *substitution* of a hydrogen by a bromine atom to form an alkyl bromide, could be represented as:

$$Br_2 + RH \xrightarrow{h\nu} RBr + HBr$$

This is somewhat oversimplified and misleading, however, because the reaction normally leads not to a single alkyl bromide but to a mixture of products. Notice that hydrobromic acid, HBr, is also always a product of the reaction.

Bromination of alkenes and alkynes is very fast at room temperature in the dark. Unlike the reactions with alkanes and cycloalkanes, these are *addition* reactions. Bromine adds to each side of the double bond or triple bond. The reaction does not involve free radicals, as in the case of alkanes, but has an *ionic* mechanism. Because hydrogen is not displaced, HBr is *not* a product of these addition reactions.

$$\begin{array}{cc} & \text{H} \quad \text{H} \\ & | \quad | \\ \text{Br}_2 + \text{R}-\text{C}=\text{C}-\text{R} \rightarrow & \text{R}-\overset{\displaystyle |}{\underset{\displaystyle |}{\text{C}}}-\overset{\displaystyle |}{\underset{\displaystyle |}{\text{C}}}-\text{R} \\ & \text{Br} \quad \text{Br} \end{array}$$

$$\begin{array}{cc} & \text{Br} \quad \text{Br} \\ & | \quad | \\ 2\text{Br}_2 + \text{R}-\text{C}\equiv\text{C}-\text{R} \rightarrow & \text{R}-\overset{\displaystyle |}{\underset{\displaystyle |}{\text{C}}}-\overset{\displaystyle |}{\underset{\displaystyle |}{\text{C}}}-\text{R} \\ & \text{Br} \quad \text{Br} \end{array}$$

Bromination of benzene, the simplest aromatic hydrocarbon, is very slow either in the dark or in the light, because of the unusual stability resulting from delocalization of the pi electrons in the ring. When benzene does react, it reacts by substitution, so HBr is a product. Bromination of other aromatic compounds, such as toluene, phenol, or nitrobenzene, depends on the substituent(s) on the aromatic ring. Some substituents speed the substitution reaction; others make it go more slowly.

Reactions of Hydrocarbons with Potassium Permanganate

Saturated hydrocarbons are not oxidized by dilute potassium permanganate, but alkenes and alkynes react to give oxidation products. The reaction between an unsaturated compound and permanganate in neutral or slightly basic solution is called *Baeyer's test*. In the reaction, purple permanganate (MnO_4^-) ion is reduced to a brown precipitate, $MnO_2(s)$. Alkenes are oxidized to diols (glycols); and alkynes, to acid salts..

$$\begin{array}{c} \text{H} \quad \text{H} \qquad\qquad\qquad\qquad \text{OH} \ \text{OH} \\ | \quad | \qquad\qquad\qquad\qquad\qquad | \quad | \\ 3\text{R}-\text{C}=\text{C}-\text{R}' + 2\text{KMnO}_4 + 4\text{H}_2\text{O} \rightarrow 3\text{R}-\text{C}-\text{C}-\text{R}' + 2\text{MnO}_2(s) + 2\text{KOH} \\ | \quad | \\ \text{H} \quad \text{H} \end{array}$$

$$\text{R}-\text{C}\equiv\text{C}-\text{R} + 2\text{KMnO}_4 \rightarrow 2(\text{R}-\text{COO}^- \text{K}^+) + 2\text{MnO}_2(s)$$

Benzene itself is not oxidized by dilute permanganate, but the aromatic ring makes side chains attached to the ring susceptible to oxidation. For unbranched alkyl side chains, such as methyl, ethyl, or *n*-propyl groups, the primary oxidation product (for one side chain attached to the aromatic ring) is always benzoic acid, C_6H_5COOH.

Reactions of Unsaturated Hydrocarbons with Concentrated H_2SO_4

Sulfuric acid will not react with saturated hydrocarbons, but it will add across a double bond in an alkene to give a product called an alkyl hydrogen sulfate.

$$\begin{array}{c} \text{H} \quad \text{H} \\ | \quad\; | \\ \text{R}-\text{C}=\text{C}-\text{R}' \end{array} + \text{H}_2\text{SO}_4 \rightarrow \begin{array}{c} \text{H} \quad\; \text{OSO}_3\text{H} \\ | \quad\quad | \\ \text{R}-\text{C}\text{——}\text{C}-\text{R} \\ | \quad\quad | \\ \text{H} \quad\; \text{H} \end{array}$$

The alkyl hydrogen sulfate product is insoluble in hydrocarbons but soluble in sulfuric acid. If the hydrocarbon reacts to give an alkyl hydrogen sulfate, the hydrocarbon layer should disappear as the reaction proceeds; if there is *no* reaction, the hydrocarbon will form a separate layer (since sulfuric acid is very dense compared to hydrocarbons, it will be the bottom layer). It must be noted that the reaction between an alkene and H_2SO_4 is quite exothermic, enough so that other reactions often take place in addition to the one producing the colorless alkyl hydrogen sulfate. One side reaction that often occurs under these conditions is a chain reaction of the alkene to produce colored polymeric substances. These colored compounds tend to be soluble in neither the hydrocarbon nor the sulfuric acid layers.

Procedure

(a) ### Reaction of Hydrocarbons with Bromine
Note: Please discard all of your reaction products for this part of the experiment in the waste bottle labeled "Waste bromination solutions—Exp. O-3—Hydrocarbons."

Place 3 mL of each of the following hydrocarbons in separate small test tubes: *n*-heptane, cyclohexane, 1-pentene, cyclohexene, and toluene.

■ CAUTION

Bromine is toxic and can cause painful burns. If you get it on your skin, immediately wash with soap and water; then rinse with sodium thiosulfate solution and wash again. Notify your instructor.

■ **WEAR YOUR
SAFETY
GOGGLES**

To each tube add 1 mL (15 drops) of fresh bromine water; stopper and shake until the bromine color has left the water layer. Note what happens in the hydrocarbon layer in each test tube. (Bromine dissolved in a hydrocarbon has a red-brown color; alkyl bromides are colorless.)

If the bromine color was *not immediately* decolorized (indicating a slow reaction or no reaction) in one or more of the test tubes, set *those* tubes in separate beakers, *remove the stoppers from the tubes,* and set the beakers under a sunlamp in the hood for 5 min. The tops of the test tubes should be approximately 15 cm from the sunlamp.

■ **CAUTION**

Avoid looking at the light from the sunlamp any more than necessary; the ultraviolet light is not good for your eyes.

If the color has not gone after 5 min, put the tubes back for another 5 min, repeating until the reaction has taken place. Then take the test tubes back to your desk.

(b)   **Reaction of Hydrocarbons with Potassium Permanganate**
Note: Please discard all of your reaction products for this part of the experiment in the waste bottle labeled "Waste from KMnO$_4$ + Hydrocarbons—Exp. O-3—Hydrocarbons."

■ **WEAR YOUR
SAFETY
GOGGLES**

Place 2 mL of each of the following hydrocarbons in separate small test tubes: *n*-heptane, cyclohexane, 1-pentene, cyclohexene, and toluene.

To each hydrocarbon sample, add 10 drops of 1% KMnO$_4$ solution, stopper, and shake. Note any evidence for reaction, and, in test tubes where a reaction does take place, note how long it takes relative to the others. [If reaction occurs, the purple color of KMnO$_4$ should be replaced by the brown color of MnO$_2(s)$.]

(c)   **Reactions of Acetylene (Demonstration by Instructor)**
Acetylene (gas) will be prepared in a 16 × 150 mm test tube fitted with a one-hole stopper through which a bent piece of fire-polished glass tubing has been inserted. The glass tubing should be approximately 30 cm in length and should contain two right-angle bends, one at 5 cm and one at 15 cm from the end to be inserted through the stopper, giving it the shape of an inverted J. The test tube should be supported on a ring stand, high enough so that the open end of the glass tubing

may be inserted into test tubes containing liquids. Acetylene is produced by placing three or four small chunks of calcium carbide in the bottom of the test tube, adding enough water to nearly cover the chunks, and inserting the stopper. The instructor should repeat this demonstration several times for groups of seven or eight students. After the demonstrations, the water should be decanted off the calcium carbide chunks, the chunks should be placed in the large beaker in the hood labeled "Used Calcium Carbide—Exp. O-3—Hydrocarbons," and the test tube should be cleaned for the next lab section.

Acetylene + Bromine   Place 3 mL of bromine water in a test tube, and bubble acetylene gas through the liquid. Students should note any results and record their observations.

Acetylene + KMnO$_4$ Place 3 mL of 1% KMnO$_4$ solution in a test tube, and bubble acetylene gas through the liquid. Students should record any observations.

(d)   **Reaction of Unsaturated Hydrocarbons with H$_2$SO$_4$** (Demonstration by Instructor)

■ CAUTION   **Because concentrated H$_2$SO$_4$ is extremely corrosive, it is recommended that these reactions be carried out by the instructor, with students observing and taking notes.**

Place 3 mL of concentrated sulfuric acid in each of four different *dry* 16 × 150 mm test tubes. To each test tube add 5 drops, one drop at a time, of one of the following hydrocarbons: *n*-heptane, cyclohexane, 1-pentene, or cyclohexene. Agitate the mixture in each test tube *after the addition of each drop* by rapping it briskly on the side, being very careful not to splash any of the liquid out of the test tube. Note if any heat is evolved. If there is no reaction, two separate phases should remain; if reaction occurs, there should be only one phase after reaction. If no reaction is observed after all 5 drops have been added, even after agitation, try mixing the liquids with a glass stirring rod. Students should record all observations. Following these demonstrations, the instructor should pour the waste solutions into a large beaker of water, *slowly*, with lots of stirring, and then down the drain, flushing with plenty of water.

Data and Results

Record each of your observations, and write out equations for any reactions that occurred. Compare and explain reactivity of the different hydrocarbons with each reagent.

(a)   Observations, Equations, and Explanations for Observations on Reactivity of Hydrocarbons with Bromine
n-Heptane

Cyclohexane

1-Pentene

Cyclohexene

Toluene

Compare and explain reactivity.

(b)   **Observations, Equations, and Explanations for Observations on Reactivity of Hydrocarbons with KMnO$_4$**
n-Heptane

Cyclohexane

1-Pentene

Cyclohexene

Toluene

Compare and explain reactivity.

(c) **Observations, Equations, and Explanations for Observations on Reactivity of Acetylene**
With bromine

With KMnO$_4$

(d) **Observations, Equations, and Explanations for Observations on Reactivity of Hydrocarbons with Concentrated H$_2$SO$_4$**
n-Heptane

Cyclohexane

1-Pentene

Cyclohexene

Toluene

Compare and explain reactivity.

Questions

O.3.1   Write out formulas for and name all of the monobromination products produced by treating 2-methylpropane with bromine in the presence of light.

O.3.2   From what you observed in this experiment, what can you say about the solubility of bromine in water compared to its solubility in hydrocarbons? Explain.

O.3.3   Cyclopropane reacts in the dark with bromine, just as if it were an unsaturated hydrocarbon. How might you account for this?

O.3.4   From the way in which toluene (methylbenzene) reacted with bromine, even if you had had a way of detecting whether or not HBr had been produced, you could not tell whether the bromine reacted at a site on the aromatic ring (maybe the methyl group "activated" the ring) or the bromine reacted to form a bond with the carbon in the methyl group. Explain.

O.3.5   Suppose you had a sample of 3-methylheptane contaminated by a tiny amount of 2-pentene. Explain how you could purify the sample by using concentrated sulfuric acid. Write out an equation for the reaction.

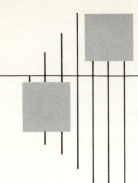

Experiment O.4

Alkyl Halides and Alcohols

Some of the most important reactions of alkyl halides and alcohols are nucleophilic substitution reactions, in which some other group is exchanged for the halogen (Cl, Br, or I) of an alkyl halide or the $-OH$ group of an alcohol. The mechanism for a particular reaction depends on the structure of the alkyl halide or alcohol, the nature of the leaving group and of the entering group, and reaction conditions, such as the polarity of the solvent. If the reaction rate is solely dependent on the concentration of a *single* reactant, such as an alkyl halide, the substitution reaction will have an S_N1 mechanism; if the rate depends on the concentration of *two* reactants, such as an alkyl halide *and* a nucleophilic entering group, the reaction will have an S_N2 mechanism. There is often the possibility for competing side reactions, such as elimination reactions; these usually can be controlled by judicious choice of reaction conditions. In this experiment you will investigate a number of substitution reactions of alkyl halides and alcohols. In addition, you will examine the oxidation of alcohols with dichromate as well as reactions of alcohols with sodium metal.

(a) **Reactions of Alkyl Halides with Sodium Iodide in Anhydrous Acetone**   Iodide ion is a good nucleophile and, under the right conditions, will react with an alkyl chloride or alkyl bromide to give the corresponding alkyl iodide. Anhydrous acetone is polar enough to dissolve many polar and weakly ionic solutes, but not polar enough to dissolve strongly ionic solutes. Sodium chloride and sodium bromide are somewhat more ionic than sodium iodide, so sodium chloride and sodium bromide are considerably less soluble in anhydrous acetone than is sodium iodide. Therefore, if an alkyl chloride or bromide reacts with NaI in acetone, $NaCl(s)$ or $NaBr(s)$ will precipitate from the reaction mixture. Formation of the precipitate both favors the forward reaction and provides evidence of reaction. The reaction rate depends on the concentrations of both the iodide ion and the alkyl chloride or alkyl bromide; therefore the mechanism is S_N2. Because the car-

bon-bromine bond in an alkyl bromide is weaker than a car-
bon-chlorine bond in an alkyl chloride, and because bromide
ion is more stable than chloride ion, bromide is a better leaving group than chloride. The reaction is fastest for primary
bromides and slowest for tertiary alkyl chlorides.

(b) **Reactions of Alkyl Halides with Silver Nitrate in Anhydrous Ethanol**   When an alkyl (or aryl) halide is treated with
anhydrous ethanol containing dissolved silver nitrate, an ethyl
ether and insoluble silver halide [$AgCl(s)$, $AgBr(s)$, or
$AgI(s)$] are formed. The mechanism of the reaction is S_N1,
the rate being dependent solely on the concentration of the
alkyl halide. The first (slowest) step of the mechanism involves formation of a carbonium ion. Then, in very fast steps,
the halide ion (produced along with the carbonium ion in the
first step) reacts with Ag^+ to give a silver halide precipitate,
and the carbonium ion reacts with the ethanol to give the ethyl
ether and hydrogen ion. Tertiary iodides react fastest and primary chlorides slowest.

(c) **Lucas Test for Alcohols**   This reaction is of use in classifying alcohols as primary, secondary, or tertiary, providing
the alcohols being tested are soluble in aqueous HCl. The Lucas test reagent is anhydrous $ZnCl_2$ dissolved in concentrated
HCl. In the net reaction, an alcohol reacts with HCl to give
an alkyl chloride; $ZnCl_2$ is added as a catalyst. Alkyl chlorides are less soluble than the corresponding alcohols; if reaction occurs, the alkyl chloride will separate out as a second
layer or the mixture will become cloudy. The first step of the
reaction is fast and involves formation of a coordination complex between the alcohol and $ZnCl_2$.

$$ROH + ZnCl_2 \rightarrow R-\overset{\overset{\displaystyle H}{|}}{O}-ZnCl_2$$

The second step is rate-determining and involves formation of
a carbonium ion, R^+, accompanied by loss of $[HOZnCl_2]^-$.

$$R-\overset{\overset{\displaystyle H}{|}}{O}-ZnCl_2 \rightarrow R^+ + [HOZnCl_2]^-$$

The $ZnCl_2$ catalyst is regenerated by combining with H^+ from
the HCl in a subsequent fast step.

$$[HOZnCl_2]^- + H^+ \rightarrow H_2O + ZnCl_2$$

The final step, formation of the alkyl chloride, is also fast.

$$R^+ + Cl^- \rightarrow RCl$$

Because the rate-determining step involves only one substance, the substitution reaction provides an example of an S_N1 mechanism. The rate depends on how hard it is to form the carbonium ion, R^+, so it is found that the rate is fastest for tertiary and slowest for primary alcohols.

(d)   **Oxidation of Alcohols with Potassium Dichromate Solution**   Acidified dilute $K_2Cr_2O_7$ will oxidize a primary alcohol first to an aldehyde and then to a carboxylic acid. It will oxidize a secondary alcohol to a ketone, but it will not react with a tertiary alcohol. In the reaction, orange dichromate ion, $Cr_2O_7^{2-}$, is reduced to green chromic ion, Cr^{3+}.

(e)   **Reaction of Alcohols with Metallic Sodium**   Water reacts vigorously with metallic sodium to give sodium hydroxide and hydrogen gas.

$$2Na(s) + 2H_2O \rightarrow 2NaOH + H_2(g)$$

Alcohols react in an analogous manner with sodium to give sodium alkoxides and hydrogen gas.

$$2Na(s) + 2ROH \rightarrow 2NaOR + H_2(g)$$

The vigor of the reaction between an alcohol and sodium depends on the strength of the O—H bond; the weaker the O—H bond, the more vigorous the reaction. Because the strength of this same bond determines the acidity of an alcohol (relative tendency to release H^+), the reaction with sodium provides a measure of the acidity of an alcohol; the more vigorous the reaction, the more acidic the alcohol.

Procedure

■ **WEAR YOUR SAFETY GOGGLES**

Perform the test reactions as directed below. Upon the completion of each part, dispose of your waste solutions as directed by your instructor.

(a)   **Alkyl Halides + NaI**   Support your largest beaker on a ring stand, fill it half full of tap water, heat the water to approximately 60°C, and *shut off your burner.*

■ **CAUTION**

Flammable reagents; keep away from flames!

Number six small test tubes 1 to 6, and, in each tube place four drops of a different alkyl halide, as follows: (1) 1-chlorobutane, (2) 2-chlorobutane, (3) 2-bromobutane, (4) 2-

chloro-2-methylpropane (*t*-butyl chloride), (5) benzyl chloride, and (6) chlorocyclohexane. Into each of six other clean, dry test tubes, measure 2 mL of sodium iodide reagent (15% NaI in anhydrous acetone). Moving quickly, add the 2-mL volumes of sodium iodide reagent to the numbered test tubes containing the different alkyl halides, noting the time of addition. If there is a reaction, a precipitate of either sodium chloride, NaCl(s), or sodium bromide, NaBr(s), both of which are insoluble in anhydrous acetone, should form. If you observe a precipitate in any of the test tubes, record your observation and the time. If there are any tubes in which no reaction occurred, set those tubes in the beaker of hot water you set up earlier. After placing the test tubes in the beaker, check the temperature of the water in the beaker; if it has fallen below 50°C, *remove any flammable reagents from the vicinity of the burner*, heat the water to 50°C, and shut off the burner. Allow the test tubes to remain in the hot water bath for about 6 or 7 min, remove them, check for any sign of reaction, and record your observations.

(b)   **Alkyl Halides + AgNO₃**   As you did in part (*a*), number six small test tubes 1 to 6 and, in each tube, place 4 drops of a different alkyl halide, as follows: (1) 1-chlorobutane, (2) 2-chlorobutane, (3) 2-bromobutane, (4) 2-chloro-2-methylpropane (*t*-butyl chloride), (5) benzyl chloride, and (6) chlorocyclohexane. Measure 2 mL of ethanolic silver nitrate reagent (1.5% AgNO₃ in absolute ethanol) into each of six different clean, dry test tubes. Within as short a time span as possible, add the 2-mL volumes of silver nitrate reagent to each of the six test tubes containing the different alkyl halides, and record the time of addition. Look for the formation of any precipitates [AgBr(s), AgCl(s)], recording the time you first observe them. After about 5 min, *after removing any flammable reagents from the vicinity of your burner*, place those test tubes in which no precipitate was observed in the beaker of hot water you used in part (*a*), heat the water to boiling, shut off the burner, and allow the test tubes to stand in the hot water for several minutes. Using a test tube holder, remove each tube. Look for any signs of reaction, and record your observations.

■ WEAR YOUR SAFETY GOGGLES

(c)   **Lucas Test for Alcohols**   Label three 16 × 150 mm test tubes with numbers 1 to 3, and place 5 mL of Lucas reagent (ZnCl₂ in concentrated HCl) in each test tube. Also number three small 13 × 100 mm test tubes, and to each of them add 1 mL of one of the following alcohols: (1) 1-butanol, (2) 2-butanol, and (3) 2-methyl-2-propanol (*t*-butyl alcohol). Transfer each alcohol as quickly as possible from the smaller

■ WEAR YOUR SAFETY GOGGLES

test tube into the correspondingly numbered larger test tube containing Lucas reagent; then stopper each test tube, mix well, and note the time. Note any immediate reactions, and record your observations. Then set the solutions aside, checking for any subsequent reactions at 5-min intervals for about $\frac{1}{2}$ hour. These alcohols are soluble in the test solution, but the alkyl chlorides are insoluble; thus, the formation of two phases, as evidenced by cloudiness, is a positive test.

(d)   **Alcohols + $K_2Cr_2O_7$ Solution**   Number four small test tubes, and to each add 1 mL of one of the following alcohols: (1) 1-butanol, (2) 2-butanol, (3) 2-methyl-2-propanol, and (4) cyclohexanol. Then add 10 drops of 6 M HCl, followed by 1 drop of saturated $K_2Cr_2O_7$ (in water); stopper each tube, and shake to mix the contents. Note any evidence of reactions, and record your observations. A color change from yellow-orange to green is a positive test.

■ CAUTION   **Potassium dichromate is quite toxic; wash your hands after using this chemical.**

(e)   **Alcohols + Sodium Metal (Demonstration by Instructor)**   Your instructor will place a few milliliters of each of the following alcohols in large beakers and cover each beaker with a large watch glass: (1) methanol, (2) 1-butanol, (3) 2-butanol, and (4) 2-methyl-2-propanol. Then he or she will add a small (pea-sized) chunk of freshly cut sodium metal to each beaker and will quickly replace the watch glass after each addition. Students should observe what happens in each case and record their observations.

Data and Results

(a)   **Alkyl Halides + NaI**   Write out a structural formula for each of the following, and record your observations in the space provided:

Compound	Formula	Observations
1-Chlorobutane		

Compound	Formula	Observations
2-Chlorobutane		
2-Bromobutane		
2-Chloro-2-methylpropane		
Benzyl chloride (ϕ-CH$_2$Cl)		
Chlorocyclohexane		

Conclusions:

(b)   Alkyl Halides + AgNO₃   Record your observations for the reaction of $AgNO_3$ with each of the listed alkyl halides.

Compound	Observations
1-Chlorobutane	
2-Chlorobutane	
2-Bromobutane	
2-Chloro-2-methylpropane	
Benzyl chloride (ϕ-CH₂Cl)	
Chlorocyclohexane	

Conclusions:

(c)   **Lucas Test**   Write out a structural formula for each of the following alcohols, and record your observations in the space provided:

Compound	Formula	Observations
1-Butanol		
2-Butanol		
2-Methyl-2-propanol		

(d)   **Alcohols + $K_2Cr_2O_7$**   Record your observations for the reactions with the following alcohols with acidic potassium dichromate solution.

Compound	Observations
1-Butanol	
2-Butanol	
2-Methyl-2-propanol	

Compound	Observations
Cyclohexanol	

Conclusions:

(e)   **Alcohols + Sodium Metal**   Record your observations for the reactions with the following alcohols with sodium metal.

Compound	Observations
Methanol	
1-Butanol	
2-Butanol	

Compound   **Observations**

2-Methyl-2-propanol

Conclusions:

Questions

O.4.1   In reacting sodium iodide in acetone with alkyl chlorides and
bromides,
   (*a*) Why must the acetone be anhydrous (not contain water)?
   (*b*) Do primary bromides react faster than tertiary bromides?
   (*c*) Do secondary bromides react faster than secondary chlorides?

O.4.2   Suggest one or more reasons for the observation that tertiary
alkyl iodides react faster than primary alkyl chlorides with
alcoholic silver nitrate.

O.4.3   In the Lucas test, why must the alcohol being tested be soluble
in the Lucas test reagent?

O.4.4   Why does acidified dilute $K_2Cr_2O_7$ *not* oxidize tertiary alcohols at room temperature?

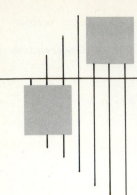

Experiment O.5

Aldehydes and Ketones

Aldehydes and ketones exhibit very similar chemical behavior, due to the fact that they both contain a carbonyl group. The differences result from the fact that in aldehydes the carbonyl group is bonded to a carbon on one side and a hydrogen on the other, whereas in ketones the carbonyl carbon is bonded to two other carbons. To oxidize a ketone is quite difficult, because it means breaking a strong carbon-carbon bond; oxidation of an aldehyde, on the other hand, is relatively easy

$$\diagdown \atop \diagup \!\!\! C=O \qquad\qquad {H \diagdown \atop R \diagup} C=O \qquad\qquad {R \diagdown \atop R' \diagup} C=O$$

Carbonyl group   Aldehyde   Ketone

with mild oxidizing agents. Special oxidizing agents have been developed that are strong enough to oxidize aldehydes but not strong enough to oxidize ketones, thereby allowing one to distinguish between aldehydes and ketones.

Because the oxygen in the carbonyl group is substantially more electronegative than the carbon, the oxygen atom draws electron density away from the carbon toward itself, giving the oxygen a slightly negative charge. The carbon takes on a slightly positive charge, making the carbon a center of attack for nucleophiles (electron-rich groups). The withdrawal of electron density from the carbonyl carbon is also felt, to a lesser extent, by α carbons, which are carbon atoms adjacent to the carbonyl group. Hydrogen atoms bonded to the α carbons, called α hydrogens, are easier to remove (more acidic) than hydrogens bonded to other carbon atoms. In this experiment you will investigate the oxidation of aldehydes and ketones, a reaction involving α hydrogens, and the reactivity of the carbonyl group with a nucleophile.

(a) **Oxidation with Potassium Dichromate Solution**   Primary alcohols are oxidized to aldehydes, and then to carboxylic acids, by $K_2Cr_2O_7$ in acidic solution. Aldehydes, of course, are oxidized directly to acids. Secondary alcohols are oxidized to ketones. Ketones are not ordinarily oxidized by this reagent. If a reaction occurs, orange dichromate ion, $K_2Cr_2O_7{}^{2-}$, is reduced to green chromic ion, Cr^{3+}.

(b)   Oxidation with Fehling's Solution   Fehling's solution is a milder oxidizing agent than dichromate. Its reaction with alcohols and aromatic aldehydes (such as benzaldehyde) is much slower than its reaction with aliphatic aldehydes; it does not react at all with most ketones. Fehling's solution is a freshly prepared basic tartrate complex of Cu^{2+} ion, which is blue. If Fehling's solution is reduced by an aldehyde, the Cu^{2+} ion is reduced to Cu^+ ion, which forms a red precipitate of $Cu_2O(s)$ in the basic solution. With a positive test, the blue of the tartrate complex often combines with the red of the precipitate to give a greenish mixture. If, after getting a positive test, one continues to heat the mixture, often the $Cu_2O(s)$ will be further reduced to copper metal, which will plate out on the inside of the test tube.

(c)   Oxidation with Tollen's Reagent   The strength of Tollen's reagent as an oxidizing agent lies between that of dichromate and Fehling's solution. As with the other two, it will not react with ordinary ketones, but, unlike Fehling's solution, it will oxidize aromatic aldehydes in addition to aliphatic aldehydes. Tollen's reagent is a basic solution of $Ag(NH_3)_2{}^+$ ion. In a very clean test tube, Tollen's reagent will react with an aldehyde to produce metallic silver, which forms as a silver mirror on the inside of the test tube.

(d)   The Iodoform Reaction   Besides its primary function, which is testing for the presence of a terminal methyl group bonded to a carbonyl group, this reaction serves as an illustration of the acidity of the α-hydrogen atoms of a carbonyl compound. The iodoform reagent is prepared by dissolving 5 g of iodine and 10 g of potassium iodide in 80 mL of water (a solution of KI and I_2 has been prepared for you and is available on the side shelf). To carry out the reaction, a small amount of this solution is mixed with the organic compound. Then sodium hydroxide is added to the mixture until the dark iodine color just barely disappears. In the basic solution, a hydrogen on the α-carbon atom is easily removed and replaced by an iodine atom. Because iodine, like oxygen, is more electronegative than carbon, the remaining C—H bonds on the α carbon are weakened even more than before, leading to their rapid replacement by iodine. The resulting triiodomethyl group is cleaved by the base, giving iodoform [$CHI_3(s)$, a yellow solid with a characteristic medicinal odor] together with a carboxylic acid salt. The iodoform test is positive for acetaldehyde

and ketones with the $\diagdown$ $C{=}O$ group. It also gives a positive
$\diagup$
CH_3

test for the $\diagdown$ $C{-}OH$ group, which is first oxidized by iodine
$\diagup$ $\diagdown$
CH_3 H

in the basic solution to give the required terminal methyl-carbonyl group; then the carbonyl compound reacts as usual.

(e)   **Reaction with 2,4-Dinitrophenylhydrazine**   A convenient way of purifying an aldehyde or ketone is to react it with something that will give a solid crystalline product that is easy to recrystallize. After purification of the solid derivative, the aldehyde or ketone may be recovered in a subsequent reaction. Besides their use in purification, the crystalline derivatives are of great value in identification of aldehydes and ketones, because the derivatives often have characteristic melting points. The reagents used to prepare derivatives are basic compounds that donate electrons to the slightly positive carbonyl carbon atom. In the reaction of a carbonyl compound with 2,4-dinitrophenylhydrazine, a pair of electrons from the terminal amine group are donated to the carbonyl carbon, forming an intermediate that immediately loses a water molecule to give a 2,4-dinitrophenylhydrazone derivative (Figure O.5.1).

Procedure

(a)   **Oxidation with $K_2Cr_2O_7$ Solution (Chromic Acid Test)**   Number five small (13×100 mm) test tubes, and place 1 mL of one of the following compounds in each tube: (1) ethanol, (2) 3-pentanone, (3) 20% acetaldehyde (aqueous) (4) 1-propanal (propionaldehyde), and (5) benzaldehyde. Add 10 drops of 6 M HCl and 1 drop of saturated $K_2Cr_2O_7(aq)$ to each test tube, stopper, and shake. A positive test is a color change from yellow-orange to green. (*Note:* Since benzalde-

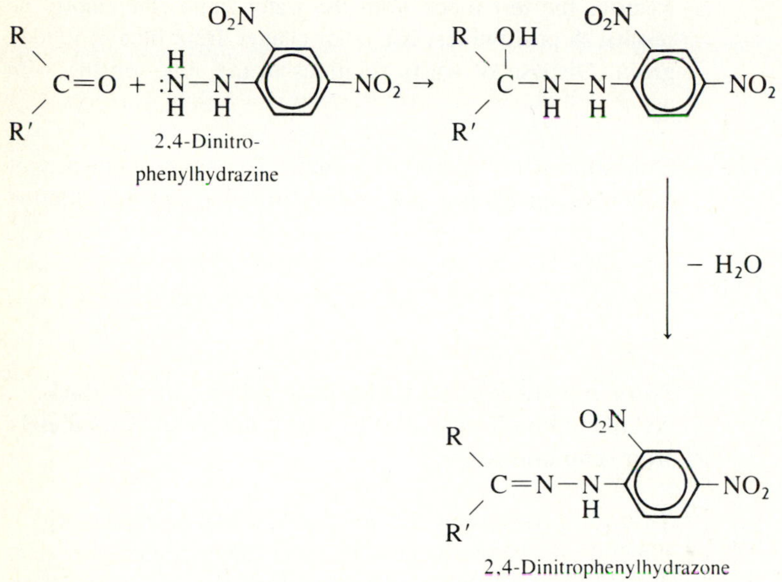

Figure O.5.1

hyde is quite insoluble in water, you should shake the benz-aldehyde reaction mixture vigorously enough to get a fine emulsion.) Record the results of your tests. Waste solutions may be rinsed down the sink with water.

(b) **Oxidation with Fehling's Solution**   Label two clean 16×150 mm test tubes, one with an A, the other with a B. Place 5 mL (approximately one-fourth full) of Fehling's solution A in test tube A and 5 mL of Fehling's solution B in test tube B. Thoroughly clean and label five small test tubes with numbers 1 to 5. Place 1 mL of Fehling's solution A in each of the five test tubes. Add 1 mL of Fehling's solution B and 1 mL of water to each of the five small test tubes, and mix until all of the light-blue $Cu(OH)_2(s)$ dissolves to form a dark-blue cupric tartrate complex. Then add the following quantities of these compounds to the combined Fehling's solution in the numbered test tubes: (1) 4 drops of 1-propanal, (2) 4 drops of 3-pentanone, (3) 4 drops of 1-propanol, (4) 10 drops of 20% glucose (aqueous), and (5) 5 drops of benzaldehyde. Mix by rapping the test tubes with the side of your finger. Set up a large beaker half full of tap water on your ring stand, place the test tubes in the beaker, and heat the water until it boils. Boil *gently* for 2 min (no longer), and then shut off the burner.

Remove the test tubes from the water; note and record the results. A positive test is a color change from blue to muddy green. Dispose of waste solutions in the sink, rinsing with water.

(c)   **Oxidation with Tollen's Reagent**   To prepare a quantity of Tollen's reagent, first place 6 mL of a 5% aqueous solution of silver nitrate ($AgNO_3$) in a *scrupulously* clean 16 × 150 mm test tube, and add 3 drops of 10% aqueous NaOH solution.

■ CAUTION   **Silver nitrate will react with your skin to turn it black. Avoid spilling it on yourself; if you do, wash immediately with soap and water.**

This will give a brownish precipitate of silver oxide, Ag_2O. Finally, add 10% aqueous ammonia (ammonium hydroxide) drop by drop, shaking after every few drops, until the silver oxide just barely dissolves. This should take about 3 mL of the aqueous ammonia. Don't use any more ammonia than necessary. To get good results with Tollen's test, you must use *very* clean test tubes.

■ CAUTION   **Tollen's reagent must not be stored, because it can form silver fulminate, which is explosive, if allowed to evaporate to dryness. Dispose of unused Tollen's reagent and any reaction products before you leave today by rinsing them down the sink with plenty of water.**

Pour the Tollen's reagent into five very clean test tubes, which you have numbered beforehand, dividing it equally among them (about 2 mL in each). Then add one of the following compounds to each test tube in the quantity specified: (1) 4 drops of 1-propanal, (2) 4 drops of 3-pentanone, (3) 4 drops of 1-propanol, (4) 10 drops of 20% glucose (aqueous), (5) 5 drops of benzaldehyde. Mix the contents of each test tube by rapping briskly on the side of the test tube with your finger (in the case of benzaldehyde, stopper with a very clean stopper, and shake to form an emulsion), and allow the reaction mixtures to stand for about 10 min. Note and record any observations. Formation of a silver mirror indicates a positive test. If there is no reaction after 10 min, warm the test tubes in which no reaction took place by placing them in a water bath (60°C, *no higher*) for another 5 min. Record your observations. Any silver mirrors should be destroyed by treatment with aqua regia (mixture of HNO_3 and HCl) in the hood. Ask your instructor for assistance with this.

(d)   **Iodoform Reaction**   Place 1 mL of water in each of four numbered test tubes, and then add one of the following compounds to each test tube as directed: (1) 10 drops of 20% aqueous acetaldehyde, (2) 2 drops of 2-propanol (isopropyl alcohol), (3) 2 drops of acetone, and (4) 2 drops of 3-pentanone. Add 2 drops of iodoform reagent (iodine–potassium iodide solution) to each test tube, and mix the contents of each test tube by rapping on the outside of the tube. Finally, add 10% NaOH solution, drop by drop, mixing after each addition, until the dark color of the iodine just disappears. If reaction is immediate, note the results and record; if not, place the test tubes in which there was no reaction in a 60°C water bath for 10 min. Observe and record results. Dispose of reaction products by rinsing them down the sink with plenty of water.

(e)   **Reaction with 2,4-Dinitrophenylhydrazine**   Obtain 8 mL of 2,4-dinitrophenylhydrazine reagent in a graduated cylinder and 16 mL of 95% ethanol in a small beaker.

■ CAUTION

2,4-Dinitrophenylhydrazine is toxic and will react with your skin to turn it yellow. If you spill it on yourself, wash immediately and thoroughly with soap and water.

Number four 16 × 150 mm test tubes, and place 4 mL of 95% ethanol in each of them. To each of the test tubes add 5 drops of one of the following compounds, and shake to mix: (1) acetone, (2) benzaldehyde, (3) butanone, and (4) cyclohexanone. Finally, add 2 mL of 2,4-dinitrophenylhydrazine reagent to each test tube, and allow reaction to take place for about 10 min at room temperature. Observe, and record your results. Discard waste solids and solutions as directed by your instructor.

[*Optional:* If there is time, your instructor may direct you to filter (vacuum filtration, Büchner funnel) one or more of the crystalline products, dry them, and determine their melting points.]

Data and Results

(a) Oxidation with $K_2Cr_2O_7$ Solution   Write out a structural formula for each of the following compounds and record your observations in the space given:

Compound	Formula	Observations
Ethanol		
3-Pentanone		
20% Acetaldehyde		
1-Propanal		
Benzaldehyde		

Conclusions:

(b)   **Oxidation with Fehling's Solution**   Write out structural formulas for each organic reactant *and* each organic product, and give your observations in the space provided. For no reaction, write "NR."

Compound   Formula   Product Formula   Observations

1-Propanal

3-Pentanone

1-Propanol

Glucose

Benzaldehyde

Conclusions:

(c)   **Oxidation with Tollen's Reagent**   Give your observations for each of the following compounds:

Compound	Observations
1-Propanal	
3-Pentanone	
1-Propanol	
Glucose	
Benzaldehyde	

Conclusions:

(d)   **Iodoform Reaction**   Write out a structural formula for each of the following compounds, and enter your observations in the space given:

Compound	Formula	Observations
Acetaldehyde		

2-Propanol

Acetone

3-Pentanone

Conclusions:

(e)   **Reactions with 2,4-Dinitrohydrazine**   Write out a structural formula for the organic *product* (2,4-dinitrophenylhydrazone) formed with each of the following, and give your observations:

Compound	Formula of Product	Observations
Acetone		
Benzaldehyde		
Butanone		
Cyclohexanone		

Conclusions:

Questions

O.5.1   Write out a structural formula for 3-heptanone. Circle all of the α hydrogens. Using a lowercase Greek delta, δ, for "slightly," label the polarity of the carbon atom and that of the oxygen atom in the carbonyl group as "slightly positive" (δ^+) or "slightly negative" (δ^-)

O.5.2   Consider the following groups: alkanes, 1° alcohols, 2° alcohols, 3° alcohols, aliphatic aldehydes, aromatic aldehydes, and ketones. Which of these groups could be distinguished from each other on the basis of their reactions with:
(*a*) Acidic dichromate
(*b*) Fehling's solution
(*c*) Tollen's reagent

O.5.3   Write out a stepwise reaction sequence for what occurs when iodoform reagent is added to 2-pentanol.

O.5.4   Write out a stepwise reaction sequence for what occurs when 2,4-dinitrohydrazine reagent is added to 3-heptanone.

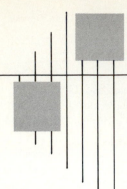

Experiment O.6

Carboxylic Acids and Esters

One of the most important reactions of carboxylic acids, which contain the carboxyl, $\underset{/}{\overset{\displaystyle OH}{\diagdown}} C=O$, group, often written as —COOH, is their reaction with bases such as hydroxide ion to form carboxylate salts. In this reaction, the acidic hydrogen atom of the carboxyl group is donated to the base:

$$RCOOH + K^+ + OH^- \longrightarrow RCOO^- + K^+ + H_2O$$

Carboxylic acids are considerably more soluble in water than many other organic compounds because they can ionize to some extent and also because they can interact with the water by forming hydrogen bonds. Carboxylate salts, formed by the reaction of a carboxylic acid with a base, as described above, are even more soluble than their parent acids, because they are much more highly ionized and therefore can interact more strongly with the polar water molecules.

Carboxylic acids can also react by replacement of the hydroxyl (—OH) group in the carboxyl group by some other group. If, for example, the hydroxyl group is replaced by an alkoxide group, RO^-, the result is an ester. An ester can be prepared by reacting an alcohol with an acid in the presence of an acidic catalyst.

$$ROH + R'-\underset{\underset{OH}{|}}{C}=O \overset{H^+}{\rightleftharpoons} R'-\underset{\underset{OR}{|}}{C}=O + H_2O$$

It is important to note that the oxygen in the —OR group in the ester comes *not* from the carboxylic acid but from the alcohol. Also note that the reaction is reversible. The forward reaction is called *esterification*, and the reverse reaction is called *saponification*. Just as esterification is catalyzed by acid, saponification is catalyzed by base. The products of the saponification reaction are the alcohol and the salt of the carboxylic acid.

Organic molecules are not limited to having a single carboxyl group; there are, for example, many *dicarboxylic* acids, molecules that contain two carboxyl groups. Maleic and fumaric acids are isomeric dicarboxylic acids, both having the formula $HOOC-CH=CH-COOH$. They are alike in that each has the same structural formula, but they are different in the spatial arrangement of their carboxyl groups and therefore are *geometric* isomers. Fumaric acid is the *trans* isomer, and maleic acid the *cis* isomer. Because of their different geometries, these isomers have considerably different physical and chemical properties. In this experiment you will examine chemical and physical properties of a variety of carboxylic acids, including maleic and fumaric acids, and you will also investigate some reactions and physical properties of esters.

Procedure

(a)   **Saponification of an Ester**   [Students should work in groups, half on this part, and half on part (*b*).] Assemble a reflux condenser as shown in Figure O.6.1, using a 300-mL Florence flask (or 250-mL Erlenmeyer flask). Connect the rubber tubing to the water tap and condenser such that the water goes in at the bottom and out at the top. Put in three or four boiling chips, and then add 15 mL of 5 *M* NaOH and 5 mL of ethyl benzoate to the flask.

■ WEAR YOUR
SAFETY
GOGGLES

■ CAUTION

Sodium hydroxide is very caustic and *must* be kept out of the eyes, or severe damage, even blindness, may result. If you should get any NaOH solution into your eyes, flush them immediately with copious amounts of water and then notify your instructor.

Before heating, ask your instructor to check your setup. Turn on the water so that it runs *slowly* through the condenser and into the sink.

Heat the mixture so that it boils *gently* for a minimum of 1 h 30 min, or until the ester layer has disappeared. From time to time, check to see that the vapors are not passing through the top of the condenser; if you can see the vapors condensing any higher than halfway up the tube, slow down the heating. While you are waiting, proceed with the rest of

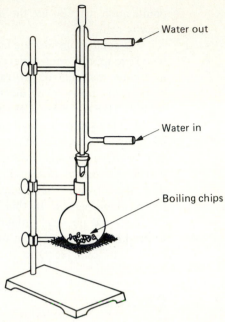

Water out

Water in

Boiling chips

Figure O.6.1

the experiment. When the saponification is complete, shut off your gas burner, and allow the flask to cool enough so that you can handle it comfortably. Then remove it, cool it under the tap, and add 50 mL of cold wtaer. Add 3 *M* HCl in 5-mL increments until the solution turns blue litmus red. What is the purpose of adding the acid? After each addition, swirl the flask and, using a glass stirring rod, remove a drop of the solution and touch the drop to a piece of litmus paper. After the excess NaOH has been completely neutralized, add an additional 10 mL of 3 *M* HCl to the flask in order to convert the soluble sodium benzoate to insoluble benzoic acid.

Filter out the crystalline benzoic acid using vacuum filtration. Before transferring the product to the Büchner funnel, decant off most of the supernatant liquid into the sink, rinsing it down the drain with plenty of water. Then, using a rubber policeman attached to a glass stirring rod, transfer the crystals to the filter paper in the funnel. Remove any boiling chips, and wash the benzoic acid with a few milliliters of cold water. Then transfer the solid to a 250-mL Erlenmeyer flask, cover the flask with a watch glass, and recrystallize the benzoic acid by dissolving the solid in a minimum amount of hot water (80° C) and then allowing the solution to cool slowly. After crystallization appears to be complete, decant off most of the supernatant liquid, and filter by vacuum filtration as before. Allow air to be drawn through the crystals several minutes to dry them; then remove them from the funnel and allow them to finish drying. Divide your crystals among the members of your group for further testing. Describe the shape and color of the benzoic acid crystals, and write equations both for the saponification reaction and for the conversion of sodium benzoate to benzoic acid.

(b)   **Maleic and Fumaric Acids**   (Students should work in groups.) Set up a reflux condenser as in part (*a*), and ask your instructor to check it over. Place three boiling chips in the flask. Weigh out 10 g of maleic acid, place it in the flask, and add 10 mL of water. Without attaching the flask to the condenser, warm the mixture, swirling the flask, just enough to dissolve the maleic acid. Then add 25 mL of concentrated HCl to the flask, and attach it to the condenser. Turn on the water so it runs slowly through the condenser. Heat the mixture, refluxing *gently* for 30 min. (Crystals of fumaric acid should begin forming soon after the refluxing starts.) Remove the heat, and allow the flask to cool enough so you can handle it; then remove the flask from the condenser and cool it thoroughly by running tap water over it.

Carefully decant most of the supernatant liquid into the sink, and then filter the crystals using vacuum filtration. Remove the boiling chips from the fumaric acid crystals. Turn the aspirator off, add 10 mL of cold water to the filter cake, stir the crystals around in the rinse water, and then turn the aspirator back on until the crystals are dry. Then shut off the aspirator and repeat the washing procedure with another 10 mL of cold water. Remove the fumaric acid crystals from the Büchner funnel, and divide them among the members of your group for further analysis. Record any observations, such as the time required to dissolve the maleic acid in the warm water, how long it took before fumaric acid crystals began forming, and the appearance of the two acids. Also write out formulas for the two isomeric acids.

(c)   **Solubility of Carboxylic Acids in Water and in Aqueous 2.5 *M* Sodium Hydroxide**   In classifying compounds with respect to their solubility in different solvents, we will say that if less than 2 drops (or 0.1 g of solid) of solute will dissolve in 3 mL of solvent, the substance is "insoluble"; if 2 drops dissolve in 3 mL, we will call it "slightly soluble"; if between 3 and 6 drops dissolve in 3 mL, we will say that the solute is "moderately soluble"; if more than 6 drops (or

■ **WEAR YOUR SAFETY GOGGLES**

equivalent amount of solid) dissolve in 3 mL of solvent, we will say that the compound is "soluble." Your instructor will show you how much solid material constitutes 0.1 g. To determine the solubility of a carboxylic acid, first place 3 mL of either water or 2.5 M NaOH in a 16 × 150 mm test tube. Then add a certain number of drops or grams of solute (start with 2 drops or 0.1 g) to the test tube, stopper, and mix the contents by rapping the side of the test tube briskly with your finger.

Perform solubility tests on each of the following carboxylic acids [use the products prepared in parts (*a*) and (*b*), supplemented by materials from the side shelf if necessary], both in water and in 2.5 M NaOH: (1) benzoic acid, (2) propanoic acid, (3) fumaric acid, (4) maleic acid, and (5) stearic acid. Then place 1 g of benzoic acid in 50 mL of water in a beaker, support it on a ring stand, and heat the mixture to about 80°C. Cover the beaker with a watch glass, and allow the solution to cool to room temperature. Observe, and record your observations.

(d)   Preparation of Esters   Your instructor will demonstrate the reaction between ethyl alcohol and butyric acid. Students should note the odor of butyric acid (cautiously waft the vapor toward the nose with the hand) and also the odor of the product. To produce the ester, place 3 mL of absolute ethanol in a dry 16 × 150 mm test tube, add 1 mL of butyric acid, and mix the contents. Slowly and carefully add, drop by drop, a total of 10 drops of concentrated H_2SO_4, and heat the mixture in a water bath (70°C) for 10 min. Finally, pour the contents of the test tube into a beaker containing 20 mL of hot water. Students should note the odor and record their observations. The remaining esters should be produced by students, working in pairs.

Methyl Alcohol + Salicylic Acid   Place 3 mL of methanol in a dry test tube, add 0.5 g of salicylic acid, and mix. Slowly add, drop by drop, 10 drops of concentrated H_2SO_4. Heat in a 70°C water bath for 10 min, pour the product into a beaker containing 20 mL of hot water, and note the odor. Record your observations.

Isoamyl Alcohol + Acetic Acid   Place 3 mL of isoamyl alcohol (3-methyl-1-butanol) in a dry test tube, add 1 mL of acetic acid, and mix. Slowly add, drop by drop, 10 drops of concentrated H_2SO_4, heat in a 70°C water bath for 10 min, and then pour the product into a beaker containing 20 mL of hot water. Note the odor, and record your observations. Dispose of waste solutions by pouring them into the sink, rinsing them down the drain with plenty of water.

Data and Results

(a) Saponification of an Ester
Observations

Saponification reaction

Reaction for conversion of benzoate to benzoic acid

(b) Maleic and Fumaric Acids
Observations

Formulas

(c)   Solubility of Carboxylic Acids in Water and in 2.5 *M* NaOH

Compound	Solubility in H_2O	Solubility in 2.5 *M* NaOH
Benzoic acid	_____	_____
Propanoic acid	_____	_____
Fumaric acid	_____	_____
Maleic acid	_____	_____
($C_{17}H_{35}COOH$)	_____	_____

Observations on heating and cooling benzoic acid solution:

(d)   Esters
Odor of butyric acid

Odors of compounds and equations for esterification reactions.

Compound	Odor	Equation
Ethyl butyrate		

Compound	Odor	Equation
Methyl salicylate		
Isoamyl acetate		

Questions

O.6.1   In recrystallizing the benzoic acid you prepare in part (*a*), you are directed to cover the flask with a watch glass. Suggest a reason for this.

O.6.2   Like esters, acid anhydrides are considered to be derivatives of carboxylic acids. In the case of acid anhydrides, the group replacing the $-OH$ group of the carboxyl group is an $RCOO-$ group. For example, the acid anhydride of acetic acid, CH_3COOH, has the following formula:

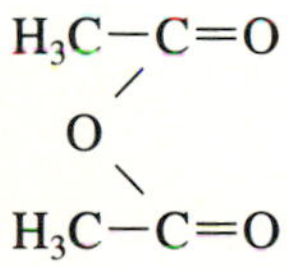

$$H_3C-C=O$$
$$\diagup$$
$$O$$
$$\diagdown$$
$$H_3C-C=O$$

When a mole of water is added to a mole of acetic anhydride, 2 mol of acetic acid are produced. When maleic acid, $HOOC-CH=CH-COOH$, is heated under reduced pressure just above its melting point, it forms an "internal" anhydride having the molecular formula $C_4H_2O_3$. Write out a structural formula for maleic anhydride. How many moles of water would convert 1 mol of maleic anhydride to maleic acid? Why is fumaric acid *not* observed to form an anhydride having the same molecular formula as that of maleic anhydride when it is heated just above its (fumaric acid's) melting point?

O.6.3   How can you explain the great difference in solubility observed for benzoic acid in sodium hydroxide and in water?

0.6.4 Suggest one or more reasons for the differences in solubility observed for maleic and fumaric acids.

0.6.5 Why is esterification favored by an acidic solution and saponification favored by a basic solution? Give a mechanism for each reaction.

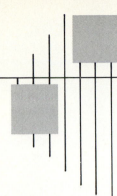

Experiment O.7

Preparation of Aspirin

Special Items

250 mL distilled water
1 g salicylic acid
3 mL acetic anhydride
   (density = 1.08 g/mL)
Few drops 85% phosphoric
   acid

Aspirin can be made by treating salicyclic acid with acetic anhydride in the presence of an acid catalyst.

Salicylic acid $+$ Acetic anhydride $\longrightarrow$

Aspirin $+$ Acetic acid

Procedure

■ WEAR YOUR SAFETY GOGGLES

Weigh 1.0 g of salicylic acid into a 100-mL beaker, and add 3 mL of acetic anhydride followed by a few drops of 85% phosphoric acid. In the hood, warm the mixture gently for 5 min. Remove the burner; while the mixture is still warm, slowly add dropwise about 1 mL of water to decompose excess acetic anhydride.

■ CAUTION

The reaction is vigorous and may spatter.

   Add 15 mL of water to the solution, heat until clear, and then allow to cool. As soon as cloudiness appears, scratch the inside of the beaker with a stirring rod until the aspirin starts to crystallize. When the mixture has cooled to room temperature, filter and wash the crystals on the filter paper thoroughly with several small portions of cold water. Allow to drain completely, and press dry between several layers of filter paper. When dry, weigh the product, and calculate the percent yield based on the original mass of salicylic acid.

Data and Results

Questions

O.7.1   List all safety precautions that must be observed in this experiment regarding the handling of toxic and/or flammable materials.

O.7.2   Write a mechanism for the acid-catalyzed reaction between salicylic acid and acetic anhydride to form aspirin.

O.7.3   In this experiment an acid catalyst is used. What chemical serves as the catalyst?

O.7.4   "Aspirin" is a trivial name. Give a more "systematic" name (or names) for aspirin.

O.7.5   After the reaction of salicylic acid with acetic anhydride, 1 mL of water is added. Write the equation for the reaction that should occur between water and acetic anhydride.

O.7.6   Explain, in general terms, how to calculate percent yield.

O.7.7   What is produced in this experiment besides aspirin? How is this by-product removed?

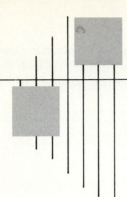

Experiment O.8

Organic Polymers

Special Items

5 mL methyl methacrylate
25 mg benzoyl peroxide
2 mL adipoyl chloride
50 mL cyclohexane
3 mL 70%
    hexamethylenediamine
1.5 g NaOH
2 mL methyl red (0.1 g in 100
    mL 95% ethanol)
2 mL methyl orange (0.1 g in
    100 mL 95% ethanol)
2 mL bromcresol green (0.1 g
    in 100 mL 95% ethanol)

Lucite (polymethylmethacrylate) is prepared from the monomer methyl methacrylate. The reaction is initiated by free radicals formed when benzoyl peroxide is heated in the presence of the monomer.

$$\text{Benzoyl peroxide} \quad \xrightarrow[\text{heat}]{\Delta} \quad 2 \quad \text{Free radical}$$

The reaction is *initiated* when the free-radical fragments attack the methyl methacrylate, creating new free radicals.

New free radical

Polymerization is *propagated* by the new free radicals attacking more of the monomer, creating longer chains, which are also new free radicals.

New free radical

The process continues until a *termination* step, which involves two free radicals coming together, ends the chain formation. Each chain may contain hundreds of repeating units.

Polymethylmethacrylate (Lucite)
($n = 100$ or more)

Procedure

(a)   **Lucite**   Place three or four boiling chips in a 600-mL beaker, add about 200 mL of tap water, set the beaker on a ring stand equipped with a wire gauze, and heat the water to boiling. Place 5 mL of methyl methacrylate in a clean, dry 15-cm test tube. Ask your lab instructor to add about 25 mg of benzoyl peroxide. Being careful to avoid the formation of bubbles in the liquid, stir the mixture in the test tube with a stirring rod until the benzoyl peroxide is uniformly dispersed throughout the liquid. Using your test-tube holder, place the test tube in the boiling water and stir the mixture gently until the benzoyl peroxide completely dissolves. Remove the test tube, and stir the mixture to try to remove any bubbles that may have formed. Allow the test tube to remain in the hot water for 5 min without stirring. Cool the water to about 70°C by adding, with stirring, about 100 mL of cold water to the beaker. At this point you may wish to add a small object that you wish to immortalize to the viscous plastic. Allow the beaker to cool, with the test tube in it, until the end of the laboratory period. Then put the test tube in your desk until the next laboratory period.

At the beginning of the next laboratory period, wrap your test tube in a cloth and break the tube by striking it with the edge of a ring stand. Put the broken glass in the waste glass bucket and examine the polymer. Heat the piece of plastic in hot water, and note whether or not it softens.

(b)   **Nylon**   Nylon may be prepared by reacting hexamethylenediamine with adipoyl chloride. For each reaction between the two starting materials, one HCl molecule is split out and an amide bond is formed. Because each of the reactants has two functional groups, one at each end, both molecules react in both directions, and the reaction repeats until a long-chain polyamide results.

$$n \quad \overset{\displaystyle H}{\underset{\displaystyle H}{N}}\!-\!(CH_2)_6\!-\!\overset{\displaystyle H}{\underset{\displaystyle (H}{N}} \quad + \quad n \quad \overset{\displaystyle O}{C}\!-\!(CH_2)_4\!-\!\overset{\displaystyle O}{\underset{\displaystyle Cl}{C}} \quad \longrightarrow$$

Hexamethylenediamine   Adipoyl chloride

$$n\text{HCl} \;+\; \left[\!-\!\overset{\displaystyle H}{N}\!-\!\overset{\displaystyle O}{C}\!-\!(CH_2)_4\!-\!\underset{\displaystyle O}{C}\!-\!\overset{\displaystyle H}{N}\!-\!(CH_2)_6\!-\!\right]_n$$

Nylon 66

■ CAUTION

Before starting this portion of the experiment, make sure that all gas burners have been turned off. Mix all solutions in the open hood. Do not work at your regular bench space until the solutions have been mixed and poured together.

Place 2 mL of adipoyl chloride in 50 mL of cyclohexane in a dry 250-mL Erlenmeyer flask, and swirl the flask until the contents are well mixed. In a dry 200-mL beaker, dissolve 3 mL of 70% hexamethylenediamine in 50 mL of water, and then add 1.5 g of sodium hydroxide pellets. Stir the mixture until the sodium hydroxide pellets dissolve. Also add 2 mL of methyl red, methyl orange, or bromcresol green to the water solution in the beaker.

The density of the water solution in the beaker is greater than that of the cyclohexane solution in the flask. Grasping the beaker in one hand, tilt it; and taking the flask in your other hand, carefully pour the cyclohexane solution of adipoyl chloride down the side of the tilted beaker such that it ends up floating on top of the water solution of hexamethylenediamine with as little mixture as possible. At the interface between the two solutions you should see nylon begin to form immediately. See Figure O.8.1.

Being careful not to disturb the solutions in the beaker, carry the beaker from the open hood to your regular working space. Place a paper towel in the sink and another on the bench top. Set the beaker containing the reacting mixture on the paper towel in the sink. Using your stirring rod, free the walls of the beaker of any strands or film of nylon, and push the material toward the center. Then reach into the beaker with your test-tube holder and grasp the center of the nylon film. Carefully lift up on the test-tube holder and, twisting as you pull, lift a rope of nylon out of the solution. Wind the strand onto a wooden splint until you have pulled out a yard or more

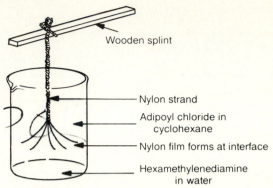

Wooden splint

Nylon strand

Adipoyl chloride in cyclohexane

Nylon film forms at interface

Hexamethylenediamine in water

Figure O.8.1

of nylon. Try to keep the fresh nylon off the bench top by keeping it above the paper towel.

Wash the nylon rope with water, and then wash it with about 5 mL of acetone in a beaker. Decant the used acetone into the sink, and wash it down with water. Pour the solutions remaining in the reaction beaker into the waste bottle provided, *not* down the sink. (Why not?)

Data and Results

Questions

O.8.1   List all safety precautions that must be observed in this experiment regarding the handling of toxic and/or flammable materials.

O.8.2   What is the purpose of heating the mixture of benzoyl peroxide and methyl methacrylate in part (*a*)?

O.8.3   Find the molecular weight of
   (*a*) A Lucite molecule made from 200 methyl methacrylate molecules and one benzoyl peroxide molecule.
   (*b*) A Nylon-66 molecule containing 200 of the smallest repeating units.

O.8.4   The product formed when adipoyl chloride and hexamethylenediamine react is a straight-chain polymer. Draw a structure of a molecule that could be used in place of adipoyl chloride that would react with hexamethylenediamine to give a branched-chain polymer.

O.8.5   Why is it very important *not* to pour your reaction solutions into the sink drain after the nylon synthesis?

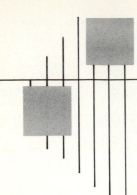

Experiment O.9

Soap

Special Items

Burette
1 cm³ lard
10 mL 5 M NaOH
5 mL glycol
5 g NaCl
3 g castile soap
25 mL 1 M HCl
50 mL standard 1 M NaOH
Phenolphthalein

Ordinary soap is usually the sodium salt of a long-chain organic acid such as stearic acid ($C_{17}H_{35}COOH$) or oleic acid ($C_{17}H_{33}COOH$). In this experiment, you will make some soap by the reaction of NaOH with lard, a mixture of high-molecular-weight esters. For simplicity, we can consider lard to be the triester glyceryl stearate [$(C_{17}H_{35}COO)_3C_3H_5$], which on boiling with NaOH forms glycerine and soap.

$$
\begin{array}{ccc}
& O & H \\
& \| & | \\
C_{17}H_{35}-C-O-C-H & & H-O-C-H \\
& O & | \\
& \| & | \\
C_{17}H_{35}-C-O-C-H & + 3OH^- \rightarrow & H-O-C-H + 3C_{17}H_{35}COO^- \\
& O & | \\
& \| & | \\
C_{17}H_{35}-C-O-C-H & & H-O-C-H \\
& | & | \\
& H & H
\end{array}
$$

Triester   Glycerine

In the second pard of this experiment, you will analyze a sample of soap, assumed to contain sodium stearate ($NaOOCC_{17}H_{35}$), in order to estimate the water content. You will do this by mixing a known mass of soap with a known volume of standard acid, allowing the stearate ion to react with H_3O^+ to precipitate stearic acid, and then titrating the excess H_3O^+ with known base. Since 1 mol of stearate ion reacts with 1 mol of H_3O^+, the number of moles of H_3O^+ used up by the soap is equal to the number of moles of sodium stearate in the original sample. From the formula weight of $NaOOCC_{17}H_{35}$ and the original mass of sample, you can calculate the percentage of water in the sample.

Procedure

■ **WEAR YOUR SAFETY GOGGLES**

(a) Place about 1 cm³ of lard in your 100-mL beaker. Add 10 mL of 5 M NaOH and 5 mL of ethylene glycol. The glycol ($HOCH_2CH_2OH$) serves to raise the boiling point and to help dissolve the lard. Boil gently over a low flame.

■ CAUTION

Hot NaOH is corrosive to the skin.

From time to time, add water to replace that lost by evaporation. Continue boiling until reaction is complete, as shown by the disappearance of the oil droplets in the mixture. [Meanwhile, proceed with part (*b*).] Allow the mixture to cool, and then add 10 mL of water and 5 g of NaCl. Stir well. Skim off the product, and test it by shaking a bit of it in a test tube with water.

(b)   Weigh out on the platform balance 3.0 g of the soap provided on the reagent shelf. Dissolve it with heating in about 25 mL of distilled water. Let cool. Add to the solution dropwise with constant stirring 25.0 mL of 1.0 *M* HCl from your graduated cylinder. Stir the mixture well until precipitation is complete. Set up a suction filter (such as is shown in Part A, Figure A.14), and filter the liquid into a clean bottle. Rinse the solid on the filter several times with distilled H_2O. Add a few drops of phenolphthalein to the filtrate, and titrate the unreacted acid with 1.0 *M* NaOH from a burette.

Data

Results

Calculate the percentage water content of the soap sample.

Questions

O.9.1 Assuming that the lard in part (*a*) is 1.00 g of pure glyceryl stearate, write a balanced equation for the reaction, and find the number of moles of the reagent in excess left over after complete reaction has taken place.

O.9.2 Soluble sodium stearate (soap) precipitates as insoluble calcium stearate, $(C_{17}H_{35}COO^-)_2Ca^{2+}$, in the presence of calcium ion. How many milliliters of 1.0 M $CaCl_2$ solution would be required to precipitate totally the amount of soap you analyzed in part (*b*)?

O.9.3 How would each of the following affect your calculated percentage of water in your soap sample?
(*a*) The soap contains some Na_2CO_3.
(*b*) Not all the soap was dissolved in water before adding the HCl, so that the stearic acid in the filter paper was contaminated with some sodium stearate.

O.9.4 A typical bar of soap weighs about 200 g. Find the number of grams of stearic acid and of NaOH required to make a bar of soap having the same percentage of water as the soap you analyzed in part (*b*).

O.9.5 Obtain from your classmates at least eight values for the percentage of water in the soap sample, and find an average value for the percentage, together with an average deviation.

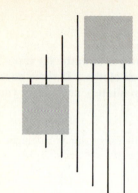

Preparation of Sulfanilamide and Melting-Point Determination

Special Items

10 g *p*-acetaminobenzene-
 sulfonyl chloride
15 mL concentrated NH_3
15 mL 6 *M* HCl
10 g charcoal
8 g $NaHCO_3$
50 mL mineral oil
1 g hydroquinone
funnel
Stirrer, stopper, 250°C
 thermometer assembly
Melting point capillary

Sulfanilamide, for which the structure is

$$H_2N - \underset{}{\bigcirc} - \overset{\displaystyle O}{\underset{\displaystyle O}{S}} - NH_2$$

is an example of a sulfa drug, which probably owes its extreme toxicity for some forms of bacteria to its molecular resemblance to *para*-aminobenzoic acid,

$$H_2N - \underset{}{\bigcirc} - C \overset{\displaystyle O}{\underset{\displaystyle OH}{}}$$

This latter molecule is used to synthesize an important vitamin, folic acid (also called vitamin B_c), and it is presumed that when sulfanilamide is present the bacteria mistakenly try to synthesize folic acid from it instead. When this fails, they die. Human beings, however, are immune to sulfanilamide, since they do not synthesize folic acid but generally get it from the bacteria that normally reside in their intestinal tracts.

In this experiment you will synthesize sulfanilamide from *para*-acetaminobenzenesulfonyl chloride, for which the structure is

$$CH_3 - \overset{\displaystyle O}{\underset{}{C}} - \overset{\displaystyle H}{\underset{}{N}} - \underset{}{\bigcirc} - \overset{\displaystyle O}{\underset{\displaystyle O}{S}} - Cl$$

■ CAUTION

This substance is toxic. Try not to breathe its vapor, and do not let it touch your skin or clothes. If some does get on your skin, wash it off immediately and notify your instructor.

The sequence of steps you will carry out is indicated schematically as follows:

$$CH_3CONH \langle \bigcirc \rangle SO_2Cl \rightarrow CH_3CONH \langle \bigcirc \rangle SO_2NH_2$$

$$\rightarrow H_2N \langle \bigcirc \rangle SO_2NH_2 \cdot 2HCl$$

$$\rightarrow H_2N \langle \bigcirc \rangle SO_2NH_2$$

At the conclusion of the synthesis, you will purify your material and determine its melting point so as to identify it. The normal melting point of sulfanilamide is 163°C. You will also investigate the general effect of an impurity on melting point.

Procedure

(a)

■ WEAR YOUR
SAFETY
GOGGLES

Preparation of Sulfanilamide Get from your instructor a glass vial containing about 10 g of *para*-acetaminobenzene-sulfonyl chloride. Remember that this substance is toxic, so handle it carefully. Weigh the vial and contents, pour the contents into a 250-mL beaker, cover the beaker with an evaporating dish, and weigh the vial again.

Mix together (in the hood) 15 mL of concentrated aqueous ammonia and 15 mL of distilled water in a 100-mL beaker. Add the mixture slowly, with stirring, to the weighed *para*-acetaminobenzenesulfonyl chloride. Heat the mixture in the hood on a wire gauze over a flame, with occasional stirring. Maintain the temperature just below boiling for about 5 min. A change should be observed as the sulfonyl chloride changes to a more pasty suspension of the amide.

Set up a filtration apparatus. Cool the above suspension in an ice bath and filter off the *para*-acetaminobenzenesulfonamide. Allow it to drain as thoroughly as possible. (If you have available a Büchner funnel for the filtration, as described in Part A, Figure A.14, so much the better.) Scrape the moist amide back into the original 250-mL beaker, and slowly add

15 mL of 6 N HCl. White fumes will be evolved—but this is normal.

Boil the mixture gently until all the solid has dissolved. Note the level of the liquid in the beaker and boil gently for 15 min, adding water as required to maintain the original volume. (If an evaporating dish is placed on top of the beaker, it should not be necessary to add much water.) Cooling of the solution should now give you no precipitate of unhydrolyzed amide.

If the solution is not colorless, add a small amount of finely divided charcoal and filter. (Charcoal is often able to remove color from a solution by adsorbing the molecules that produce the color.) Prepare a solution of sodium bicarbonate in a 100-mL beaker by dissolving about 8 g of $NaHCO_3$ in 90 mL of H_2O. Return the decolorized solution to the 250-mL beaker and, *very slowly and carefully*, with stirring, add the sodium bicarbonate solution to it. After the foam subsides, test the suspension with litmus paper; if it is still acidic, add more bicarbonate solution until it has become neutralized. Cool the solution in an ice bath and filter. This product is impure sulfanilamide.

To purify a solid, one can dissolve it in a solvent at the boiling point, filter the hot solution to remove any suspended particles, and then crystallize. Note, however, that if too much solvent is used, the yield will be reduced. Place the impure sulfanilamide in a clean 250-mL beaker and add enough water to make the final volume of suspension about 50 to 60 mL. Heat this to boiling, and filter while hot into a 100-mL beaker. Wash the funnel with a bit of hot water. Evaporate the solution to a volume of about 50 mL, and allow the beaker to cool, eventually standing it undisturbed for a while in an ice bath.

Collect the crystals by filtration. Let them dry on a filter paper in your desk. Weigh the product, and determine the percentage of theoretical yield you actually obtained.

(b) **Determination of Melting Point**   Following the procedure given in Section 2g in Part A of this manual, determine the melting range of the following samples:

1   Your pure, *dry* sulfanilamide. To dry, press a small amount between two pieces of dry filter paper until the sample appears powdery.

2   Pure hydroquinone, $C_6H_4(OH)_2$.

3   A mixture of sulfanilamide and hydroquinone. Prepare the mixture by thoroughly powdering together on a watch glass approximately equal small quantities of the two substances.

Cleanup   Clean the thermometer by wiping thoroughly with a dry towel. If the used oil is dirty or dark in color, pour it into the waste bottle provided. If it is in the same condition as at the start of the experiment, pour it back into one of the oil bottles. Using a brush, wash out the test tube with soap and water.

Questions

Prelaboratory

O.10.1   Safety precautions are always important, but especially in this experiment. Examine the procedure and make a list of the precautions to be taken at each stage.

O.10.2   What procedural steps are taken in order to obtain the maximum amount of pure, colorless sulfanilamide product?

O.10.3   Make up a flowchart of this experiment showing reagents added and species formed at each stage of the experiment.

O.10.4   What is the main purpose for adding the $NaHCO_3$ in the procedure in part (a)? Calculate the minimum amount needed for this purpose, and compare with the amount you are directed to use.

Postlaboratory

O.10.5   Organic chemistry is largely the study of reactions involving functional groups. List the organic compounds used in this experiment, and identify all the important functional groups in each.

O.10.6   Which is more soluble in water, sulfanilamide or sulfanilamide hydrochloride? Suggest a reason.

O.10.7   What is the rationale behind saying that an observed melting point lower than the accepted true value probably means an impure product?

O.10.8   Just like sulfanilamide, phenylbenzoic acid has a melting point of 163°C. A student has made a product that she knows is either sulfanilamide or phenylbenzoic acid. She has an authentic sample of sulfanilamide. How can she use melting-point procedure to identify what her product is?

O.10.9   Why, in melting-point determination, do you use a *fresh* sample of unknown for your second run after you have made a rough first-run determination?

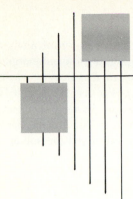

Experiment O.11

Paper Chromatography of Amino Acids

Special Items

1000-mL beaker
Oven

When a small spot containing a mixture of amino acids is placed near the bottom of a piece of filter paper, and the filter paper is placed (vertically) in a covered beaker containing a small amount of suitable solvent, the solvent moves up the filter paper, carrying each different amino acid in the mixture up the paper to a different extent. This results in a series of spots (a chromatogram) on the paper, each spot corresponding to a different compound. The paper is called the *stationary phase*, and the solvent is called the *mobile phase*. An amino acid that interacts with (binds to) the surface of the paper more strongly than it interacts with (dissolves in) the solvent will not move very far from the origin compared to an amino acid that interacts more strongly with the solvent than with the paper. Filter paper is nearly pure cellulose, a carbohydrate. The surface of the paper is normally covered with water molecules, attracted there by the many —OH (hydroxyl) groups of the cellulose molecules. Both water and the hydroxyl groups are polar and can form hydrogen bonds. Amino acids differ in their polarities and vary in their capacity to form hydrogen bonds, so each amino acid will bind to the paper more or less strongly than the others. Each will also differ in its behavior toward the solvent. By choosing just the right solvent system, one can achieve maximum separation of the components of a mixture of amino acids.

The reason that this separation technique is called chromatography is that it was first used to separate colored substances (*chroma* is the Greek word for color). As you will find in this experiment, however, chromatography may also be used to separate colorless compounds, such as amino acids. The compound ninhydrin reacts with all amino acids to produce purple products. After you develop your chromatogram, your instructor will spray it with a solution of ninhydrin so that you can see the spots corresponding to the different amino acids. You will then determine an R_f value for each compound. The R_f value is the ratio of the distance traveled by the compound to the distance traveled by the solvent in a specified time period. If the center of the spot for a particular amino acid were halfway between the origin and the solvent front, for example, its R_f value would be 0.50.

Procedure

■ **WEAR YOUR SAFETY GOGGLES**

You will need at least five capillary tubes for use in placing the amino acid spots on the filter paper. To get best results, these tubes should be very thin, so thin that you couldn't slide a common pin inside. To make them, heat a piece of soft glass tubing (6 mm) in the flame of a gas burner, rotating the glass constantly, until the glass is yellow and soft. Then remove the tubing from the flame and pull out, from side to side, until the capillary tubing you have produced is of the desired diameter. Set the hot glass on several transite mats to cool. Then break the capillary tubing into 10-cm lengths, making certain that the ends are squared off. Place 20 mL of 2-propanol (isopropyl alcohol) in a 1000-mL beaker, add 10 mL of 2% aqueous ammonia (NH_4OH), and swirl the beaker to mix. Use a piece of aluminum foil to cover the beaker.

Making sure to touch it only along the top edge (see Figure O.11.1), place a precut (12×25 cm) sheet of Whatman No. 1 filter paper on a clean sheet of notebook paper, with the long (25 cm) way aligned to the left and right. Using a straightedge, draw a light pencil (*not ink!*) line from left to right, parallel to and up from the bottom edge by 2.0 cm. Place 10 small pencil marks at 2.0-cm intervals along this line—five to the left, five to the right, of the center—being careful not to touch the paper with your fingers. Label the five

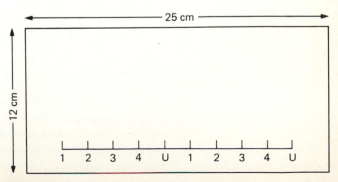

Figure O.11.1

marks on the left 1, 2, 3, 4, U, and also the five marks on the right (1, 2, 3, 4, U), just under the marks.

Label five small test tubes 1, 2, 3, 4, U. Obtain 5 drops of an unknown (either a single compound or a mixture of two amino acids) from your instructor in test tube U and 5 drops of each of the following amino acids from the side shelf in test tubes 1 to 4: (1) aspartic acid, (2) glycine, (3) leucine, and (4) tyrosine. All of the amino acids are 0.05 M, dissolved in aqueous 1.5% HCl. Place a capillary tube in each of the five test tubes. Using a spare sheet of ordinary filter paper, practice using a capillary tube to make a spot between 0.4 and 0.5 cm in diameter; check to see just how much solution in the capillary gives the right-sized spot when the capillary tube is touched to the paper. Then dip the capillary tubes into the amino acid solutions and make the spots on the large sheet, applying them at the positions marked earlier. Allow the paper to dry for a few minutes, and then make a second application at the same positions as the first. Allow the paper to dry for 10 min. While you are waiting, roll the paper into a cylinder, such that the amino acid spots are on the outside, and staple the paper twice, once near the top and once near the middle. It is important that when the staples are put in, the edges of the paper *not* be allowed to touch; there must be a gap.

Being certain not to allow it to touch the inside wall of the beaker and not to splash the solvent up onto the paper, carefully place the cylinder, with the pencil line near the bottom, in the 1000-mL beaker, and cover the beaker tightly with aluminum foil. Allow the chromatogram to develop for at least an hour, preferably longer. When development is complete, remove the chromatogram, and immediately, before it has a chance to dry, mark the location of the solvent front, the highest position reached by the solvent. Then, before removing the staples, set the chromatogram, top side down, on a piece of paper and allow it to dry for 10 min. Then remove the staples and take it to the spray box in the hood, where your instructor will hang it up and spray it with ninhydrin solution for you.

■ CAUTION   **Avoid getting ninhydrin on your hands or clothing. It will cause stains that are difficult or impossible to remove.**

Using a folded piece of paper to carry your chromatogram, put it in an oven held at 105°C, and leave it there for 10 min. Remove the chromatogram, and take it back to your desk. Circle each spot, and measure the distance from the origin to the center of each spot as well as the distance from the origin to the solvent front. Calculate all R_f values, and determine the identity of the components of your unknown. Since you have two chromatograms for each substance, use average distances in making your calculations of R_f values. Submit your developed chromatogram to your instructor with your report.

Data

Unknown No. _____

Distance from base line (origin) to solvent front   _____

Average measured distance from origin to center of:

    Aspartic acid spot   _____

    Glycine spot   _____

    Leucine spot   _____

    Tyrosine spot   _____

    Top unknown spot   _____

    Bottom unknown spot   _____

Results

R_f value calculated for:

    Aspartic acid   _____

    Glycine   _____

    Leucine   _____

    Tyrosine   _____

    Top unknown compound   _____

    Bottom unknown compound   _____

Identities of amino acids in unknown:

Give supporting evidence for your conclusions.

Questions

O.11.1   Draw Fischer projections of each of the amino acids studied in this experiment.

O.11.2   The R_f value for a particular amino acid in a particular solvent depends on properties of the amino acid, the paper, and the solvent. What properties do you think might be important here? Would changing the pH of the solvent system have any effect? Explain.

O.11.3   Why does one use pencil, and not ink, in marking the paper?

O.11.4   In stapling the ends of the paper together, you are cautioned *not* to allow the edges of the paper to come into contact. Suggest a reason for this.

O.11.5   When placing the amino acid solutions on the paper, why is it important not to produce large spots, that is, why are smaller spots preferred?

O.11.6   The baseline is the line upon which the spots of amino acids are applied. Why is it important that the depth of solvent in the bottom of the large beaker *not* be greater than the distance from the bottom of the filter paper to the baseline?

O.11.7   If you start with a mixture of amino acids, the R_f values that you observe for each amino acid are usually somewhat different from the R_f values obtained when you start with pure samples of each amino acid. Suggest a reason for this observation.

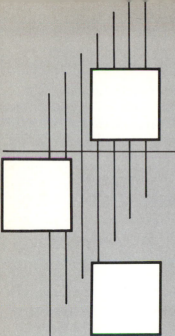

PART D

Qualitative Analysis

Introduction

The term "qualitative analysis" means the detection of the presence of chemical elements in an unknown sample. It is to be distinguished from quantitative analysis, which means the determination of the relative amounts of the elements present in an unknown. In this part of the lab manual, you will study qualitative analysis, not so much to be able to test for the presence of elements but because it is one of the most effective ways to learn the chemistry of the elements and become familiar with equilibria in aqueous solutions. The chemistry that you will learn from qualitative analysis will help you to understand nature, not only because many of the reactions you will observe occur commonly in the environment but also because the principles that underlie the scheme of analysis are general and may be applied to the understanding of many other materials and systems.

In testing a sample for the presence of a particular element, other elements may be present that interfere with the test. Thus, in searching for Hg^{2+} by adding H_2S to see whether black HgS is precipitated, an erroneous conclusion could be drawn if Ag^+ were present, since it also forms a black insoluble sulfide. To avoid such problems, it is necessary to attack the analysis systematically. In the case mentioned, prior addition of HCl would remove Ag^+ as white insoluble AgCl so that any subsequent black sulfide precipitate could not be caused by Ag^+.

One of the most successful schemes for systematic qualitative analysis depends on separating the elements into "groups" based on solubility behavior. This is done by consecutive addition of precipitating agents and filtration or centrifugation of precipitates formed at each stage. In many cases an element would precipitate with more than one of the precipitating agents and hence would seem to belong to several groups. However, since the reagents are added consecutively, in most cases one of the reagents will precipitate the ion in question before others get a chance to react. Thus, Ag^+ is classified with the insoluble chlorides even though it also forms an insoluble sulfide.

The usual groups for qualitative analysis in the order in which the ions react are:

1 Ions that form chlorides insoluble in acid
2 Ions that form sulfides insoluble in 0.3 N acid
3 Ions that form sulfides and hydroxides when base is added to the filtrate from 2
4 Ions that form insoluble carbonates and phosphates
5 Ions that remain in residual solution as unprecipitated ions

Refer to pages 411–413 for an overview of how the above separation can be done.

Once the ions are separated into the various groups, the analysis is not finished, because each group consists of several ions. To identify the ions within a group, it is necessary to make use of characteristic chemical reactions such as complex-ion formation, amphoteric behavior, etc. It is very important to observe carefully, to take good notes, and to follow up on your observations by referring to your textbook and to other reference books. You are not interested so much in what *should* occur as in what *does* occur. If for example, a precipitate theoretically should dissolve under certain conditions but doesn't dissolve in practice, it could be that the dissolving occurs, but only very slowly. Observations need not be positive to be valuable; it is just as useful to know that a precipitate does *not* result as to observe that a precipitate is formed upon addition of a certain reagent. In earlier experiments, you will be directed to perform many tests that give negative results. As you build up the scheme in later experiments, you will see the value in those earlier observations.

The precipitation of the sulfides can be done by adding H_2S gas to the solution; however, use of H_2S gas is both unpleasant and dangerous. To avoid it, H_2S is generated directly in the solution as needed by reaction of thioacetamide (CH_3CSNH_2) with water. When an aqueous solution of thioacetamide is heated, the following reaction slowly occurs to produce a saturated (0.1 M) H_2S solution:

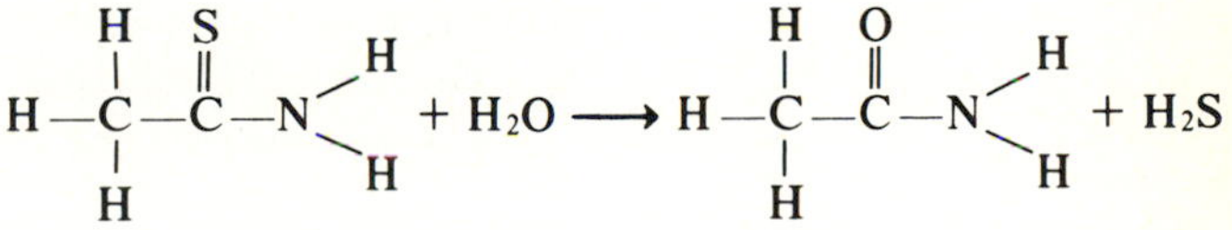

$$H-\overset{\overset{\displaystyle H}{|}}{\underset{\underset{\displaystyle H}{|}}{C}}-\overset{\overset{\displaystyle S}{||}}{C}-N\overset{\diagup H}{\diagdown_H} + H_2O \longrightarrow H-\overset{\overset{\displaystyle H}{|}}{\underset{\underset{\displaystyle H}{|}}{C}}-\overset{\overset{\displaystyle O}{||}}{C}-N\overset{\diagup H}{\diagdown_H} + H_2S$$

■ CAUTION

Thioacetamide may be toxic when ingested. If you spill some on your hands, wash them immediately so as to reduce the risk of getting thioacetamide into your mouth.

In the following experiments, you will investigate the reactions of 23 frequently encountered elements (Na, K, Mg, Ca, Ba, Cr, Mn, Fe, Co, Cu, Ag, Zn, Hg, Al, C, Sn, Pb, N, S, F, Cl, Br, and I). They will be taken up, as listed, in the same order as they occur from left to right in the periodic table. As each new element is studied, it will be added to the scheme of analysis containing all the elements previously studied. Thus, you will gradually build up a systematic scheme of analysis and a working knowledge of the chemistry of the common elements.

When separating a precipitate from a solution, it is imperative that you both wash the precipitate with distilled water and check the filtrate for completeness of precipitation.

1   To wash the precipitate, first decant off the supernatant solution, add about 10 drops of water to the remaining precipitate, and stir well. Centrifuge the solution, discard the supernatant solution, and save the precipitate. Repeat. This removes all traces of contaminating ions from the precipitate.

2   To test for completeness of precipitation, add one drop of the precipitating reagent to the clear supernatant solution. If a precipitate appears, centrifuge to remove the precipitate, decant off the solution, and add another drop of the precipitating reagent. If another precipitate forms, repeat. If no precipitate appears, proceed with the required test on the solution.

If the supernatant solution is murky but the precipitate is too finely divided to be adequately removed by centrifugation, then you may filter the solution to remove this finely divided precipitate.

When you are in doubt about a particular qualitative analysis test, it is good practice to test simultaneously a known solution and your unknown solution for a particular ion, so that you can directly compare the results (e.g., flame test).

■ CAUTION

Most of the chemicals you will work with are poisonous to some degree. Immediately scrub your hands thoroughly after exposure to *any* chemical, and get into the habit of always washing your hands before leaving the laboratory.

Equipment and Chemicals

In addition to the side-shelf reagents listed below, you should have in your desk dropping bottles of each of the following reagents:

6 M HCl, 15 M NH$_3$, 1.7 M thioacetamide, 3 M H$_2$SO$_4$, 3 M NH$_4$C$_2$H$_3$O$_2$, 2 M NaOH, 6 M NH$_4$Cl, saturated Na$_2$SO$_4$, 2 M NaHSO$_4$, 6 M HNO$_3$

15-cm piece nichrome wire

Double-thickness cobalt glass

Side-Shelf Reagents

0.1 M NaF

0.1 M NaCl

0.1 M NaBr

0.1 M NaI

0.5 M Ca(NO$_3$)$_2$

0.1 M FeSO$_4$ (with iron wire and 0.1 M H$_2$SO$_4$)

18 M H$_2$SO$_4$

0.1 M BaCl$_2$

0.1 M Fe(NO$_3$)$_3$ (0.1 M HNO$_3$)

0.1 M Co(NO$_3$)$_2$

0.5 M KNCS

0.1 M Cr(NO$_3$)$_3$ (0.1 M HNO$_3$)

0.1 M MnCl$_2$

0.8 M NaIO$_4$ (2 M H$_2$SO$_4$)

0.5 M Pb (C$_2$H$_3$O$_2$)$_2$

0.1 M MgCl$_2$

0.1 M CaCl$_2$

0.1 M SrCl$_2$

0.1 M SnCl$_2$ (2 M HCl plus excess tin)

0.1 M Pb(NO$_3$)$_2$

6 M HC$_2$H$_3$O$_2$

1 M K$_2$CrO$_4$

0.1 M HgCl$_2$

0.1 M Al(NO$_3$)$_3$ (0.1 M HNO$_3$)

0.1 M Cu(NO$_3$)$_2$

0.1 M Zn(NO$_3$)$_2$

0.1 M AgNO$_3$ (in amber bottle)

3 M NH$_3$

Chlorine water

Cyclohexane

0.1 M Ba(OH)$_2$

0.1 M Hg$_2$(NO$_3$)$_2$ (1 M HNO$_3$)

0.1 M Hg(NO$_3$)$_2$ (0.5 M HNO$_3$)

15 M HNO$_3$

12 M HCl

0.5 M (NH$_4$)$_2$CO$_3$ (1 M NH$_3$)

1 M (NH$_4$)$_2$HPO$_4$

3 M HC$_2$H$_3$O$_2$

Ethyl alcohol

0.25 M (NH$_4$)$_2$C$_2$O$_4$

Solid Na$_2$SO$_4$

Solid Na$_2$CO$_3$

Solid NaNO$_3$

Solid KNO$_3$

Iron wire

Side-Shelf Reagents (cont.)

3% H_2O_2	Iron filings
6 M NH_3	Solid sodium bismuthate

■ CAUTION

Guard your reagents well, and do not let them get contaminated as, for example, by putting the wrong dropper back into a reagent bottle.

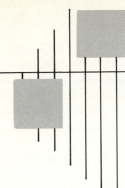

Experiment Q.1

Alkali and Alkaline-Earth Elements

It is characteristic of the alkali elements that their cations do not form insoluble chlorides, sulfides, carbonates, or phosphates. Therefore, in the qualitative analysis scheme, they do not precipitate when successive reagents are added but are left in the residual solution as unprecipitated ions. They are usually identified by their flame spectra in the presence of HCl as a volatilizing agent. Because the flame test is made after separation into all the groups and the test is sensitive to the presence of even trace amounts of element, it is necessary to guard against the introduction of contaminating amounts of alkali element when adding the precipitating reagents. For example, not only is it inadmissible to add NaCl to precipitate the insoluble chlorides, but the HCl used instead must be checked for possible presence of Na contamination. In the present scheme only the most common alkali elements (Na and K) are included.

It is characteristic of the alkaline-earth elements that their cations do not form insoluble chlorides or sulfides. Except for beryllium, which is not included in the present scheme, and magnesium, they do form insoluble carbonates and can thus be separated from the alkali elements. Once the carbonates have been precipitated, they can be redissolved in an acetic acid buffer and the alkaline-earth elements precipitated out consecutively, making use of differences in the solubilities of the chromates and oxalates. Because the difference with strontium is so slight, it is not included in the present scheme.

Procedure

(a) Be sure to record all observations in your book.

Tests for Alkali Elements Na and K Get a piece of ni-chrome wire about 15 cm long, and fashion the end of it into tiny loop. Stick the other end into a cork, which will serve as a handle. Prepare a solution by dissolving a few milligrams of Na_2CO_3 in a few milliliters of water and adding 1 mL of 6 M HCl. Dip the loop into the solution, and hold it in the blue edge of a burner flame. Note the color.

Clean the wire by heating it strongly in the flame and dipping it in a few milliliters of 6 M HCl in a test tube. Heat again. Repeat until the flame coloration reaches a low, con-stant level. Test a fresh portion of 6 M HCl to see how much Na contamination it has. In future tests, before reporting pos-itive presence of sodium, compare the flame test of the un-known with that given by reagent HCl.

Observe the flame test for K by making up a solution of a potassium salt, such as KNO_3, with added HCl. Look at the flame through a double thickness of cobalt glass. The glass transmits the K-flame color but filters out the Na color. Note that the K flame is very transient and can easily be over-looked.

Make up a solution by dissolving a very small amount of Na_2CO_3 and of KNO_3 in a few milliliters of water. Add 1 mL of 6 M HCl, and test for the flame coloration. Repeat with a fresh portion, looking through the cobalt glass at the very moment the loop is introduced into the flame.

(b) **Reactions of Alkaline-Earth Elements Mg, Ca, and Ba** (Because it is difficult to boil solutions in test tubes over open flames without blowing out the contents, it is recommended that at the start of each lab you put some water to boil in a beaker. You can use this throughout as a boiling-water bath into which you can place your test tubes whenever directed to heat or boil a solution.)

Barium ion is toxic. Wash your hands after handling.

In each of three labeled test tubes, place $\frac{1}{2}$ mL (about 10 drops) of one of the solutions $MgCl_2$, $CaCl_2$, or $BaCl_2$. If the solution is acidic, add 1 drop of concentrated NH_3 to make it basic. Add 2 drops of the ammonium carbonate solution [which is 0.5 M $(NH_4)_2CO_3$ and 1 M NH_3]. Note any car-bonate precipitation.

To any of the above solutions in which a precipitate does not appear, add 4 drops of 1 M $(NH_4)_2HPO_4$. The precipitate that forms is an ammonium phosphate, having the formula MNH_4PO_4, where M is the appropriate alkaline-earth cation.

Take each solution in which a carbonate has formed and centrifuge (or filter). Test the supernatant liquid (or filtrate) for completeness of precipitation by adding a few drops of the ammonium carbonate solution. Combine any precipitate formed with that originally obtained. Discard the supernatant liquid. Add 3 M acetic acid dropwise to the precipitates so as to dissolve them. Add half as many drops of 3 M ammonium acetate to the solutions. Warm gently, and then add 1 M K_2CrO_4 dropwise until any precipitation of $MCrO_4$ is complete. Centrifuge and separate. Dissolve any $MCrO_4$ formed in 6 M HCl, and test the resulting solution for flame coloration. [If a white $MCl_2(s)$ precipitate forms in the 6 M HCl, centrifuge, separate, and dissolve the $MCl_2(s)$ in distilled water. Test the MCl_2 solution for flame coloration.]

In any of the above cases where $MCrO_4$ did not precipitate, add 3 M NH_3 dropwise until the solution is neutral or the color has changed from orange to yellow, and then add about 5 drops of 0.25 M $(NH_4)_2C_2O_4$. Heat to boiling, and look for a white precipitate of MC_2O_4.

(c)   **Analysis of Unknown Possibly Containing Na, K, Mg, Ca, Ba**   On the basis of observations made in (*a*) and (*b*), fill out the scheme of analysis that follows by writing formulas and colors of all species formed. Also, fill in the same information in the complete analysis scheme shown in Experiment Q.9.

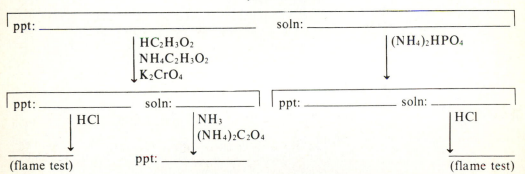

When you have mastered the chemistry of the above scheme, report to your lab instructor for oral examination. Get 5 mL of an unknown solution, which may contain any or all of the ions above. Analyze 1 mL of it by using the reactions indicated. Where precipitation occurs, separate the solid and solution by centrifugation (or filtration) before adding further reagents. Be extremely careful with your unknown sample;

you will not be given any refill. In case any test comes out to be doubtful, repeat the test on a known solution and compare with your unknown. Report final results to your instructor.

Questions

Q.1.1   What could you do to minimize each of the following errors commonly made by students in analyzing unknowns containing alkali and alkaline-earth elements?
(a) Reporting sodium present when it is there only as an impurity
(b) Missing potassium when it is actually present
(c) Obtaining an MNH_4PO_4 precipitate when Ba^{2+} and/or Ca^{2+} are present but Mg^{2+} is absent

Q.1.2   Given that the K_{sp} for $Mg(OH)_2 = 8.9 \times 10^{-12}$ and K_{diss} for $NH_3 = 1.8 \times 10^{-5}$, show whether or not $Mg(OH)_2(s)$ should precipitate from a buffer solution containing equal concentrations of NH_3 and NH_4^+, if $[Mg^{2+}] = 0.010\ M$.

Q.1.3   For the following reaction in aqueous solution,

$$Cr_2O_7^{2-} + 3H_2O \rightleftharpoons 2CrO_4^{2-} + 2H_3O^+$$

$$K = [CrO_4^{2-}]^2[H_3O^+]^2/[Cr_2O_7^{2-}] = 2.3 \times 10^{-15}$$

consider several aqueous solutions of $Na_2Cr_2O_7$. Find the ratio of $[CrO_4^{2-}]^2$ to $[Cr_2O_7^{2-}]$:
(a) In a neutral solution (that is, pH = 7)
(b) In a basic solution in which pH = 12.0
(c) In an acetic acid–acetate buffer solution (K_{diss} of acetic acid = 1.8×10^{-5})

Q.1.4   Given that K_{sp} of $BaCrO_4 = 8.5 \times 10^{-11}$, show whether or not $BaCrO_4(s)$ should precipitate from a solution in which $[Ba^{2+}] = 0.010\ M$ and $[Cr_2O_7^{2-}] = 0.10\ M$ and in which there is also an acetic acid–acetate buffer, that is, $[HC_2H_3O_2] = [C_2H_3O_2^-] = 1\ M$. See question Q.1.3 above.

Q.1.5   In preparation for a flame test on the solution resulting from dissolving the $MCrO_4$ precipitate in 6 M HCl, it is sometimes found that the yellow chromate solid is simply replaced by a white MCl_2 precipitate. In the instructions for Experiment Q.1, it is suggested that if this happens, the mixture should be centrifuged and the separated $MCl_2(s)$ dissolved in water. Why should the MCl_2 be more soluble in water than in 6 M HCl?

Q.1.6   In attempting to precipitate Ca^{2+} and Ba^{2+} as carbonates, leaving Mg^{2+}, Na^+, and K^+ in solution, why would selection of NaHCO$_3$ solution as the precipitating agent *not* give satisfactory results?

Q.1.7   Using the model in part (*c*) of this experiment, give a complete flow diagram for the analysis of a solution possibly containing Na$^+$, K$^+$, Mg^{2+}, Ca^{2+}, and Ba^{2+}. Write out an equation for each reaction that occurs.

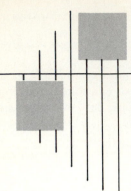

Experiment Q.2

Chromium and Manganese

Chromium is frequently encountered in solution as the chromic ion Cr^{3+} (purple), chromate ion CrO_4^{2-} (yellow), or dichromate ion $Cr_2O_7^{2-}$ (orange). The latter two are reduced to the $3+$ state by H_2S in the usual scheme of qualitative analysis. In this reaction colloidal sulfur frequently results.

Manganese is encountered in solution usually as the colorless manganous cation Mn^{2+} or as the deep-violet permanganate anion MnO_4^-. On addition of H_2S, MnO_4^- is reduced to Mn^{2+} in acid solution, but the color of MnO_4^- is such a giveaway of its presence that it is rarely included in a qualitative analysis unknown.

In this experiment you will for the first time go through the full scheme of qualitative analysis. Take careful notes on *all* your observations, noting both *what happens* and *what does not happen* when reagents are added. In part (*a*) you will do this for separate portions of chromic and manganous solutions. In part (*b*), you will use the entire scheme to analyze an unknown that may contain chromium and/or manganese plus any of the other elements studied so far.

The solubility of some compounds, particularly hydroxides and sulfides, depends on the pH of the solution. In this experiment, chromium and manganese do not precipitate when the solution contains H_3O^+ greater than 0.3 N but do precipitate when the solution is made basic. The precipitates are then redissolved with a solution buffered at pH 2. The buffer in this case is prepared from an equimolar mixture of $NaHSO_4$ and Na_2SO_4, which involves the equilibrium dissociation of HSO_4^-, $K = 1.3 \times 10^{-2}$.

Procedure

Be sure to record all observations.

(a) **Reactions of Chromium and Manganese** Into each of two labeled test tubes place $\frac{1}{2}$ mL (about 10 drops) of one of the standard solutions of Cr^{3+} and of Mn^{2+}. Add to each 2 drops of 6 M HCl. (Any insoluble chlorides would precipitate at this point.) Add $\frac{1}{2}$ mL of 1.7 M thioacetamide. Boil gently for at least 5 min by placing the test tubes in a beaker of boiling water. (Any acid-insoluble sulfides would precipitate here.) Add 1 mL of water, 2 drops of 6 M NH_4Cl, and enough 15 M NH_3 to make the solution basic to litmus.

■ CAUTION

H_2S and thioacetamide are very poisonous. Heat solutions containing sulfides, H_2S, or thioacetamide in the hood in a beaker of water provided. Wash hands thoroughly after using.

Centrifuge, note the color of the precipitate, and then, with a stirring rod, mix the precipitate back into the supernatant liquid. To the suspension in the test tube add 5 drops of 1.7 M thioacetamide and boil gently (in a water bath in the hood) for several minutes. Centrifuge the mixture. (Precipitation of insoluble hydroxides and any basic-insoluble sulfides is now complete.) The supernatant liquid, which would contain alkali metal and alkaline-earth ions in the complete scheme, may be discarded.

Wash any precipitate with a few drops of water, discarding the wash water. Add $\frac{1}{2}$ mL of saturated Na_2SO_4 solution and $\frac{1}{2}$ mL of 2 M $NaHSO_4$. Stir for 2 min, and note what happens. Add 10 drops of 3 M H_2SO_4, and boil to expel H_2S. Make basic with 2 M NaOH. Add $\frac{1}{2}$ mL of 3% H_2O_2. Heat. A solid residue remains in one case but not in the other. Centrifuge this residue off, and add to it a few drops of 3 M H_2SO_4 and 4 drops of acid 0.8 M $NaIO_4$ or a pea-sized chunk of sodium bismuthate. Heat, and note any color. To see the color of the solution in the presence of the bismuthate, which remains as a yellow solid, it may be necessary to centrifuge again.

■ CAUTION

Lead compounds are toxic. Wash.

In the case where a solid residue is not present, take the solution, which should have a characteristic color, boil it to destroy any excess H_2O_2, and acidify with 6 M $HC_2H_3O_2$. Add a couple of drops of 0.5 M $Pb(C_2H_3O_2)_2$ dropwise. [It is important to note the color of the solution before the addition of the acetate. If chromium is present as CrO_4^{2-}, the solution should be yellow. Addition of one drop of

$Pb(C_2H_3O_2)_2$ to this yellow solution will result in the precipitation of yellow $PbCrO_4$. If excess $Pb(C_2H_3O_2)_2$ is added, white $PbSO_4$ will coprecipitate with the $PbCrO_4$ and mask its yellow color. The SO_4^{2-} is derived from H_2SO_4 that was added earlier in the qualitative analysis scheme.]

(b) **Analysis of Unknown Possibly Containing Na, K, Mg, Ca, Ba, Mn, Cr** On the basis of the observations made in (a), draw up a scheme of analysis for chromium and manganese similar to the one in Experiment Q.1(c). Consult your textbook for any information required. You will find it useful to know that chromium precipitates as $Cr(OH)_3$ and manganese as MnS, which is subsequently oxidized to MnO_2 and then to MnO_4^-. Indicate how this scheme connects with that of Experiment Q.1(c). Enter as much information as possible into the general outline shown in Experiment Q.9.

When you understand the full scheme, report to your instructor for an oral quiz, and obtain an unknown solution. Analyze. (Your instructor may choose to omit the alkali and alkaline-earth elements, in which case you will be so informed.)

Questions

Q.2.1 From what you have observed in this experiment, what can you say about each of the following?
(a) Solubility of $CrCl_3$ and $MnCl_2$ in dilute HCl
(b) Solubility of MnS and $Cr(OH)_3$ in dilute HCl (and 0.1 M H_2S)
(c) Solubility of MnS and $Cr(OH)_3$ in NH_3/NH_4^+ buffer (in the presence of thioacetamide)
(d) Colors of Mn^{2+}, Cr^{3+}, MnS(s), $Cr(OH)_3(s)$, $Cr(OH)_4^-$, $Mn(OH)_2(s)$, $Mn(OH)_3(s)$, CrO_4^{2-}, $MnO_2(s)$, MnO_4^-, and $PbCrO_4(s)$.

Q.2.2 Given, K_{diss} for $NH_3 = 1.8 \times 10^{-5}$, K_I for H_2S (that is, $H_2S \rightleftharpoons H^+ + HS^-$) $= 1.1 \times 10^{-7}$, and $K_w = 1.0 \times 10^{-14}$ find K for the reaction, $NH_3 + H_2S \rightleftharpoons NH_4^+ + HS^-$.

Q.2.3 Write a balanced equation for the disproportionation reaction that occurs when $Mn(OH)_3(s)$ is treated with acid. What is the oxidation state of manganese in $Mn(OH)_3$ and in each of the products?

Q.2.4 Write an equation for the reactions that occur when a $Na_2SO_4/NaHSO_4$ buffer solution is added to the precipitate formed when Mn^{2+} and Cr^{3+} are treated with thioacetamide in the presence of NH_4^+/NH_3 buffer. Explain why it is necessary to treat the resulting solution with 3 M H_2SO_4 and boil before making the solution basic with 2 M NaOH.

Q.2.5   Give a complete flowchart for analysis of a solution possibly containing Na^+, K^+, Mg^{2+}, Ca^{2+}, Ba^{2+}, Mn^{2+}, and Cr^{3+}. Show the formulas and colors of species present at each stage.

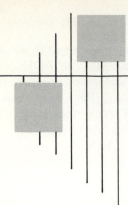

Experiment Q.3

Iron and Cobalt

Unknown solutions generally contain iron as ferrous ion (Fe^{2+}) or ferric ion (Fe^{3+}) and cobalt as cobaltous ion (Co^{2+}). In the scheme of analysis, Fe^{3+} is reduced by H_2S in acid to Fe^{2+} (forming colloidal sulfur in the process). In subsequent steps, Fe^{2+} is oxidized to Fe^{3+}, so it is usually impossible to tell the oxidation state of any iron in the original unknown solution.

The separation of Co^{2+} and Fe^{2+} as sulfides depends on the fact that one of these sulfides is slow to dissolve in slightly acid solution.

The final test for cobalt ion depends on its conversion to the complex ion $CoCl_4^{2-}$, which has a characteristic blue color but may show up as green if certain other species are present.

The final test for ferric ion depends on its forming a characteristically colored soluble complex ($FeNCS^{2+}$).

Procedure

Be sure to record all observations.

(a) **Reactions of Iron and Cobalt**   Starting with separate $\frac{1}{2}$-mL samples containing (1) Fe^{2+}, (2) Fe^{3+}, and (3) Co^{2+}, carry out the procedure outlined in the first paragraph of Experiment Q.2(a) to see where iron and cobalt first precipitate.

■ WEAR YOUR SAFETY GOGGLES

Wash the precipitated sulfides with water, and transfer them to clean test tubes. Add $\frac{1}{2}$ mL of saturated Na_2SO_4 solution and $\frac{1}{2}$ mL of 2 M $NaHSO_4$, in that order. Stir for 2 min. Note the difference in behavior of the CoS and FeS. Separate the insoluble one by centrifugation, wash with water, and transfer to an evaporating dish. Add $\frac{1}{2}$ mL of 6 M HCl and $\frac{1}{2}$ mL of 6 M HNO_3. Heat to boiling. Add 1 mL of water. Collect any sulfur on a stirring rod, and discard. (All black precipitate should have dissolved. If not, repeat the aqua regia treatment.) Evaporate to dryness gently. Note the color. Dissolve the residue in $\frac{1}{2}$ mL of water, transfer to a test tube, and add excess 2 M NaOH (ca. 10 drops) and 2 drops of H_2O_2. Heat to destroy excess H_2O_2. A brown-black precipitate of $Co(OH)_3$ should form. [In subsequent experiments the supernatant solution might contain $Zn(OH)_4{}^{2-}$.] The precipitate should be washed and as much of the water removed as possible. Then add excess 12 M HCl to the precipitate, and look for a bright blue due to $CoCl_4{}^{2-}$. Frequently, a bright-green solution will be obtained. This green solution is also indicative of the presence of cobalt.

To the test tube containing the sulfide soluble in Na_2SO_4-$NaHSO_4$, add 10 drops of 3 M H_2SO_4, and boil to expel H_2S. Make basic with 2 M NaOH. Add $\frac{1}{2}$ mL of 3% H_2O_2. Heat. Acidify with 3 M H_2SO_4. Add 2 drops of 0.5 M KNCS. Note the color.

(b)   **Analysis of Unknown Possibly Containing Na, K, Mg, Ca, Ba, Mn, Cr, Fe, Co**   (Your instructor may choose to omit the alkaline-earth and alkali groups from this unknown, in which case you will be so informed.) On the basis of the observations made in (a), draw up a scheme of analysis that includes iron and cobalt along with manganese and chromium. Also enter the information into the general scheme given in Experiment Q.9.

When you understand the chemistry, report to your instructor for an oral quiz, and obtain an unknown solution. Analyze.

Questions

Q.3.1   From what you have observed in this experiment, what can you say about each of the following?

(d)   Solubility of $FeCl_3$, $FeCl_2$, and $CoCl_2$ in dilute HCl

(b)   Solubility of FeS and CoS in dilute HCl (and $0.1\ M\ H_2S$)

(c)   Solubility of FeS and CoS in NH_3/NH_4^+ buffer (in the presence of thioacetamide)

(d)   Colors of Fe^{2+}, Fe^{3+}, Co^{2+}, $FeS(s)$, $CoS(s)$, $CoCl_4^{2-}$, $CoCl_2(s)$, $Co(OH)_3(s)$, $Fe(OH)_2(s)$, $Fe(OH)_3(s)$, and $FeNCS^{2+}$

(e)   Which precipitate dissolved faster in $Na_2SO_4/NaHSO_4$ buffer, FeS or CoS?

(f)   Amphoterism of $Co(OH)_3(s)$, if any

Q.3.2   Heating a thioacetamide solution in the presence of NH_3/NH_4^+ buffer produces a slightly different result with Fe^{3+} than with Fe^{2+}. Explain. (*Hint:* You get an additional product with Fe^{3+}.)

Q.3.3   Write a balanced equation for the reaction that occurs when $CoS(s)$ is treated with aqua regia, assuming the products are $CoCl_4^{2-}$, $NO_2(g)$, and $S(s)$.

Q.3.4   In the final test for iron, why must H_2O_2 be added before adding the KNCS solution?

Q.3.5   Make a flowchart showing the reagents used and the species present, together with their colors at each stage of the analysis, of a solution possibly containing Na^+, K^+, Mg^{2+}, Ca^{2+}, Ba^{2+}, Mn^{2+}, Cr^{3+}, Fe^{3+}, and Co^{2+}.

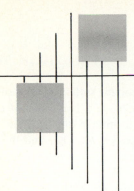

Experiment Q.4

Copper, Silver, Zinc, Mercury

Of the four elements copper, silver, zinc, and mercury, only mercury may occur in your unknown in two oxidation states, 1+ (mercurous, Hg_2^{2+}) and 2+ (mercuric, Hg^{2+}). These can be distinguished in the usual course of analysis because they precipitate at different points in the scheme.

The identification of Cu^{2+} depends on the formation of the colored complex $Cu(NH_3)_4^{2+}$.

The identification of Ag^+ is made on the basis that AgCl dissolves in NH_3 through formation of $Ag(NH_3)_2^+$ but reprecipitates when the solution is acidified. Although Hg_2Cl_2 is also insoluble, it does not dissolve in NH_3 but is converted to mercuric ammonobasic chloride $HgNH_2Cl$ and mercury.

Zn^{2+} can be identified because in the scheme it goes along with Co^{2+}, except that Zn^{2+} is soluble in excess NaOH, from which it can be precipitated as white ZnS.

Final identification of mercuric ion depends upon reduction by Sn^{2+} to Hg_2Cl_2 and Hg.

Procedure

(a) **Reactions of Cu, Ag, Zn, and Hg** Starting with separate $\frac{1}{2}$-mL samples containing (1) Cu^{2+}, (2) Ag^+, (3) Zn^{2+}, (4) Hg^{2+}, and (5) Hg_2^{2+}, carry out the successive group separations [by adding HCl, thioacetamide, base, etc., as in Experiment Q.2(a)] until each of these ions has precipitated. Make sure you heat the thioacetamide long enough to ensure hydrolysis and precipitation.

■ WEAR YOUR
SAFETY
GOGGLES

■ CAUTION

These chemicals are toxic when ingested. Use care in handling them.

Filter off the insoluble chlorides. (It is better here to use filtration than centrifugation.) With the precipitate still on the filter paper, place a test tube under the funnel and add 5 drops of 15 M NH_3 and 10 drops of water to the solid. Loosen the filter paper slightly to allow the funnel stem to drain. To each filtrate add 6 M HNO_3 dropwise.

Centrifuge off the acid-insoluble sulfides, and wash them with no more than 1 mL of water. Add 1 mL of 6 M HNO_3, and heat to boiling for at least 2 min. In one case the sulfide precipitate will react to give sulfur, $NO(g)$, and the metal ion; in the other case the sulfide precipitate will not react with the HNO_3. Cool each test tube. With a stirring rod, remove the sulfur residue from the test tube in which sulfur is formed, and to that solution add 15 M NH_3 drop by drop until basic. Note any color. To the undissolved sulfide in the other test tube, add 2 drops of 15 M HNO_3 and 10 drops of 12 M HCl. In the hood, heat to boiling, and stir to dissolve. Evaporate until only a few drops remain. Add 1 mL of water and a couple of drops of 0.1 M $SnCl_2$.

Centrifuge off the base-insoluble sulfide, and test its solubility in the Na_2SO_4–$NaHSO_4$ buffer ($\frac{1}{2}$ mL each of saturated Na_2SO_4 and 2 M $NaHSO_4$). [If it dissolves, your buffer solution is too acidic. Try using a 5:3 mixture of Na_2SO_4–$NaHSO_4$ solution (total volume 1 mL).] Centrifuge, and wash the precipitate with a few milliliters of water. Transfer it to an evaporating dish, add $\frac{1}{2}$ mL of 6 M HCl and $\frac{1}{2}$ mL of 6 M HNO_3, and heat to boiling. Add 1 mL of water; collect any sulfur on a stirring rod, and discard it. Evaporate to dryness *gently*. Dissolve the residue in $\frac{1}{2}$ mL of water. Transfer to a test tube, add excess 2 M NaOH and 2 drops of H_2O_2. Heat to destroy any excess H_2O_2. Add 10 drops of 1.7 M thioacetamide. Heat just to boiling and let cool, while watching for a precipitate.

(b)   **Analysis of Unknown Possibly Containing Mn, Cr, Fe, Co, Cu, Ag, Zn, Hg**   Draw up a scheme of analysis for these elements. Also enter the information into the general scheme outlined in Experiment Q.9.

Report to your instructor for oral quiz and an unknown solution. Analyze.

Questions

Q.4.1   From what you have observed in this experiment, what can you say about each of the following?

    (*a*)   Solubility of $CuCl_2$, $AgCl$, $ZnCl_2$, Hg_2Cl_2, and $HgCl_2$ in dilute HCl

    (*b*)   Solubility of CuS, ZnS, and HgS in dilute HCl (and 0.1 M H_2S)

    (*c*)   Of these ions, Cu^{2+}, Ag^+, Zn^{2+}, and Hg^{2+}, which precipitate in the analysis scheme as base-insoluble sulfides?

    (*d*)   Colors of Cu^{2+}, Ag^+, Zn^{2+}, Hg^{2+}, Hg_2^{2+}, mixture of $HgNH_2Cl(s)$ + $Hg(l)$, $Ag(NH_3)_2^+$, $CuS(s)$, $HgS(s)$, $Cu(NH_3)_4^{2+}$, $AgCl(s)$, $ZnCl_2(s)$, $Hg_2Cl_2(s)$, $HgCl_4^{2-}$, $Zn(OH)_4^{2-}$, $ZnS(s)$

Q.4.2   What problem will develop in subsequent stages of the analysis of a solution containing Ag^+ if the Ag^+ is not completely removed as a chloride in the first step?

Q.4.3   Given K_1 for $H_2S = 1.1 \times 10^{-7}$ and K_2 for $H_2S = 1 \times 10^{-14}$, explain why $[S^{2-}]$ is greater in a solution of higher pH than in one of lower pH.

Q.4.4   Write balanced equations for each of the following reactions:

    (*a*)   Conversion of $HgS(s)$ to $NO_2(g)$, $HgCl_4^{2-}$, $S(s)$, and H_2O by reaction with aqua regia

    (*b*)   Reaction of $HgCl_4^{2-}$ with $SnCl_2$ to give $Hg_2Cl_2(s)$, $SnCl_4$, and Cl^-

    (*c*)   Reaction of $Hg_2Cl_2(s)$ with $SnCl_2$ to give $Hg(l)$ and $SnCl_4$.

    (*d*)   Reaction between $AgCl(s)$ and excess NH_3

    (*e*)   Reaction between Cu^{2+} and excess NH_3

    (*f*)   Conversion of $CuS(s)$ to Cu^{2+}, $NO(g)$, H_2O, and $S(s)$ by reaction with concentrated HNO_3

Q.4.5   Make a flowchart showing reagents to be used, species formed at each stage, and reactions that occur in the analysis of a solution containing Mn^{2+}, Cr^{3+}, Fe^{2+}, Co^{2+}, Cu^{2+}, Ag^+, Zn^{2+}, Hg_2^{2+}, and Hg^{2+}.

Q.4.6   Addition of a small amount of aqueous NH_3 to either Cu^{2+} or Ag^+ results in formation of a precipitate that does not contain ammonia. What are these precipitates, and why are they formed?

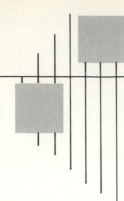

Experiment Q.5

Aluminum

Aluminum precipitates as the hydroxide with the base-insoluble sulfides. Its subsequent identification depends on the fact that aluminum hydroxide is soluble in excess base but, being amphoteric, reprecipitates when the pH of the basic solution is lowered.

Procedure

(a) **Reactions of Aluminum** Start with $\frac{1}{2}$ mL of solution containing Al^{3+}. Add 2 drops of 6 M NH_4Cl and 1 drop of 15 M NH_3. Look for a gelatinous precipitate (which may be hard to see). Centrifuge off the precipitate, and discard the liquid. Add $\frac{1}{2}$ mL of saturated Na_2SO_4 and $\frac{1}{2}$ mL of 2 M $NaHSO_4$. Stir for 2 min until the precipitate dissolves. Add 2 M NaOH dropwise until a precipitate forms and then redissolves. Add an equal number of drops of 6 M NH_4Cl, and watch for a reprecipitate.

■ WEAR YOUR SAFETY GOGGLES

(b) **Analysis of Unknown Possibly Containing Mn, Cr, Fe, Co, Cu, Ag, Zn, Hg, Al** Extend your previous scheme to include aluminum. Enter the information into the scheme outlined in Experiment Q.9. Learn it. See your lab instructor for an oral quiz and an unknown solution.

Questions

Q.5.1 From what you have observed in this experiment, what can you say about each of the following?
(a) Color and appearance of Al^{3+}, $Al(OH)_3(s)$, and $Al(OH)_4^-$
(b) Solubility of $Al(OH)_3(s)$ in NH_3/NH_4^+ buffer
(c) Solubility of $Al(OH)_3(s)$ in SO_4^{2-}/HSO_4^- buffer

Q.5.2 Complete and balance the equation for each of the following:
(a) $AlCl_3 + H_2O \rightarrow Al(OH)^{2+}$
(b) $Al(OH)^{2+} + OH^- \rightarrow Al(OH)_3(s)$
(c) $Al(OH)_3(s) + OH^- \rightarrow$

Q.5.3   By means of balanced equations, show how each of the following separations is made in the standard scheme of analysis:
(*a*) Al^{3+} from Ag^+
(*b*) Al^{3+} from Cu^{2+}
(*c*) $Al(OH)_3(s)$ from $ZnS(s)$
(*d*) $Al(OH)_4^-$ from CrO_4^{2-}
(*e*) Al^{3+} from Mn^{2+}

Q.5.4   Write a complete flowchart showing reagents to be used, species formed at each stage, and reactions that occur in the analysis of a solution containing the entire set of ions encountered in Experiments Q.1 to Q.5.

Q.5.5   You are given a solid sample that may contain one or more of the following: $KAl(SO_4)_2$, $MnCl_2$, $Ba(NO_3)_2$, $FeCl_2$, $AgNO_3$, $ZnCl_2$, $Na_2Cr_2O_7$. Addition of water to the unknown gives an acidic orange solution and a white precipitate, which is filtered out and washed. The precipitate partially dissolves when treated with 6 *M* NH_3. Progressive addition of 1 *M* NaOH to the orange solution produces a white precipitate, which dissolves in excess NaOH, and a yellow solution. What can you conclude about the makeup of the original unknown?

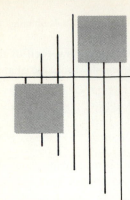

Experiment Q.6

Tin and Lead

Tin may occur in both the 2+ and the 4+ states, whereas lead is usually found only in the 2+ state. Both stannous (Sn^{2+}) and stannic (Sn^{4+}) precipitate as insoluble sulfides. To avoid the difficulty of having to dissolve SnS, it is convenient to oxidize Sn^{2+} to Sn^{4+} with H_2O_2 prior to sulfide precipitation. Yellow SnS_2 can then be converted to soluble thiostannate ion (SnS_3^{2-}), which makes possible separation of the tin from CuS, HgS, and PbS. The final test for tin is the reduction to Sn^{2+} with Fe filings followed by reaction with $HgCl_2$. In the last step, Sn^{2+} reduces $HgCl_2$ to a mixture of Hg and Hg_2Cl_2.

Lead forms an insoluble chloride, which, however, is somewhat more soluble than AgCl and Hg_2Cl_2. For this reason and because PbS is less soluble than $PbCl_2$, lead shows up in the qualitative analysis scheme in two places. In the insoluble chloride group, it can be distinguished by its solubility in hot water and subsequent precipitation as lead chromate. In the insoluble sulfide group, lead differs from mercury in that PbS, unlike HgS, dissolves in nitric acid and differs from copper in forming an insoluble sulfate. The presence of lead can be confirmed by a conversion of $PbSO_4$ to $PbCrO_4$.

Procedure

(a) **Reactions of Tin and Lead** Start with separate $\frac{1}{2}$-mL samples of (1) Sn^{2+} and (2) Pb^{2+}. Add to each 2 drops of 6 M HCl. Filter off the precipitate, and *save the filtrate* for sulfide treatment. (It is better here to use filtration than centrifugation.) Gently lift the filter paper out of the funnel, and pour hot water through the funnel to heat it. Replace the filter paper, put a test tube under the stem to catch the filtrate, and pour about 1 mL of hot water on the filter paper. To the solution that has come through, add 1 drop of 6 M $HC_2H_3O_2$ and 1 drop of 1 M K_2CrO_4.

■ **WEAR YOUR SAFETY GOGGLES**

■ CAUTION   **Lead is toxic. Do not ingest.**

To the test tube containing the filtrate from the insoluble chloride saved for sulfide treatment, and to the test tube containing the soluble chloride, add 3 drops of 3% H_2O_2. Heat gently for several minutes. Add 1 drop of 6 M HCl and $\frac{1}{2}$ mL of 1.7 M thioacetamide. Boil gently for several minutes. Centrifuge, and wash the solid with a few drops of water. Add 10 drops of 15 M NH_3 and 10 drops of 1.7 M thioacetamide. Heat gently with stirring for several minutes. In the case where the precipitate has dissolved, add 10 drops of 6 M HCl, and transfer to an evaporating dish. Heat over a low flame in the hood until most of the solution is gone. Add some iron filings, a few more drops of 6 M HCl, and reheat at least until all traces of yellow are gone. Decant the solution to a test tube, add 2 drops of 0.1 M $HgCl_2$, and stir.

Centrifuge off the previously precipitated PbS, and transfer it to a test tube using about 1 mL of water. Add 1 mL of 6 M HNO_3, and heat to boiling for about a minute. Cool, and transfer to an evaporating dish. Carefully add 10 drops of 3 M H_2SO_4, and evaporate in a hood until dense white SO_3 fumes appear. (*Note:* It is easy to confuse HNO_3 fumes.) Let cool. Add $\frac{1}{2}$ mL of 3 M $NH_4C_2H_3O_2$, and heat to dissolve. Centrifuge, and discard any solid. To the solution add a drop of 1 M K_2CrO_4.

(b)   **Analysis of Unknown Possibly Containing Mn, Cr, Fe, Co, Cu, Ag, Zn, Hg, Al, Sn, Pb**   Draw up a scheme of analysis for all these elements, noting that H_2O_2 must be added prior to any sulfide precipitation. Enter the information into the scheme outlined in Experiment Q.9. Check with your instructor, and analyze the given unknown.

Questions

Q.6.1   From what you have observed in this experiment, what can you say about each of the following?
(*a*) Solubility of $SnCl_2$ and $PbCl_2$ in dilute HCl
(*b*) Solubility of SnS_2 and PbS in dilute HCl (and 0.1 M H_2S)
(*c*) Solubility of SnS_2 and PbS in NH_4^+/NH_3 buffer (and saturated thioacetamide)
(*d*) Solubility of $PbSO_4$ and $PbCrO_4$ in 3 M $NH_4C_2H_3O_2$
(*e*) Colors of Sn^{2+}, Sn^{4+}, Pb^{2+}, $PbCl_2(s)$, PbS (s), $SnS_2(s)$, SnS_3^{2-}, $PbSO_4(s)$, $PbCrO_4(s)$, mixture of $Hg_2Cl_2(s)$ + Hg (l)

Q.6.2 Complete and write balanced equations for each of the following reactions:

(a) $Sn^{2+} + H_2O_2 + H_3O^+ \rightarrow$

(b) $SnS_3^{2-} + H_3O^+ \rightarrow SnS_2(s) + H_2S$

(c) $Fe(s) + Sn^{4+} \rightarrow$

(d) $Sn^{2+} + HgCl_2 \rightarrow Hg_2Cl_2(s) + \cdots$

(e) $Sn^{2+} + Hg_2Cl_2(s) \rightarrow Sn^{4+} + \cdots$

(f) $PbS(s) + HNO_3 \rightarrow NO(g) + S(s) + \cdots$

Q.6.3 Before pouring hot water through the filter paper containing the insoluble chloride, you are directed to heat the funnel with hot water. Explain.

Q.6.4 Make a flowchart for the analysis of an unknown possibly containing Mn^{2+}, Cr^{3+}, Fe^{2+}, Co^{2+}, Cu^{2+}, Ag^+, Zn^{2+}, Hg_2^{2+}, Hg^{2+}, Al^{3+}, Sn^{4+}, Pb^{2+}. Show the species present at each stage of the analysis.

Q.6.5 Using as few reagents as possible and chemical reactions you have studied in developing the qualitative analysis scheme, tell how to separate

(a) Ag^+ from Pb^{2+}

(b) Zn^{2+} from Pb^{2+}

(c) Sn^{2+} from Cu^{2+}

(d) Sn^{4+} from Ba^{2+}

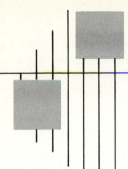

Experiment Q.7

Carbonate, Nitrate, Sulfate

Although it is possible to set up a systematic group separation for anions as we have done for cations, it is common practice to analyze specifically for each anion. In this course we consider the possible anions to be carbonate, nitrate, sulfate (Experiment Q.7), and the halides (Experiment Q.8), and will study specific tests for each.

The test for carbonate depends on the conversion of CO_3^{2-} to CO_2 by acid; it is confirmed by reaction of CO_2 with aqueous $Ba(OH)_2$ to form insoluble $BaCO_3$.

The test for nitrate depends on the reduction of NO_3^- to NO by Fe^{2+} in the presence of concentrated H_2SO_4. A brown complex $(FeNO^{2+})$ comes from the combination of NO and Fe^{2+} and is detected as a brown ring at the interface between the H_2SO_4 and the solution.

Sulfate forms an insoluble white precipitate with Ba^{2+} that is not soluble in acid solution.

Procedure

(a) **Reactions of Carbonate, Nitrate, and Sulfate**   Place in a test tube about 10 mg of solid Na_2CO_3. (Do not weigh it out! Ten mg is about the volume of one-fifth of a drop of water.) Prepare a film of $Ba(OH)_2$ solution by dipping a nichrome wire loop into your own supply of 0.10 M $Ba(OH)_2$ solution, which you have obtained from the reagent shelf. Add a few drops of dilute HNO_3 to the Na_2CO_3, hold the film in the escaping gas, and note what happens.

■ WEAR YOUR
SAFETY
GOGGLES

■ CAUTION   **Barium is toxic. Do not ingest.**

Dissolve about 10 mg of $NaNO_3$ in about $\frac{1}{2}$ mL of water in a test tube. Add a drop or two of 3 M H_2SO_4 (to make the solution acid) and 5 drops of 0.1 M $FeSO_4$. Holding the tube at an angle, slowly pour about 10 drops of 18 M H_2SO_4 *so* *that it runs down along the wall of the tube to form a separate layer beneath the solution.* Note the change at the interface.* (I^- and Br^- will interfere with the brown-ring test. See Experiment Q.8.)

■ CAUTION   **Concentrated H_2SO_4 is highly corrosive. Flush immediately with large amounts of water if you get it on your skin.**

Dissolve about 10 mg of Na_2SO_4 in $\frac{1}{2}$ mL of water. Add a few drops of 6 M HCl and a few drops of 0.10 M $BaCl_2$. Note any precipitate.

To check whether these ions will interfere with each other, make up a solid mixture consisting of Na_2CO_3 and $NaNO_3$. Test it for sulfate.

Make up a solid mixture of Na_2CO_3 and Na_2SO_4, and perform the brown-ring test on it.

Finally, test a solid mixture of $NaNO_3$ and Na_2SO_4 for carbonate.

(b)   **Analysis of a Soluble Solid Unknown for Presence of CO_3^{2-}, NO_3^-, and SO_4^{2-}**   Devise a method of attack, and check it with your instructor. Enter the information into the general scheme in Experiment Q.9. Get a solid sample, and analyze for the above anions.

Questions

Q.7.1   How many milliliters of $CO_2(g)$ at 1 atm pressure and 20.0°C will be produced by reaction of 10.0 mg of Na_2CO_3 with excess HNO_3, if the fact that some CO_2 dissolves in the solution is ignored?

Q.7.2   It is sometimes suggested that in the test for carbonate, two $Ba(OH)_2$ films be prepared, one to be held over the test tube and one to be held away from the test tube and used for comparison. What advantages might the use of such a "blank" provide?

Q.7.3   A student decided that it should be possible to test for car-

---

*Note: If you have trouble getting a positive test with this procedure, hold the test tube upright and introduce the 18 M H_2SO_4 directly at the bottom by means of a disposable pipette.

bonate by replacing the $Ba(OH)_2$ film on a wire loop with a piece of wet blue litmus paper. Should this work? Explain.

Q.7.4   Write balanced equations for the reactions that occur in the brown-ring test for nitrate.

Q.7.5   Clearly show, using flowcharts, how one could analyze an unknown possibly containing CO_3^{2-}, NO_3^-, and SO_4^{2-}.

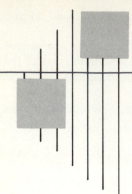

Experiment Q.8

The Halides

The halides (F^-, Cl^-, Br^-, and I^-) can be distinguished from each other in that one of them forms an insoluble calcium salt in a slightly alkaline solution whereas the other three form insoluble silver salts. The latter three differ in the extent of their dissolving in aqueous ammonia caused by the formation of the silver-ammonia complex, $Ag(NH_3)_2^+$. Additional distinguishing information can be obtained from oxidation of the halide ions to the elemental forms, which give distinctive colors in cyclohexane or other nonpolar solvents.

Procedure

(a) Reactions of Fluoride, Chloride, Bromide, and Iodide   To $\frac{1}{2}$-mL samples of (1) NaF, (2) NaCl, (3) NaBr, and (4) NaI solutions, first add enough concentrated ammonia (a few drops) to make the solutions just basic, then add several drops of 1 M $Ca(NO_3)_2$ solution. Centrifuge. To fresh $\frac{1}{2} =$ mL samples of NaF, NaCl, NaBr, and NaI solutions add several drops of 0.1 M $AgNO_3$. To each insoluble silver halide, add 5 drops of 3 M NH_3. Stir to see if the precipitate dissolves. If it does, add a few drops of dilute HNO_3 to see whether it reprecipitates.

Mix $\frac{1}{2}$ mL of chlorine water with $\frac{1}{2}$ mL of cyclohexane. Tap the side of the test tube briskly with the finger, and note color. Add $\frac{1}{2}$ mL of NaBr solution. Tap briskly, and note color.

Cyclohexane is flammable; keep away from open flames.

Mix $\frac{1}{2}$ mL of cyclohexane and $\frac{1}{2}$ mL of NaI solution. Add *dropwise*, looking for consecutive reactions, no more than 20 drops of *fresh* chlorine water, shaking between added drops. If the chlorine water is weak, it helps to add a drop of sulfuric or acetic acid. (Chlorine can oxidize I^- to I_2 and also I_2 to IO_3^-.)

Mix $\frac{1}{2}$ mL of NaI solution with $\frac{1}{2}$ mL of NaBr solution

■ WEAR YOUR SAFETY GOGGLES

■ CAUTION

and $\frac{1}{2}$ mL of cyclohexane. Note behavior on dropwise addition of chlorine water.

(b)   **Analysis of an Unknown Possibly Containing CO_3^{2-}, NO_3^-, SO_4^{2-}, F^-, CI^-, Br^-, and I^-**   Obtain a solid unknown, and to avoid interferences analyze it as follows: Split your unknown into at least four 10-mg portions. With portion 1, test for sulfate as in Experiment Q.7. [*Note:* If the test for sulfate is positive, you should also use portion 1 to test for fluoride. But first remove the sulfate by adding a slight excess of $BaCl_2$, centrifuging, neutralizing the solution with NH_3, and then adding several drops of 1 M $Ca(NO_3)_2$.] Dissolve portion 2 in $\frac{1}{2}$ mL of water, and test for carbonate as in Experiment Q.7. To the resulting solution add NH_3 until slightly basic, then add several drops of 1 M $Ca(NO_3)_2$. Centrifuge if a precipitate forms. Then add to the solution several drops of 0.1 M $AgNO_3$. If any silver salts precipitate, centrifuge them off, and wash with water. By threefold dilution of the stock 3 M NH_3, prepare 1 M NH_3, and add this to the precipitate. Stir, centrifuge off any solid, and test the separated solution by adding a few drops of 3 M HNO_3.

(You should know by this stage if CO_3^{2-}, SO_4^{2-}, F^-, and Cl^- are present in your unknown, and might know that Br^- and I^- are absent.)

Dissolve portion 3 in water, and test for Br^- and I^- as in part (*a*) of this experiment. If I^- is present, make sure that you add enough chlorine water so that any iodine color obtained is discharged.

Dissolve portion 4 in $\frac{1}{2}$ mL of water to test for nitrate. If Br^- and I^- are absent, simply proceed as in Experiment Q.7. If Br^- or I^- are present, they will be oxidized by the concentrated H_2SO_4 to produce Br_2 or I_2, thus confusing the brown-ring test. To remove Br^- or I^-, add 0.10 M Ag_2SO_4 until precipitation is complete. Filter off the precipitate, and proceed with the nitrate test as usual.

Enter the information you have learned in this experiment into the flowchart given in Experiment Q.9.

Questions

Q.8.1   Hydrofluoric acid, HF, is a weak acid. Should $CaF_2(s)$ be more soluble in an alkaline or in an acidic solution? Explain.

Q.8.2   From what you have observed in this experiment, what can you say about each of the following?

(*a*) Solubility of $CaF_2(s)$, $CaCl_2(s)$, $CaBr_2(s)$, $CaI_2(s)$ in slightly alkaline aqueous solution

(b) Solubility of $AgF(s)$, $AgCl(s)$, $AgBr(s)$, $AgI(s)$ in water

(c) Solubility of $AgF(s)$, $AgCl(s)$, $AgBr(s)$, $AgI(s)$ in dilute NH_3

(d) Colors of Cl_2, Br_2, I_2 in water and in cyclohexane

Q.8.3 How do you test for

(a) I^- in the presence of Br^-?

(b) NO_3^- in the presence of Br^-?

(c) F^- in the presence of Br^-?

Q.8.4 Write balanced equations for the reactions that occur when

(a) NH_3 is added to $AgCl(s)$

(b) Cl_2 is added to $Br^-(aq)$

(c) A small amount of Cl_2 is added to $I^-(aq)$

(d) An excess of Cl_2 is added to $I^-(aq)$

Q.8.5 Prepare flowcharts detailing how, by taking several samples, you could systematically analyze an unknown possibly containing CO_3^{2-}, NO_3^-, SO_4^{2-}, F^-, Cl^-, Br^-, I^-.

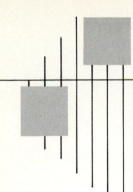

Experiment Q.9

General Unknown Solution

In the preceding experiments, you have studied the analytical reactions of some common cations and anions. In this experiment you will analyze a solution for the possible presence of all these ions. To prepare yourself for the search, it will be necessary to study the previous experiments and complete the following outline. Since it is advisable to test for anions and for cations on separate portions of the unknown solution, the anion and cation schemes are given as independent procedures. When performing the analysis, you may find shortcuts. If you find first that certain anions are present, some cations may be excluded since they would precipitate. Thus, if CO_3^{2-} is present, Ba^{2+}, for example, cannot be in the solution at the same time. Similarly, if during the group separation none of a particular group of ions precipitates, there is evidently no point in carrying out any operations to distinguish the members of that group.

Before going for your unknown, fill in the blanks on the next page and check your scheme with your instructor.

Anion Analysis (CO_3^{2-}, NO_3^-, SO_4^{2-}, F^-, Cl^-, Br^-, I^-):

Portion 1:
Add HCl and $BaCl_2$.   ppt: _____

Portion 2:
Add HNO_3. Let gas react with $Ba(OH)_2$.   ppt: _____
    ↓ Add $Ca(NO_3)_2$ to solution*

| soln: ppt: _____ |

    ↓ Add $AgNO_3$

| ppt: _____ _____ _____ |

Add NH_3 to previous precipitates

| soln: residue: _____ _____ |

    ↓ Add HNO_3
ppt: _____

Portion 3:
Add Cl_2 water dropwise. Shake with cyclohexane.

    First color: _____ caused by _____
    Later color: _____ caused by _____

Portion 4:
Add Ag_2SO_4 to remove Br^- and I^-.
Then add $FeSO_4$ and conc. H_2SO_4. Color: _____ caused by _____

*Use modified procedure with portion 1 if sulfate is present.

Cation Analysis (Na, K, Mg, Ca, Ba, Cr, Mn, Fe, Co, Cu, Ag, Zn, Hg, Al, Sn, Pb):

Unknown solution:

   ↓   Add HCl

ppt: Group A   soln:

————   Add 3% H_2O_2

————   Add HCl and thioacetamide ↓

————   ppt: Group B   soln:

        ————   Add NH_3 and thioacetamide ↓

        ————   ppt: Group C   soln:

        ————   ————   Add NH_3 and $(NH_4)_2CO_3$ ↓

        ————   ————   ppt: Group D   soln:

        ————   ————   Add $(NH_4)_2HPO_4$ ↓

        ————   ————   ppt: soln:

        ————   ————   ————   ————

        ————   ————   flame

                                               test

Group A ppt:

   ↓   Wash with hot water

soln: residue: ———— ————

   ↓   Add $HC_2H_3O_2$ and K_2CrO_4 ↓   Add NH_3

ppt: ————   soln: residue:

                                                      ————

                                         ↓   Add HNO_3 ————

                                     ppt: ————

Group B ppt:

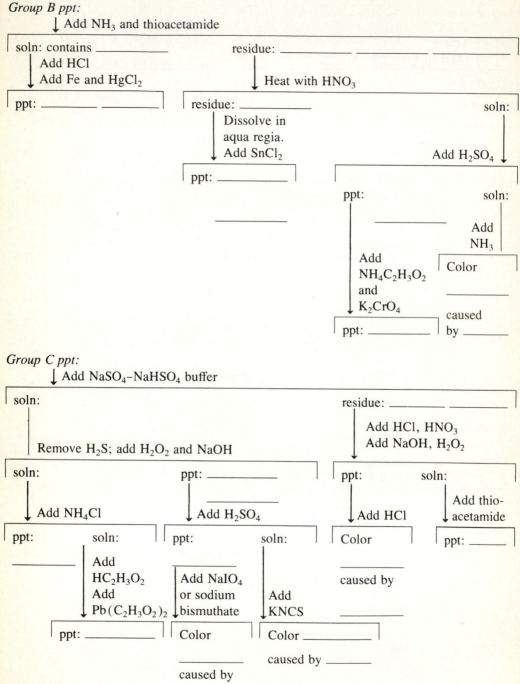

↓ Add NH₃ and thioacetamide

soln: contains _____ residue: _____ _____ _____

↓ Add HCl
↓ Add Fe and HgCl₂

ppt: _____ _____

↓ Heat with HNO₃

residue: _____ soln:

↓ Dissolve in aqua regia.
↓ Add SnCl₂

ppt: _____

Add H₂SO₄ ↓

ppt: soln:

_____

Add NH₃

Color

Add NH₄C₂H₃O₂ and K₂CrO₄ ↓

_____

ppt: _____ caused by _____

Group C ppt:

↓ Add NaSO₄–NaHSO₄ buffer

soln: residue: _____ _____

↓ Add HCl, HNO₃
↓ Add NaOH, H₂O₂

| Remove H₂S; add H₂O₂ and NaOH

soln: ppt: _____ ppt: soln:

_____

↓ Add NH₄Cl ↓ Add H₂SO₄ ↓ Add HCl ↓ Add thioacetamide

ppt: soln: ppt: soln: Color ppt: _____

_____

Add HC₂H₃O₂
Add Pb(C₂H₃O₂)₂ ↓

Add NaIO₄ or sodium bismuthate ↓

Add KNCS ↓

caused by

_____

ppt: _____ Color Color _____

_____ caused by _____

caused by

_____

Group D ppt:

  $\downarrow$ Add $HC_2H_3O_2$–$NH_4C_2H_3O_2$ buffer ,and K_2CrO_4

ppt: _____   soln:

  $\downarrow$ Dissolve in HCl $\downarrow$ Add NH_3 and $(NH_4)_2C_2O_4$

Flame test _____   ppt: _____

caused by _____

Questions

Q.9.1 Based on what you have learned in Experiments Q.1 through Q.8, what cations can you eliminate if your initial unknown is completely soluble and
(a) Cl^- is found to be present
(b) SO_4^{2-} is found to be present
(c) The solution is colorless
(d) The solution is basic

Q.9.2 Predict the specific problems that would occur in the analysis of a general cation unknown if
(a) Ag^+ were not completely removed in the first step (as an insoluble chloride)
(b) 3% H_2O_2 were not added prior to precipitation of the group B acid insoluble sulfides, and the unknown contained Sn^{2+}
(c) Ba^{2+} were not completely removed from the solution as a carbonate before the test for Mg^{2+} was made

Q.9.3 Using reactions from the qualitative analysis scheme, explain how one could separate each of the following pairs of ions with as few reagents as possible.
(a) Cu^{2+} and Al^{3+}
(b) Mn^{2+} and Fe^{2+}
(c) Ba^{2+} and Hg^{2+}
(d) Cr^{3+} and Zn^{2+}

Q.9.4 Complete and write balanced equations for each of the following reactions:
(a) $Hg_2Cl_2 + NH_3 \rightarrow$
(b) $Fe^{3+} + H_2S + H_3O^+ \rightarrow$
(c) $HgS(s) + HNO_3 + HCl \rightarrow$
(d) $Zn(OH)_2(s) + OH^- \rightarrow$
(e) $MnO_2(s) + IO_4^- + H_3O^+ \rightarrow$
(f) $Cr^{3+} + HO_2^- + OH^- \rightarrow$
(g) $HgCl_4^{2-} + Sn^{2+} \rightarrow$

Q.9.5 Suppose that your general unknown solution contains Ag^+, Mn^{2+}, Cr^{3+}, Cu^{2+}, Zn^{2+}, and Ca^{2+}. Write down the species of each metal ion present at each stage of the complete analysis.

Q.9.6 An unknown solution contains one or more of the following cations: Ag^+, Ba^{2+}, Co^{2+}, Cr^{3+}, Cu^{2+}, Hg_2^{2+}, Hg^{2+}, Mn^{2+}. The colored unknown solution, when treated with dilute HCl, gives a white precipitate (A), which is found after filtration to dissolve completely in excess NH_3. The filtrate gives a black precipitate (B) and a colored solution (C) when treated with thioacetamide. After separation, the black precipitate (B) dissolves in hot HNO_3 to give sulfur, which is removed, and a light-blue solution that gives a dark-blue solution when treated with excess NH_3. Addition of excess NH_3 to solution C and treatment with thioacetamide produces a greenish dark precipitate (D). After separation, precipitate D partially dissolves in SO_4^{2-}/HSO_4^- buffer to give a colored solution. After removal of the precipitate and of H_2S, the colored solution reacts with excess NaOH and H_2O_2 to give a clear yellow solution (E). Comment on the composition of the initial unknown solution.

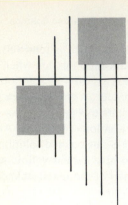

Experiment Q.10

Solid Unknowns

Solid unknowns may be either simple substances or complex mixtures. "Simple substance" is here taken to mean a pure element or a compound consisting of only one cation and one anion. Possible ions include all those previously studied plus hydroxide or oxide.

Procedure

(a) Simple Solid Unknown   Examine the unknown for color, crystal form, and any other distinguishing characteristics.

■ WEAR YOUR SAFETY GOGGLES

Test the unknown for solubility. (Use a small amount! No point in saturating the solution.) A useful sequence of solvents would be water, 6 M HNO_3, 15 M HNO_3, 15 M NH_3. Since the rate of solution may be slow, warm and wait, working in the hood where necessary. If all the above solvents fail, try heating your unknown with 1 M Na_2CO_3 for at least 5 min, decant the liquid, and treat the residue with 6 M HNO_3. Insoluble salts such as $BaSO_4$ can be thus partially converted to carbonates, which dissolve in acid. The behavior of the unknown in these solubility tests may give definite clues to the identity of the elements present as well as provide a solution sample for systematic analysis.

Use care in reporting your final results to take account of possible introduction of an ion by the reagent used to dissolve the unknown. If you find in your solution that the only anion is one introduced by the solvent, then there are three possibilities: (1) Your unknown solid contained that anion, (2) your unknown solid is an oxide or hydroxide, or (3) your unknown was a carbonate. Case 3 would be obvious from gas evolution. Cases 1 and 2 could be distinguished by changing the solvent used to dissolve the sample.

In general, successful analysis of a solid depends on ingenuity in interpreting observations and planning clever tests.

For example, a determination of the acidity or basicity of the solution of a water-soluble unknown often provides valuable information about the identity of the unknown.

(b) **Solid Mixtures** Visually examine your unknown to determine in so far as possible how many components are in it. Then dissolve and analyze. The main new problem here comes about because only part of the sample may dissolve in a particular reagent. This may actually be an advantage in picking apart the unknown, but you have to recognize that some of the sample does in fact dissolve. Sometimes the decrease in the amount of remaining solid will be easily noticeable; otherwise, you may have to evaporate some of the supernatant liquid to see if it deposits a residue.

Because you do not know how many ions you are looking for in your unknown, it will be necessary to establish the presence or absence of each of the ions previously studied. The best way to do this is by systematic analysis as in Experiment Q.9. When you solve for your unknown, the following solubility information may be useful. (Note that the generalizations apply to salts of the cations studied in previous experiments.)

Carbonates	Slightly soluble except those of the alkali metals. In dilute strong acids, all will dissolve (or be converted to an insoluble salt of the anion of the acid).
Nitrates	Soluble.
Sulfates	Soluble except those of Ca, Ba, Pb.
Fluorides	Alkali metal, Co, Cr, Al, Ag, Sn are water-soluble; others are but slightly soluble. Solubility increased by acid.
Chlorides	Soluble except AgCl and Hg_2Cl_2. $PbCl_2$ slightly soluble in cold water; dissolves in hot water.
Bromides	Soluble except AgBr, Hg_2Br_2, $HgBr_2$. $PbBr_2$ slightly soluble in cold water, dissolves in hot water.
Iodides	Soluble except AgI, Hg_2I_2, HgI_2, PbI_2.
Oxides or hydroxides	Slightly soluble except those of alkali metals and Ba. All are dissolved (or converted to insoluble salts) by strong acids. Al_2O_3, SnO_2, Cr_2O_3 are likely to be inert.

Questions

Q.10.1   The solution of cations in Experiments Q.1 to Q.9 contained
approximately 0.05 to 0.10 M cations. In dissolving your unknown in 5 mL of water, how much solid should you take to
make up a solution in which the concentrations of the cations
will be roughly in this range? Assume that your unknown is
made up of three different solids, each having a formula weight
of approximately 100 g per mol. State any other assumptions
that you would have to make.

Q.10.2   In part (*a*), it is suggested that a certain sequence of solvents be tried on your unknown solid. Why would such a sequence be useful?

Q.10.3   A solid unknown contains one or more of the following substances: NH_4NO_3, $Fe(NO_3)_3$, $NaCl$, $Cu(NO_3)_2$, $PbSO_4$, $AgNO_3$. The unknown is completely soluble in water. Addition of dilute HCl to the resulting solution gives a white precipitate. Treatment of a fresh portion of the original unknown with excess NaOH gives a clear solution having a deep-blue color. Comment on the probable composition of the original solid unknown.

Q.10.4   An unknown solution may contain appreciable amounts of one or more of the following: K^+, Ba^{2+}, Cl^-, NO_3^-, SO_4^{2-}, CO_3^{2-}. One portion of the solution turns blue litmus red and gives a white precipitate when silver nitrate solution is added. A white precipitate is also formed when dilute sulfuric acid is added to a second portion of the solution. Comment on which of the ions are definitely present, which are absent, and which are possibly present.

Q.10.5   A solid unknown possibly containing $AgNO_3$, $ZnCl_2$, NaOH, $BaCl_2$, $KHSO_4$, $CuCl_2$ is completely soluble in water. Addition of sodium carbonate to the solution produces a colorless gas and a white precipitate, which is filtered off. The precipitate is dissolved in HCl, and then NH_3 and thioacetamide are added, producing a white precipitate that is found to be insoluble in SO_4^{2-}/HSO_4^- buffer. Comment on the probable composition of the original unknown.

Q.10.6   A solid unknown contains one or more of the following substances: $Hg_2(NO_3)_2$, $FeCl_3$, $NaHCO_3$, $AlCl_3$, HgS, KNO_3. Treatment of the original solid with water gives a clear solution, a colorless gas, and a white precipitate, which is filtered out. Addition of NH_3 to the white precipitate gives a dark-gray precipitate. Addition of NH_4^+/NH_3 buffer to the clear solution gives a gelatinous precipitate that is soluble in excess NaOH. Comment on the probable composition of the original solid unknown.

Appendix A

Constants and Conversion Factors

kilo- means one thousand (10^3)
centi- means one-hundredth (10^{-2})
milli- means one-thousandth (10^{-3})
micro- means one-millionth (10^{-6})
nano- means one-billionth (10^{-9})

1 meter (m) = 100 centimeters (cm) = 39.370 inches (in.)
1 cm = 10 millimeters (mm) = 0.39370 in.

1 kilogram (kg) = 1000 grams (g) = 2.2046 pounds (lb)
1 g = 1000 milligrams (mg) = 0.035274 ounce (oz)

1 liter = 1000 milliliters (mL)
        = 1000 cubic centimeters (cm^3) = 1.0567 quarts (qt)

1 standard atmosphere (atm) = 760 mmHg (torr)
                           = 13.5 × 760 mmH$_2$O

Avogadro number = 6.0222×10^{23} molecules/mol

electron charge = 1.6022×10^{-19} coulomb
electron mass = 9.1096×10^{-28} g
Faraday constant = 9.6487×10^4 coulombs/equiv
gas constant = 8.2057×10^{-2} liter-atm/mol-deg
Planck constant = 6.6262×10^{-34} joule-sec
speed of light = 2.9979×10^{10} cm/sec

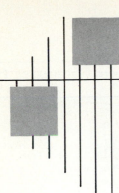

Appendix B

Vapor Pressure of Water

Temperature, °C	Pressure, atm	Pressure, mmHg
0	0.00603	4.58
1	0.00648	4.93
2	0.00697	5.29
3	0.00748	5.69
4	0.00803	6.10
5	0.00861	6.54
6	0.00923	7.01
7	0.00989	7.51
8	0.0106	8.04
9	0.0113	8.61
10	0.0121	9.21
11	0.0130	9.84
12	0.0138	10.52
13	0.0148	11.23
14	0.0158	11.99
15	0.0168	12.79
16	0.0179	13.63
17	0.0191	14.53
18	0.0204	15.48
19	0.0217	16.48
20	0.0231	17.54
21	0.0245	18.65
22	0.0261	19.83
23	0.0277	21.07
24	0.0294	22.38
25	0.0313	23.76
26	0.0332	25.21
27	0.0352	26.74
28	0.0373	28.35
29	0.0395	30.04
30	0.0419	31.82
31	0.0443	33.70
32	0.0470	35.66
33	0.0496	37.73
34	0.0525	39.90
35	0.0555	42.18

Vapor Pressure of Water (*Continued*)

Temperature, °C	Pressure, atm	Pressure, mmHg
40	0.0728	55.32
45	0.0946	71.88
50	0.1217	92.51
55	0.1553	118.04
60	0.1966	149.38
65	0.2468	187.54
70	0.3075	233.7
75	0.3804	289.1
80	0.4672	355.1
85	0.5705	433.6
90	0.6918	525.8
95	0.8341	633.9
100	1.0000	760.0
101	1.0363	787.5
105	1.1922	906.1

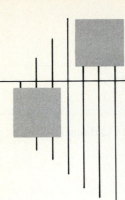

Appendix C

Commonly Encountered Ions

Simple Ions

Al^{3+}	aluminum
Ba^{2+}	barium
Br^-	bromide
Cd^{2+}	cadmium
Ca^{2+}	calcium
Cl^-	chloride
Cr^{2+}	chromous, chromium(II)
Cr^{3+}	chromic, chromium(III)
Co^{2+}	cobaltous, cobalt(II)
Co^{3+}	cobaltic, cobalt(III)
Cu^+	cuprous, copper(I)
Cu^{2+}	cupric, copper(II)
F^-	fluoride
I^-	iodide
Fe^{2+}	ferrous, iron(II)
Fe^{3+}	ferric, iron(III)
Pb^{2+}	plumbous, lead(II)
Li^+	lithium
Mg^{2+}	magnesium
Mn^{2+}	manganous, manganese(II)
Mn^{3+}	manganic, manganese(III)
Hg_2^{2+}	mercurous, mercury(I)
Hg^{2+}	mercuric, mercury(II)
Ni^{2+}	nickelous, nickel(II)
K^+	potassium
Ag^+	argentous, silver(I)
Na^+	sodium
Sr^{2+}	strontium
S^{2-}	sulfide
Sn^{2+}	stannous, tin(II)
Sn^{4+}	stannic, tin (IV)
Zn^{2+}	zinc

Complex Ions

$C_3H_2O_2{}^-$	acetate
$BrO_3{}^-$	bromate
$CO_3{}^{2-}$	carbonate
$HCO_3{}^-$	bicarbonate, hydrogen carbonate
$C_2O_4{}^{2-}$	oxalate
$ClO_2{}^-$	chlorite
ClO^-	hypochlorite
$ClO_3{}^-$	chlorate
$ClO_4{}^-$	perchlorate
$CrO_4{}^{2-}$	chromate
$CrO_2{}^-$	chromite
$Cr_2O_7{}^{2-}$	dichromate
OH^-	hydroxide
H_3O^+	hydronium
$IO_3{}^-$	iodate
$IO_4{}^-$	periodate
$MnO_4{}^{2-}$	manganate
$MnO_4{}^-$	permanganate
$NH_4{}^+$	ammonium
$NH_2{}^-$	amide
$PO_3{}^{2-}$	phosphite
$PO_4{}^{2-}$	phosphate
$HPO_4{}^{2-}$	biphosphate, hydrogen phosphate
$H_2PO_4{}^-$	dihydrogen phosphate
$SO_3{}^{2-}$	sulfite
$HSO_3{}^-$	bisulfite, hydrogen sulfite
$SO_4{}^{2-}$	sulfate
$HSO_4{}^{2-}$	bisulfate, hydrogen sulfate

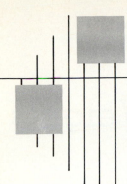

Appendix D

Equilibrium Constants

Dissociation Constants (First Step Only)

$CO_2 + H_2O$	4.2×10^{-7}	H_3AsO_4	2.5×10^{-4}
HCO_3^-	4.8×10^{-11}	$H_2ASO_4^-$	3×10^{-13}
$HC_2H_3O_2$	1.8×10^{-5}	H_2O	1.0×10^{-14}
$NH_3 + H_2O$	1.8×10^{-5}	H_2S	1.1×10^{-7}
HNO_2	4.5×10^{-4}	HS^-	1×10^{-14}
H_3PO_4	7.5×10^{-3}	H_2SO_3	1.3×10^{-2}
$H_2PO_4^-$	6.2×10^{-8}	HSO_3^-	5.6×10^{-8}
HPO_4^{2-}	$\sim 10^{-12}$	HSO_4^-	1.3×10^{-2}

Solubility Products

$Mg(OH)_2$	8.9×10^{-12}	CoS	5×10^{-22}
MgF_2	8×10^{-8}	NiS	3×10^{-21}
MgC_2O_4	8.6×10^{-5}	PtS	8×10^{-73}
$Ca(OH)_2$	1.3×10^{-6}	$Cu(OH)_2$	1.6×10^{-19}
CaF_2	1.7×10^{-10}	CuS	8×10^{-37}
$CaCO_3$	4.7×10^{-9}	$AgCl$	1.7×10^{-10}
$CaSO_4$	2.4×10^{-5}	$AgBr$	5.0×10^{-13}
CaC_2O_4	1.3×10^{-9}	AgI	8.5×10^{-17}
$Sr(OH)_2$	3.2×10^{-4}	Ag_2S	5.5×10^{-51}
$SrSO_4$	7.6×10^{-7}	ZnS	1×10^{-22}
$SrCrO_4$	3.6×10^{-5}	Hg_2Cl_2	1.1×10^{-18}
$Ba(OH)_2$	5.0×10^{-3}	Hg_2Br_2	1.3×10^{-22}
$BaSO_4$	1.5×10^{-9}	Hg_2I_2	4.5×10^{-29}
$BaCrO_4$	8.5×10^{-11}	HgS	1.6×10^{-54}
$Cr(OH)_3$	6.7×10^{-31}	$Al(OH)_3$	5×10^{-33}
$Mn(OH)_2$	2×10^{-13}	SnS	1×10^{-26}
MnS	7×10^{-16}	$Pb(OH)_2$	4.2×10^{-15}
FeS	4×10^{-19}	$PbCl_2$	1.6×10^{-5}
$Fe(OH)_3$	6×10^{-38}	PbS	7×10^{-29}

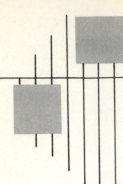

Appendix E

Atomic Weights

referred to $^{12}C = 12.000$ a.m.u.

Element	Symbol	Atomic Number	Atomic Weight	Element	Symbol	Atomic Number	Atomic Weight
Actinium	Ac	89	[227]*	Germanium	Ge	32	72.59
Aluminum	Al	13	26.98154	Gold	Au	79	196.9665
Americium	Am	95	[243]	Hafnium	Hf	72	178.49
Antimony	Sb	51	121.75	Hahnium	Ha	105	[262]
Argon	Ar	18	39.948	Helium	He	2	4.00260
Arsenic	As	33	74.9216	Holmium	Ho	67	164.9304
Astatine	At	85	[210]	Hydrogen	H	1	1.0079
Barium	Ba	56	137.33	Indium	In	49	114.82
Berkelium	Bk	97	[247]	Iodine	I	53	126.9045
Beryllium	Be	4	9.01218	Iridium	Ir	77	192.22
Bismuth	Bi	83	208.9804	Iron	Fe	26	55.847
Boron	B	5	10.81	Krypton	Kr	36	83.80
Bromine	Br	35	79.904	Kurchatovium	Ku	104	[261]
Cadmium	Cd	48	112.41	Lanthanum	La	57	138.9055
Calcium	Ca	20	40.08	Lawrencium	Lr	103	[260]
Californium	Cf	98	[251]	Lead	Pb	82	207.2
Carbon	C	6	12.011	Lithium	Li	3	6.941
Cerium	Ce	58	140.12	Lutetium	Lu	71	174.967
Cesium	Cs	55	132.9054	Magnesium	Mg	12	24.305
Chlorine	Cl	17	35.453	Manganese	Mn	25	54.9380
Chromium	Cr	24	51.996	Mendelevium	Md	101	[258]
Cobalt	Co	27	58.9332	Mercury	Hg	80	200.59
Copper	Cu	29	63.546	Molybdenum	Mo	42	95.94
Curium	Cm	96	[247]	Neodymium	Nd	60	144.24
Dysprosium	Dy	66	162.50	Neon	Ne	10	20.179
Einsteinium	Es	99	[252]	Neptunium	Np	93	237.0482
Erbium	Er	68	167.26	Nickel	Ni	28	58.69
Europium	Eu	63	151.96	Niobium	Nb	41	92.9064
Fermium	Fm	100	[257]	Nitrogen	N	7	14.0067
Fluorine	F	9	18.998403	Nobelium	No	102	[259]
Francium	Fr	87	[223]	Osmium	Os	76	190.2
Gadolinium	Gd	64	157.25	Oxygen	O	8	15.9994
Gallium	Ga	31	69.72	Palladium	Pd	46	106.42

Element	Symbol	Atomic Number	Atomic Weight	Element	Symbol	Atomic Number	Atomic Weight
Phosphorus	P	15	30.97376	Strontium	Sr	38	87.62
Platinum	Pt	78	195.08	Sulfur	S	16	32.06
Plutonium	Pu	94	[244]	Tantalum	Ta	73	180.9479
Polonium	Po	84	[209]	Technetium	Tc	43	98.9062
Potassium	K	19	39.0983	Tellurium	Te	52	127.60
Praseodymium	Pr	59	140.9077	Terbium	Tb	65	158.9254
Promethium	Pm	61	[145]	Thallium	Tl	81	204.383
Protactinium	Pa	91	231.0359	Thorium	Th	90	232.0381
Radium	Ra	88	226.0254	Thulium	Tm	69	168.9342
Radon	Rn	86	[222]	Tin	Sn	50	118.69
Rhenium	Re	75	186.207	Titanium	Ti	22	47.88
Rhodium	Rh	45	102.9055	Tungsten	W	74	183.85
Rubidium	Rb	37	85.4678	Uranium	U	92	238.0289
Ruthenium	Ru	44	101.07	Vanadium	V	23	50.9415
Samarium	Sm	62	150.36	Xenon	Xe	54	131.29
Scandium	Sc	21	44.9559	Ytterbium	Yb	70	173.04
Selenium	Se	34	78.96	Yttrium	Y	39	88.9059
Silicon	Si	14	28.0855	Zinc	Zn	30	65.38
Silver	Ag	47	107.868	Zicronium	Zr	40	91.22
Sodium	Na	11	22.98977				

*Values in brackets are mass numbers of longest-lived or best-known isotopes.